S0-CAH-130

NUCLEAR MAGNETIC RESONANCE IN BIOCHEMISTRY

NUCLEAR MAGNETIC RESONANCE IN BIOCHEMISTRY

PRINCIPLES AND APPLICATIONS

Thomas L. James

Department of Pharmaceutical Chemistry
University of California
San Francisco, California

ACADEMIC PRESS New York San Francisco London 1975

A Subsidiary of Harcourt Brace Jovanovich, Publishers

ACADEMIC PRESS, INC.
111 Fifth Avenue, New York, New York 10003

United Kingdom Edition published by
ACADEMIC PRESS, INC. (LONDON) LTD.
24/28 Oval Road, London NW1

Library of Congress Cataloging in Publication Data

James, Thomas L
Nuclear magnetic resonance in biochemistry.

Includes bibliographical references and index.
1. Nuclear magnetic resonance spectroscopy. 2. Biological chemistry. I. Title.
QD96.N8J35 538'.3 74-27782
ISBN 0-12-380950-9

PRINTED IN THE UNITED STATES OF AMERICA

To Marc and Tristan

CONTENTS

PREFACE

In the past few years, nuclear magnetic resonance (NMR) has been successfully used to unravel some knotty problems in several areas of biological and physiological chemistry. As a result of these applications, many biological and biochemical researchers are now aware of the utility of nuclear magnetic resonance. However, no text currently exists which considers NMR phenomena from the viewpoint of the life scientist. The aim of this book is to correct that condition.

The text is intended primarily for biochemists, biophysicists, and molecular biologists. It is anticipated that many chemists will also be interested in learning those aspects of nuclear magnetic resonance most useful for studying biological systems. The text should be suitable for certain graduate courses. A recent trend has been to apply more sophisticated physical techniques to biological and biochemical problems. The graduate curriculum in many schools is beginning to reflect this trend with the introduction of courses covering these newer techniques.

Several texts on NMR are available, but they generally cover material of interest to the physicist or chemist. The features of greatest utility in biological applications often differ from those presented in the chemically oriented texts. For example, the use of NMR relaxation phenomena enjoys a more prominent place in biological applications than it does in chemical applications. Appropriately, this book covers NMR relaxation and its applications in more detail than will be found in chemically oriented texts.

This book will be of interest to people with varying backgrounds. Introductory material is given in Chapter 1 for those with little previous NMR knowledge. It may be sufficient for some readers to go directly to the discussion of NMR applications in their field of interest after reading Chapter 1. It may, however, be necessary to read the appropriate sections of Chapter 2 or Chapter 3 for a better understanding of the theoretical basis for those applications. Chapter 2 presents the principles of nuclear magnetic resonance without preoccupation with derivation of equations. The emphasis is to convey an understanding of the equations, especially those which have found use in biochemical studies. In addition to a discussion of the basis for chemical shifts and spin–spin splitting, Chapters 3 and 4 include several examples of the use of these NMR parameters in studies of small molecule interactions and structure. Chapter 6 is concerned primarily with NMR spectral parameters of small molecules interacting with macromolecules, and Chapter 7 deals with the information obtainable from the spectra of biopolymers. Chapter 8 discusses NMR investigations of the state of motion of lipids in membranes and model membranes, water in macromolecular and cellular systems, and sodium ion in biological tissue.

Chapter 5 describes the experimental apparatus and procedures employed in NMR studies, with a strong emphasis on those aspects of greatest importance in biological applications, e.g., sensitivity improvement. A fairly comprehensive discussion of Fourier transform NMR is included since Fourier transform NMR is a relatively new, sophisticated technique which provides the sensitivity gain necessary for many biochemical studies as well as the capability for selective relaxation time measurements. Several promising new NMR techniques which have yet to be applied to biological problems are also briefly discussed.

I would like to express my gratitude to Daniel Buttlaire, Kenneth Gillen, Robert Hershberg, Alan McLaughlin, and James C. Orr for reviewing various chapters. I also wish to thank those individuals who kindly supplied some of the figures used in the text, and the journals for permission to reproduce published material. Last to be mentioned, but foremost in my gratitude, is my wife Joyce, who, in addition to typing the first draft and aiding with proofreading, has been exceedingly patient during the course of this writing.

Thomas L. James

NUCLEAR MAGNETIC RESONANCE IN BIOCHEMISTRY

CHAPTER 1

INTRODUCTION

The high degree of specificity for many biological reactions and processes depends on subtle differences in the structure and conformation of molecules. Nuclear magnetic resonance (NMR) spectroscopy is one of the few techniques available with the capacity to obtain detailed information about biomolecular phenomena. With NMR, an individual nucleus in a molecule can be "observed" by monitoring that nucleus' line in an NMR spectrum. The various NMR parameters of that line—frequency, splitting, linewidth, and amplitude—can be used to study the electronic and geometric structure of "simple" molecules or macromolecules, molecular motion and rate processes, and molecular interactions. Quite often the molecular information obtained is of a qualitative nature; however, in many cases NMR can provide quantitative information not obtainable by other means.

Data from many of the other physical methods used in studies of biological systems are often interpreted empirically because theory usually provides little aid in their qualitative interpretation. The usual procedure for those methods is to rely on compilations of data for comparison with present experimental results. A comparatively better understanding of theory is required for a satisfactory interpretation of NMR results. The yield for this better understanding is usually unequivocal qualitative results, or often quantitative results.

Many monographs are available that treat the NMR phenomenon with varying degrees of rigor. Some texts, in order of difficulty, are Abragam (1), Slichter (2), Carrington and McLachlan (3), and Becker (4). Reviews on various aspects of NMR studies of biological molecules have been written by pioneers in the field: Kowalsky and Cohn (5, 6), Roberts and Jardetzky (7, 8), and McDonald and Phillips (9).

1.1. Magnetic Properties of Nuclei

Roughly half of the known nuclei behave as though they were spinning like a top. The magnitude of the angular momentum

$$J' = \hbar \sqrt{I(I+1)} \tag{1-1}$$

of this spinning motion depends on the nuclear spin quantum number I, which differs for different nuclei ($\hbar$ is Planck's constant divided by 2π). The value of the nuclear spin quantum number I is determined by the mass number and atomic number according to the following tabulation.

MASS NUMBER	ATOMIC NUMBER	SPIN NUMBER, I
Odd	Odd or even	Half-integer: $\frac{1}{2}, \frac{3}{2}, \frac{5}{2}, \ldots$
Even	Even	0
Even	Odd	Integer: 1, 2, 3 . . .

Nuclei of interest having spin $\frac{1}{2}$ are ^{1}H, ^{13}C, ^{31}P, and ^{19}F. Nuclei of interest having spin 0 are ^{12}C, ^{16}O, and ^{32}S. ^{2}H (or D) and ^{14}N have $I = 1$ and ^{23}Na and ^{39}K have $I = \frac{3}{2}$. A complete list is available in Appendix 2.

Because nuclei are positively charged, a spinning nucleus gives rise to a magnetic moment

$$\vec{\mu} = \gamma \vec{J}' = \gamma \hbar \vec{I} \tag{1-2}$$

where γ is the gyromagnetic (or magnetogyric) ratio and $\vec{I}$ is a dimensionless angular momentum. As a physical picture, the nucleus with $I > 0$ is equivalent to a tiny bar magnet the axis of which is coincident with the spin axis.

Nuclei with spin $I = \frac{1}{2}$ behave as spherical entities possessing a uniform charge distribution. However, the charge distribution within a nucleus with $I \geq 1$ can be described as a prolate (cigar-shaped) or oblate (flattened) spheroid. A measure of the nonsphericity of the nuclear charge distribution is embodied in the electric quadrupole moment, which depends on I. Only nuclei with $I \geq 1$ possess an electric quadrupole moment. Therefore, when a charged species (e.g., an electron) approaches a nucleus with an electric quadrupole moment, the nucleus experiences an electric field the magnitude of which depends on the direction of approach. Possession of an electric quadrupole moment will critically affect the relaxation time of a nucleus and the coupling of that nucleus' spin with spins of neighboring nuclei. Appendix 2 lists electric quadrupole moments for those nuclei with $I \geq 1$.

1.2. Magnetic Resonance

When a nucleus with magnetic moment μ is placed in a strong, uniform magnetic field H_0 (oriented in the z direction), the magnetic dipole is quantized into a discrete set of orientations. This is referred to as "nuclear Zeeman splitting." Each one of these orientations corresponds to a nuclear energy state or level with energy

$$E = -\mu_z H_0 \tag{1-3}$$

where μ_z ($= m_I \gamma \hbar$) is the z component of the nuclear magnetic moment. The magnetic quantum number m_I, characteristic of each nuclear energy level, depends on the nuclear spin quantum number and may take on the values

$$m_I = I, (I - 1), (I - 2), \ldots, -(I - 2), -(I - 1), -I \tag{1-4}$$

Energy levels are shown in Fig. 1-1 for nuclei with spin quantum numbers

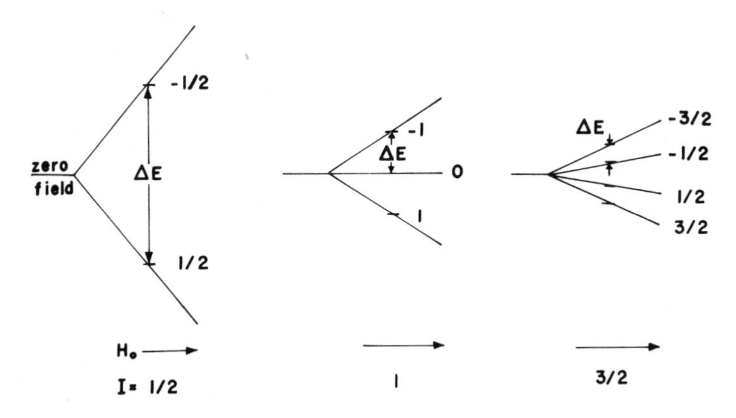

FIG. 1-1. Nuclear Zeeman splitting of energy levels in a magnetic field for various nuclear spin quantum numbers.

$I = \frac{1}{2}$, 1, and $\frac{3}{2}$. For each nuclear spin, the energy levels are equally spaced. The NMR experiment is to induce transitions between the levels by absorption or emission of a photon with the requisite energy. The selection rules of quantum mechanics prescribe that transitions between neighboring levels are the only allowable transitions. The energy of the photon required to induce transitions is just the energy separation between adjacent levels,

$$\Delta E = \gamma \hbar H_0 \tag{1-5}$$

The Bohr condition ($\Delta E = h\nu$) permits us to give the frequency of the nuclear transition

$$\nu = \gamma H_0 / 2\pi \tag{1-6}$$

which is in the radiofrequency (rf) portion (e.g., 100 MHz) of the electromagnetic spectrum. Therefore, for any particular nucleus in a given magnetic field, the NMR frequency will be characteristic, depending primarily on the gyromagnetic ratio peculiar to that particular nucleus.

The theory of electromagnetic radiation states that the probability of a photon inducing a transition from a higher energy level to a lower one is the same as the probability for induced transition from a lower to a higher energy level. Therefore, in a large ensemble of spins (such as a sample in an NMR tube), a net absorption or emission depends only on the difference between the number of nuclei in the upper and lower energy levels.

The distribution of nuclei in the various possible energy states is given, under conditions of thermal equilibrium, by the Boltzmann equation:

$$\frac{N_{\text{upper}}}{N_{\text{lower}}} = e^{-\Delta E/kT} = e^{-h\nu/Tk} \tag{1-7}$$

where N_{upper} and N_{lower} represent the population (i.e., number) of nuclei in upper and lower energy states, respectively. With radiofrequencies used in NMR, the population ratio might typically be 1.000035 (at $\nu = 220$ MHz) for hydrogen nuclei in thermal equilibrium at room temperature. That means for every 1,000,000 nuclei in the upper energy state there are 1,000,035 nuclei in the lower energy state. Without this small excess in the lower energy state, there could be no nuclear magnetic resonance phenomenon.

This small excess does permit NMR to thrive, but it also presents an unfortunate sensitivity problem. The relatively low sensitivity of NMR compared with some other spectroscopic techniques is perhaps the greatest limitation for application of NMR to biochemical systems. One of the motivating factors for use of stronger magnetic fields is the increased sensitivity. From Eqs. 1-6 and 1-7, it is seen that nuclei in a stronger magnetic field have a larger Boltzmann factor and, consequently, a larger NMR signal.

1.3. NMR Relaxation

A nuclear spin system in a stationary magnetic field H_0 may be considered. At equilibrium, the spin populations of the various Zeeman energy levels will be described by the Boltzmann distribution (Eq. 1-7), giving the lower energy levels a slightly greater spin population, as just discussed. If a radiofrequency field at the resonance frequency is applied to the system, the probability of an upward transition is equal to the probability of a downward transition. Because there is a greater spin population in the lower energy levels, there will be more upward transitions than downward transitions, resulting in a nonequilibrium spin distribution. If this process continues, the excess of nuclei in the lower energy state will continually diminish with consequent decrease in the NMR signal intensity. Under certain circumstances the two spin populations may be equal and the NMR signal may disappear completely. This phenomenon is referred to as "saturation" and, in practice, can occur if strong rf fields are applied. For an NMR signal to persist, some mechanism must be available for replenishing the number of nuclei in the lower energy state.

There are various mechanisms leading to radiationless transitions that cause the perturbed system to return to the equilibrium spin distribution. These radiationless transitions are called "relaxation processes." There are two kinds of relaxation processes: spin–lattice (or longitudinal) relaxation and spin–spin (or transverse) relaxation. The spin–lattice relaxation time is designated by T_1 and the spin–spin relaxation time is designated by T_2. T_1 is a characteristic time describing the rate at which the nonequilibrium spin distribution $(N_{\text{lower}} - N_{\text{upper}})$ exponentially approaches equilibrium $(N_{\text{lower}} - N_{\text{upper}})_{\text{equil}}$ following absorption of rf energy:

$$(N_{\text{lower}} - N_{\text{upper}}) = (N_{\text{lower}} - N_{\text{upper}})_{\text{equil}}\,(1 - e^{-t/T_1}) \qquad (1\text{-}8)$$

The lattice is the environment surrounding the nucleus—the remainder of that molecule as well as other solute and solvent molecules. Spin–lattice relaxation occurs by interaction of the nuclear spin dipole with random, fluctuating magnetic fields caused by the motion of surrounding dipoles in the lattice that happen to have components fluctuating with the same frequency as the resonance frequency described in Eq. 1-6. The energy of the radiationless transition is transferred to the various energy components of the lattice as additional rotational, translational, or vibrational energy (with total energy unchanged) until the nuclear spin system and the lattice are in thermal equilibrium. There are several possible mechanisms contributing to spin–lattice relaxation, which will be discussed in Chapter 2. In solids or viscous liquids, the T_1 relaxation may be several hours. In most nonviscous liquids and solutions, T_1 is usually on the order of 0.001–100 sec.

Spin–spin relaxation processes also have a relaxation time, T_2, which characterizes the rate of these relaxation processes. The sources of the random magnetic fields giving rise to T_1 relaxation will also lead to T_2 relaxation. However, spin–spin relaxation has other relaxation mechanisms that may contribute to T_2. Simply, the additional contribution to spin–spin (or transverse) relaxation is the result of chemical exchange or mutual exchange of spin states by two nuclei in close proximity. The distribution of energy among the spins in this manner is an adiabatic process and, although it decreases the lifetime for any particular nucleus in the higher energy state, it does not change the number of nuclei in the higher energy state.

Spin–spin relaxation is caused by random magnetic fields (usually from neighboring nuclei) in the sample that are not fluctuating. These random local fields will cause shifts in the resonance frequencies for individual nuclear spins in the sample. There will then be a distribution of resonance frequencies for any sample depending on the variation in the random local fields. The greater the variation in local fields, the greater the linewidth of the peak in the absorption spectrum will be. The spin–spin relaxation time T_2 can then be related to the resonance linewidth $W_{1/2}$ in the absorption spectrum by

$$T_2 = 1/\pi W_{1/2} \tag{1-9}$$

Equation 1-9 is valid in the absence of instrumental instability and magnetic field inhomogeneity. Variation of the stationary magnetic field H_0 over the area of the sample will cause inhomogeneity broadening just as variations in the microscopic local fields within the sample are the cause of the NMR linewidth. High-quality instruments often enable inhomogeneity and thus instrumental considerations to be overcome, with the result that Eq. 1-9 is a justifiable expression of the true T_2. However, with large values for T_2 ($\gtrsim$1.0 sec), Eq. 1-9 may not accurately relate the observed linewidth to T_2.

1.4. Chemical Shift and Shielding

Historically, NMR entered the realm of chemistry when it was discovered that all protons do not have the same resonance frequency. The application of the nuclear magnetic resonance phenomenon to chemical problems depends on the fact that the field experienced by a nucleus is not exactly the same as the applied magnetic field. The small variations in the field at the nucleus are caused by diamagnetic shielding by electrons within the molecule. The extent of this diamagnetic shielding depends on the chemical environment of the nucleus; i.e., each nucleus in a collection of chemically equivalent nuclei is shielded to the same extent, but the extent of shielding is different for any other collection of chemically equivalent nuclei. For example, the three methyl protons of ethyl bromide will form an equivalent set with resonance frequency ν_{CH_3} and the two methylene protons will form another equivalent set of nuclei with resonance frequency ν_{CH_2}.

The shielding of a nucleus is caused by the motion of electrons in the molecule induced by application of the stationary magnetic field H_0. The induced motion of those electrons sets up a local magnetic field opposed to the H_0 field. The magnitude of the effective field perceived by a set of equivalent nuclei will be proportional to the stationary field:

$$H_{\text{eff}} = H_0 - H_0\sigma = H_0(1 - \sigma) \tag{1-10}$$

where σ is a nondimensional screening or shielding constant. $H_0\sigma$ is the induced field caused by the motion of surrounding electrons. The resonance frequency will therefore be decreased by the shielding, as illustrated in Fig. 1-2.

It can be seen that the exact resonance frequencies for nuclei in a magnetic field depend on the strength of the magnetic field and on the precise frequency of the applied radiofrequency field. The nuclear magnetic resonance spectrum can be generated in either of two ways: (*1*) field sweep—the frequency of rf field is maintained at a fixed value and the strength of the applied H_0 field is slowly varied over a small range, or (*2*) frequency sweep—the strength of the applied H_0 field is maintained at a fixed value and the rf frequency is slowly varied. The positions of the resonance lines in the field sweep mode can be related to the positions in the frequency sweep mode simply by applying Eq. 1-6, $\nu = \gamma H_0/2\pi$, which expresses the resonance condition. For liquid samples, the usual practice is to describe the sweep in frequency units even if the field sweep method is being used.

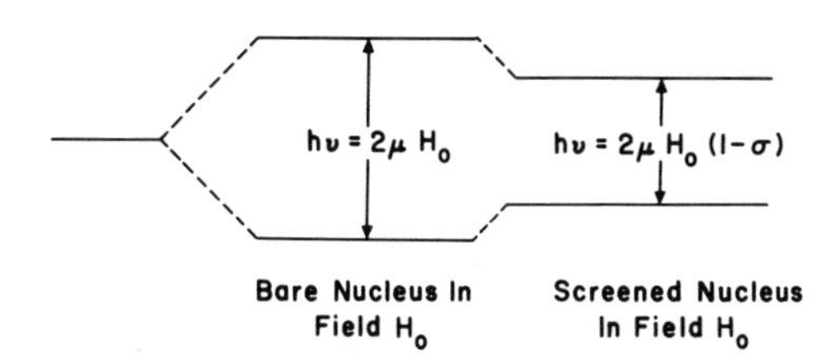

FIG. 1-2. Effect of shielding (screening) on the resonance frequency.

In most compounds, protons absorb at frequencies ranging over about 2500 Hz at a field strength of about 51,000 Gauss (or 5.1 Tesla). At this field strength, the protons resonate at a frequency of about 220×10^6 Hz. Knowledge of proton resonance frequencies provides valuable information about molecules. Therefore, it is desirable to make accurate measurements of the proton resonance positions. If the resonance position is to be determined with an accuracy of about ± 1 Hz, the 220 MHz instrument must be able to distinguish between frequencies of the order of 1 part in 10^8. The strength of the magnetic field cannot be determined with such a degree of accuracy, so the absolute resonance frequency cannot be determined to ± 1 Hz. However, the relative positions of two lines in the proton NMR spectrum can be determined to less than ± 1 Hz. The difference between a proton resonance position in the NMR spectrum and the position of some arbitrarily chosen reference is called the "chemical shift."

Typical reference compounds are 85% phosphoric acid for ^{31}P; trifluoroacetic acid or carbon tetrafluoride for ^{19}F; carbon disulfide or tetramethylsilane for ^{13}C; and cyclohexane, tetramethylammonium ion, tetramethylsilane (TMS), hexamethyldisiloxane (HMDS), or the methyl resonance of sodium 2,2-dimethyl-2-silapentane-5-sulfonate (DSS), $(CH_3)_3Si(CH_2)_3SO_3^-Na^+$, for ^1H. For proton NMR, TMS has become the ultimate reference in nonaqueous solutions and DSS the reference in aqueous solutions. The resonance frequency of the methyl protons of DSS and TMS are nearly identical.

Chemical shifts can be expressed in terms of the number of Hertz a resonance peak is from a reference peak. However, because the chemical shift is dependent on the strength of the applied magnetic field, it is advantageous to express the chemical shift as a function of field strength necessary to achieve the resonance condition:

$$\delta = \frac{H_{ref} - H_{samp}}{H_{ref}} \tag{1-11}$$

$$\delta = \frac{\Delta\nu \times 10^6}{\nu_{instr}} \tag{1-12}$$

where H_{ref} and H_{samp} are the resonance field strengths for the reference and sample nuclei, respectively; $\Delta\nu$ is the difference between the resonance frequencies of the reference and sample (in Hz); and ν_{instr} is the oscillator frequency (in Hz) characteristic of the instrument. The chemical shift is then obtained as a dimensionless number, expressed as parts per million (ppm), and chemical shifts are determined using spectrometers operating at different frequencies can be directly compared. For example, the chloroform proton resonates at 436 Hz downfield from the TMS reference on a 60 MHz (60×10^6 Hz) instrument and 1598 Hz on a 220 MHz instrument. In terms of δ units, the chemical shift is 7.25 ppm for either instrument.

Proton magnetic resonance has been more highly developed and conventions for presenting data are more uniform. In the literature, chemical shifts are given in any of three ways: (*1*) Hertz—the reference compound must be listed and the instrument frequency given, (*2*) δ (ppm)—the reference compound must be listed, and (*3*) τ (ppm)—TMS or DSS is assumed to be the reference compound with a value of $\tau = 10$ ppm. If TMS is used as the reference in (*2*), the τ and δ scales are related: $\tau = 10 - \delta$. Very few protons have resonance positions at a higher field than TMS and only a few resonances for protons in very electronegative environments or quite acidic protons are found below 10 ppm downfield from TMS. Therefore, the majority of proton chemical shifts can be found between 0 and 10 on the τ scale. Figure 1-3 shows the spectrum of adenine in D_2O using DSS as an external standard, i.e., the standard is in a capillary inside the NMR tube.

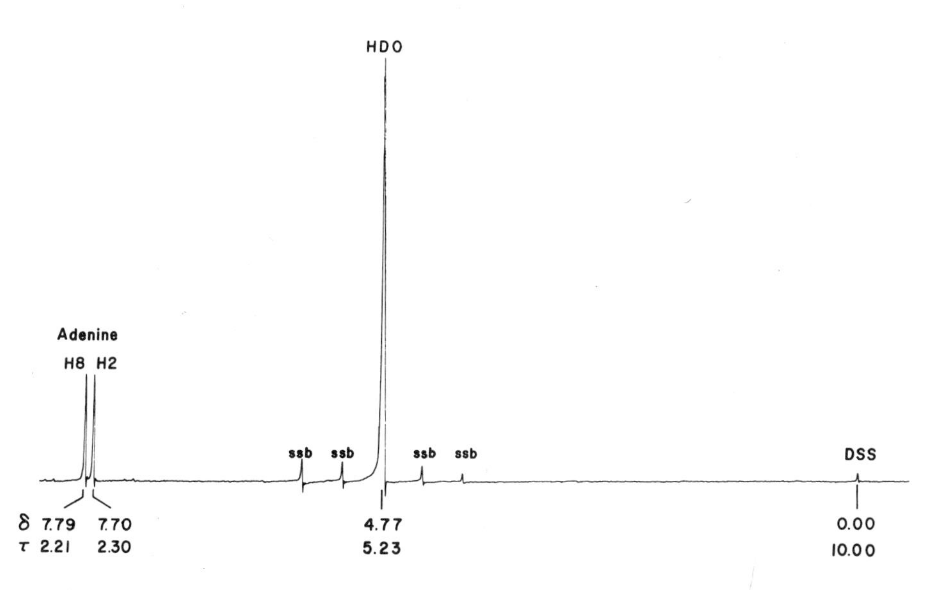

FIG. 1-3. 220 MHz proton NMR spectrum of 0.40 *M* adenine in 1 *M* NaOD with D_2O as solvent and sodium 2,2-dimethyl-2-silapentane-5-sulfonate (DSS) as an external reference present as a D_2O solution in a coaxial capillary tube. The spinning side bands (designated ssb) flanking the residual HDO peak occur at integral multiples of the sample spinning frequency. The spinning side bands result from the sample spinning through a slightly inhomogeneous field.

1.5. Spin–Spin Splitting

Nuclei with spin $I \neq 0$ can interact with other nuclei possessing magnetic moments to cause mutual splitting of the nuclear magnetic resonance peaks into multiplets. The coupling of one set of equivalent spins with another set of spins is termed "spin–spin coupling" or "spin–spin splitting." The magnitude of the splitting is designated by the spin–spin coupling constant J and is independent of the applied magnetic field strength.

The 60 MHz proton NMR spectrum of ethoxyacetic acid (EOAA) in aqueous solution is given in Fig. 1-4. The peak assignments and chemical shifts using DSS as a reference are as follows:

$$\begin{array}{llll} & CH_3\text{—}CH_2\text{—}O\text{—}CH_2\text{—}COO^- & & \\ & c & b & a \\ \delta & 1.19 & 3.56 & 3.89 \\ \tau & 8.81 & 6.44 & 6.11 \end{array}$$

The chemical shift value for any resonance that is split into a multiplet is properly determined only by mathematical analysis of the spectral data. However, the chemical shift may be approximated by the center of the multiplet when the difference in chemical shift between the two coupled resonances (in Hz) is much greater than the spin–spin coupling constant; it is usually sufficient if $\Delta\delta \geq 10J$. Roberts (10) goes into the details of the analysis of complex splitting patterns. In the ethoxyacetate spectrum, methyl resonance c is a triplet and methylene resonance b is a quartet. The spacing between adjacent components in the triplet is the same as the spacing between adjacent components in the quartet. That spacing (in Hz) is the coupling constant $J_{bc} = 7.1$ Hz.

These multiplets are caused by an interaction between neighboring nuclear spins. For liquids, direct dipole–dipole (through space) interactions are averaged to zero by molecular tumbling. The most plausible explanation

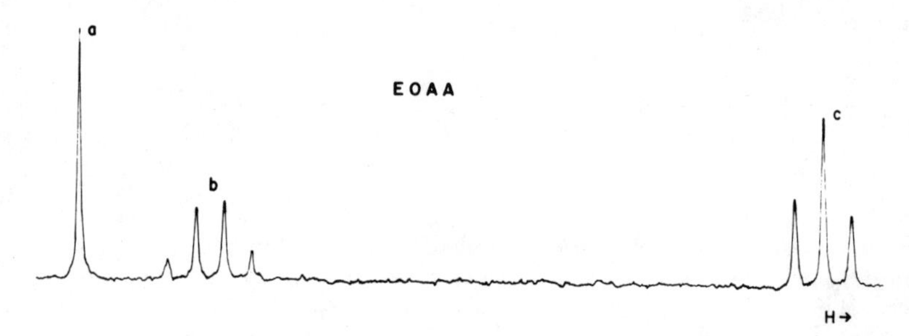

Fig. 1-4. 60 MHz proton spectrum of 0.20 M ethoxyacetic acid (EOAA) in aqueous solution.

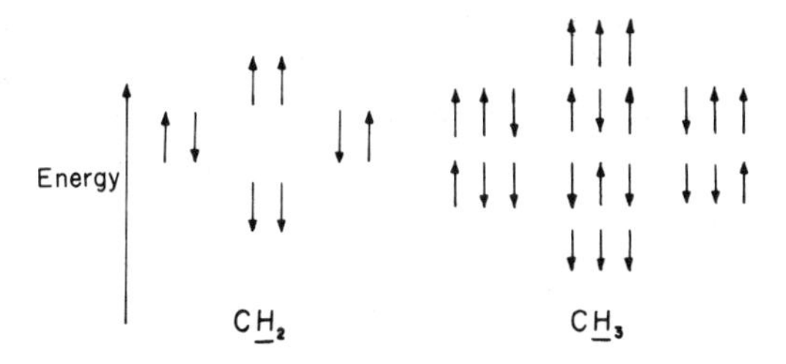

FIG. 1-5. Possible orientations of the proton spins in the ethyl group.

for the observed splittings, therefore, is that the nuclear spin interactions are mediated by bonding electrons. In this manner, a nuclear spin couples with a nearby electron spin, which then couples with other electron spins, which in turn couple with other nuclear spins. The magnitude of the coupling attenuates rapidly with distance but is also dependent on the chemical bond type, bond angle, and nuclear spin. It will be noted that the J coupling mechanism does not require a magnetic field. Therefore, the magnitude of the coupling constant will be independent of the applied magnetic field H_0. The coupling constant in a 22.6 MHz ^{13}C NMR spectrum will be the same as in a 15.1 MHz ^{13}C spectrum.

First-order multiplicities can be predicted from the following rules when the difference in chemical shift between two resonances is at least ten times as great as the coupling constant, i.e., $\Delta\delta \gtrsim 10J$.

1. An unperturbed nucleus or equivalent set of nuclei will give a singlet. An example is the *a* resonance of ethoxyacetate in Fig. 1-4. It should be noted that there is no coupling between methylene protons *a* and methylene protons *b* across the oxygen atom.

2. Multiplet splitting of a resonance from a set of equivalent nuclei will be determined by nearby sets of equivalent nuclei. Each set of equivalent nuclei will split the resonance of neighboring nuclei into $(2nI + 1)$ components, where n is the number of nuclei in the equivalent set and I is the nuclear spin of the coupling nuclei.

The possible spin orientations of the ethyl protons in ethoxyacetate, leading to the quartet and triplet in Fig. 1-4, can be examined. The possible orientations are depicted in Fig. 1-5. It can be seen that there are three possible energy configurations for the methylene protons. Each of these three configurations will couple with the methyl protons, giving rise to the methyl triplet. Likewise, the four possible energy configurations for the methyl protons will give rise to the methylene quartet.

If there are two sets of nuclei causing splitting of a third set, the number of components in the multiplet will just be $(2n_1I_1 + 1)(2n_2I_2 + 1)$.

3. The components of a multiplet are evenly spaced and have the same spacing as the multiplet resonance of the set of protons causing the splitting. The spacing is equal to J, the spin–spin coupling constant, in Hertz.

4. The intensities of the components of a multiplet are proportional to the coefficients in the expansion of

$$(1 + x + \ldots + x^{2I})^n \tag{1-13}$$

where, again, n is the number of nuclei and I is the nuclear spin of the nuclei causing the splitting. In the case of spin $\frac{1}{2}$ nuclei, possible multiplets and intensity ratios are doublet, 1:1; triplet 1:2:1; quartet 1:3:3:1; quintet, 1:4:6:4:1; sextet, 1:5:10:10:5:1, etc. It will be noticed from Fig. 1-5 that the probabilities for occurrence of the energy configurations in methylene and methyl resonances is the same as the intensity ratio predicted by Eq. 1-13, i.e., 1:2:1 and 1:3:3:1. If first-order splitting holds, then the components and their intensities will be symmetric about the midpoint of the multiplet.

The following multiplets and intensities for the following compounds in solution or liquid might be expected. $^{1}H(I = \frac{1}{2})$ spectra: acetaldehyde, a doublet (1:1) and a quartet (1:3:3:1); tetramethylammonium ion, a triplet (1:1:1) from ^{14}N ($I = 1$) splitting; and isopropyl ether, a doublet (1:1) and a septet (1:6:15:20:15:6:1). ^{13}C ($I = \frac{1}{2}$) spectra: benzene, a doublet (1:1); and dimethyl ether, a quartet (1:3:3:1). ^{31}P ($I = \frac{1}{2}$) spectra: tripolyphosphate, a doublet (1:1) and a triplet (1:2:1); and $P(OCH_3)_3$ a ten-line multiplet (1:9:36:84:126:126:84:36:9:1).

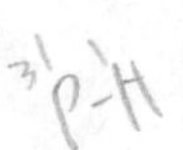

The simple rules for the number and intensity of components in a multiplet do not hold if the chemical shift differences are not much greater than the coupling constant. When $\Delta\delta \sim J$, second-order splittings can occur with the appearance of additional peaks, uneven spacings between peaks, and drastically altered peak intensity ratios. Increasing magnetic field strength, as discussed in Section 1.4, will increase the chemical shift differences and, in many cases, will remove second-order splittings.

One further complication (or simplification) may arise. One might predict that the methyl proton magnetic resonance of methanol would be a doublet resulting from coupling to the hydroxyl proton. However, in the presence of H^+ or small amounts of water, the resonances are sharp singlets. The reason for this is the rapid chemical exchange of hydroxyl protons between different hydroxyls; the H^+ or water catalyzes the exchange. If the rate of chemical exchange is slow (as with very pure methanol at low temperature) the expected multiplicity in the methyl and hydroxyl proton resonance is observed. At intermediate exchange rates, broad peaks will result. If the proton NMR resonance of an aqueous methanol solution is recorded, only two singlets are observed. One is the methyl resonance and the other is caused by the protons which are rapidly jumping among the methanol hydroxyls and the water molecules. This observation of rapid exchange is common with several groups. In general —NH_2, >N—H, —OH, —COOH, —CHO, and —SH protons will not be observed in aqueous solution because rapid exchange will bury these resonances in the water resonance. For example, aqueous serine solutions will show no separate resonances for the hydroxyl, the amino, or the carboxyl protons. This is not universally true, however. Proton resonances from certain indole NH protons of tryptophan and certain amide protons of amino acid residues hidden in the interior hydrophobic regions of proteins in aqueous solutions can be observed because these protons are unable to exchange with H_2O.

Double irradiation can be used to help unravel a complicated spectrum, as can isotopic substitution (say, deuterium for proton). With the double resonance method, a second strong rf field is applied to the sample at the resonance frequency of the nuclei that are causing the splitting. This second rf field, H_2 will saturate the one resonance and cause a collapse of the multiplet being monitored with the H_1 field. This collapse of the multiplet into a singlet is termed "spin decoupling." If the H_2 field is not too strong, the multiplet will be perturbed but not completely collapsed; this is "spin tickling." Either homonuclear (like nuclei) or heteronuclear (unlike nuclei) decoupling can be done.

1.6. General Features of the NMR Spectrum

The various nuclear magnetic resonance spectral parameters are summarized in Fig. 1-6. The NMR spectrum is composed of bell-shaped

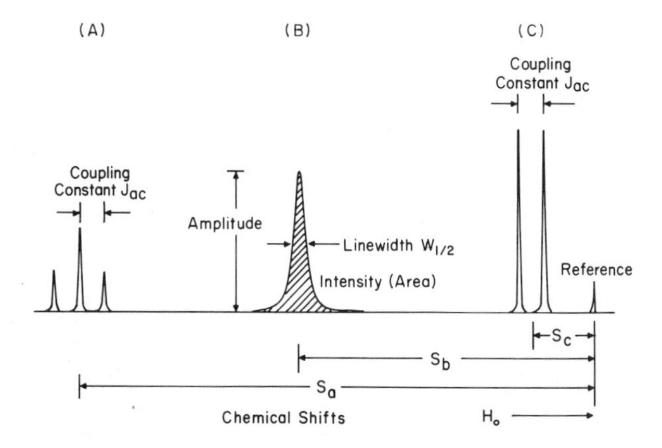

FIG. 1-6. Nuclear magnetic resonance spectral parameters.

resonance peaks with the following characteristics: (*1*) chemical shift—denotes peak position relative to a reference, (*2*) amplitude—height of peak in arbitrary units, (*3*) intensity—integrated area under the peak, (*4*) linewidth—measure of the width of the peak at half maximal amplitude (in Hz), and (*5*) spin–spin coupling constant—separation (in Hz) between the peaks in a multiplet.

References

1. A. Abragam, "Principles of Nuclear Magnetism." Oxford Univ. Press, London and New York, 1961.
2. C. P. Slichter, "Principles of Magnetic Resonance." Harper, New York, 1963.
3. A. Carrington and A. D. McLachlan, "Introduction to Magnetic Resonance." Harper, New York, 1967.
4. E. D. Becker, "High Resolution NMR." Academic Press, New York, 1969.
5. A. Kowalsky and M. Cohn, *Annu. Rev. Biochem.* **33,** 481 (1964).
6. M. Cohn, *Quart. Rev. Biophys.* **3,** 61 (1970).
7. O. Jardetzky, *Advan. Chem. Phys.* **7,** 499 (1964).
8. G. C. K. Roberts and O. Jardetzky, *Advan. Protein Chem.* **24,** 447 (1970).
9. C. C. McDonald and W. D. Phillips, *in* "Biological Macromolecules" (G. Fasman and S. N. Timasheff, eds.), Vol. IV, p. 1. Dekker, New York, 1970.
10. J. D. Roberts, "An Introduction to the Analysis of Spin-Spin Splitting in High-Resolution Nuclear Magnetic Resonance Spectra." Benjamin, New York, 1962.

CHAPTER 2

PRINCIPLES OF NUCLEAR MAGNETIC RESONANCE

There are basically two ways of looking at the nuclear magnetic resonance phenomenon. The first was the basis for the discussion in Chapter 1—essentially a quantum mechanical description of discrete nuclear energy levels, the nuclear populations of which can be described by the Boltzmann distribution (Eq. 1–7). Another way of viewing the nuclear magnetic resonance phenomenon is to consider it as a forced precession of the nuclear magnetization in the stationary magnetic field, H_0, which is perturbed by application of radiofrequency field, H_1. This classical approach is largely owing to Felix Bloch (1) and, somewhat surprisingly, describes NMR experiments quite accurately, especially for liquid samples.

2.1. Classical Description—Bloch Equations

Early in the history of NMR, Bloch (1) provided a classical description of the NMR experiment, deriving a set of very useful equations describing a group of nuclei in a magnetic field simply from phenomenological considerations. The rationale of the approach is now presented. For a more rigorous treatment, Slichter's text (2) may be consulted.

To simplify matters, all magnetic nuclei in a system will be assumed to be identical and to have spin I = ½. According to the classical theory of electromagnetism, a magnetic moment $\vec{\mu}$ in a field $\vec{H}_0$ will experience a torque

$$\vec{T} = \vec{\mu} \times \vec{H}_0 = d\vec{J}'/dt \tag{2-1}$$

where $\vec{J}'$ is the angular momentum. Because $\vec{\mu} = \gamma\vec{J}'$, the motion of the magnetic moment is given by

$$\frac{d\vec{\mu}}{dt} = \gamma\vec{\mu} \times \vec{H}_0 \tag{2-2}$$

which describes the precession of the nuclear magnetic moment about the z axis. The z axis is defined as the direction of the $\vec{H}_0$ field. This precession causes $\vec{\mu}$ to generate a cone about the z axis, as shown in Fig. 2-1. The frequency of the precessional motion

$$\nu_0 = \gamma H_0/2\pi \tag{2-3}$$

is called the "Larmor frequency"; it is, in fact, the NMR frequency for that nucleus as will be seen. The Larmor angular frequency $\omega_0 = 2\pi\nu_0$.

We consider now the application of a small rf field $\vec{H}_1$ rotating in the plane perpendicular to $\vec{H}_0$ with angular frequency ω as shown in Fig. 2-1. The $\vec{H}_1$ field will produce a torque

$$\vec{\mu} \times \vec{H}_1$$

on the moment $\vec{\mu}$, tending to tilt $\vec{\mu}$ toward the plane perpendicular to H_0. If $\omega \neq \omega_0$, only a small periodic perturbation will be exerted on $\vec{\mu}$ as $\vec{H}_1$

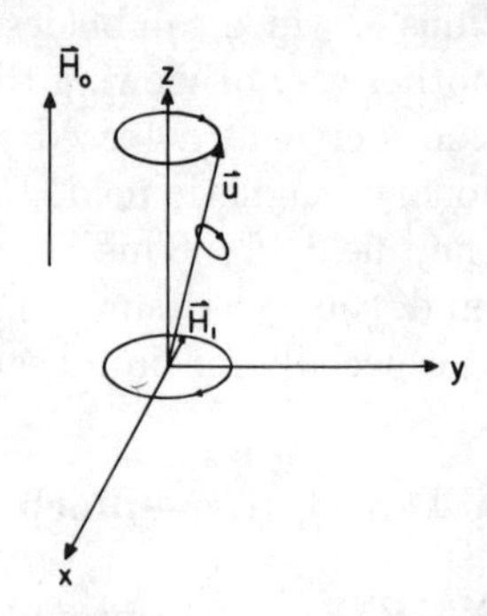

FIG. 2-1. Precession of magnetic moment $\vec{\mu}$ in a stationary magnetic field $\vec{H}_0$ rotating at the Larmor angular frequency ω_0. $\vec{H}_1$ is a small radiofrequency field rotating with angular frequency ω.

quickly "loses step" with $\vec{\mu}$. This results in a small precessing motion of $\vec{\mu}$ about the original $\vec{\mu}$ direction as shown in Fig. 2-1. As ω approaches ω_0, the torque from the $\vec{H}_1$ field tilting $\vec{\mu}$ toward the plane perpendicular to $\vec{H}_0$ will be exerted for a longer time. This will give a larger amplitude of precession. At $\omega = \omega_0$, the effect of the torque will be the greatest and the amplitude of the precession will be the largest. In this process, energy is provided to the spin at the expense of the rf field. If the frequency ω of the rotation of $\vec{H}_1$ is varied through the Larmor frequency, energy is absorbed by the spin system and the magnetic resonance phenomenon is observed. In practice an oscillating rf field is used. However, the oscillating field can be broken into two components rotating with equal angular velocities in opposite directions. Because one component will rotate in the same direction as the precession of the moment, that component will affect the moment. The component rotating in the opposite direction will have a negligible effect.

In this chapter and in Chapter 1, the properties of those nuclei of interest for NMR were discussed, namely, those possessing an angular momentum $\vec{J}'$ and a magnetic moment $\vec{\mu}$. In practice, a large number of nuclei are observed rather than a single nucleus. This ensemble of nuclei can be treated on a macroscopic scale. The effects on a macroscopic sample of a stationary magnetic field $\vec{H}_0$, a time-dependent magnetic field $\vec{H}_1$, and relaxation phenomena will be considered here.

The macroscopic nuclear magnetization

$$\vec{M} = (M_x, M_y, M_z) \tag{2-4}$$

of a sample is defined as the magnetic moment per unit volume. In the absence of an H_1 field at the resonance condition, the x and y components of the individual spins are randomly oriented in the xy plane such that $M_x + M_y = 0$. Consequently, the macroscopic magnetization is the sum of the z components of the individual nuclear spins:

$$\vec{M} = M_z\vec{k} = (\sum \mu_z)\vec{k} \tag{2-5}$$

where the z axis is defined by the direction of the stationary field $\vec{H}_0$ and the summation is over a unit volume of the sample ($\vec{k}$ is a unit vector oriented along the z axis). For spin $I = \frac{1}{2}$ nuclei, the individual moments will align themselves, as shown in Fig. 2-2, such that their z components will be either parallel or antiparallel to the direction of $\vec{H}_0$, corresponding to two energy states. An $\vec{M}$ exists because there are a greater number of spins in one state (parallel to $\vec{H}_0$), as described by the Boltzmann distribution and discussed in Chapter 1.

The equation of motion for the macroscopic magnetization in a homo-

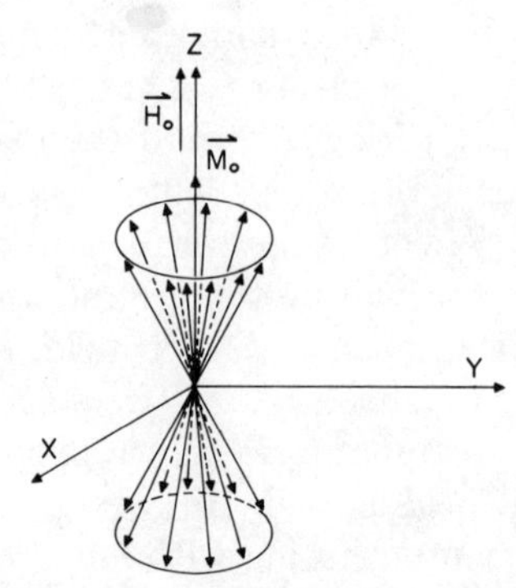

FIG. 2-2. Orientation and precession of nuclear magnetic moments in an ensemble of nuclei with spin $I = \frac{1}{2}$, at thermal equilibrium in a stationary magnetic field $\vec{H}_0$. Direction of precession indicates $\gamma > 0$.

geneous field may be written as

$$d\vec{M}/dt = \gamma(\vec{M} \times \vec{H}) \tag{2-6}$$

because the macroscopic moment is obtained as the sum of the individual moments. $\vec{H}$ is composed of $\vec{H}_0$ and $\vec{H}_1$.

Equation 2-6 describes the precession in the $\vec{H}_0$ field and the absorption of energy by application of $\vec{H}_1$. However, it does not account for the relaxation processes that tend to redistribute the energy absorbed by the system of spins. In a stationary field $H_0 = H_z$ the z component of $\vec{M}$, M_z, will try to take on its equilibrium value M_0. Analogously to Eq. 1-8, the relaxation of M_z back to its equilibrium value following absorption of rf energy can generally be described by an exponential decay:

$$dM_z/dt = -(M_z - M_0)/T_1 \tag{2-7}$$

where T_1 is the spin–lattice relaxation time. T_1 is also often referred to as the "longitudinal relaxation time," being the relaxation of the component of $\vec{M}$ in the $\vec{H}_0$ direction.

Absorption of rf energy can also give rise to x and y components of $\vec{M}$. These components will rotate about the z axis at the Larmor frequency. However, because of fluctuations and slight differences in local magnetic fields, the transverse components of the individual nuclear spins will get out of phase. Because M_x and M_y are sums of the individual nuclei's x and y components, M_x and M_y will decay to zero. This decay may be described by

$$\frac{dM_x}{dt} = \frac{-M_x}{T_2} \tag{2-8}$$

and

$$\frac{dM_y}{dt} = \frac{-M_y}{T_2} \tag{2-9}$$

where T_2 is the transverse relaxation time. As noted in Chapter 1, T_2 is also called the "spin–spin relaxation time" and may be defined as the time constant for the exponential decay to zero of the magnetization in the xy plane. For the following discussion, inhomogeneities in the magnetic field of the laboratory magnet will be considered negligible, so the decay constant will be a true T_2. The decay constant will be smaller if the magnetic field is not completely homogeneous, and the decay constant caused by magnetic field inhomogeneity will be designated T_2^*.

Combining the relaxation processes with the motions produced by torque, we have the Bloch equation:

$$\frac{d\vec{M}}{dt} = \underbrace{\gamma(\vec{M} \times \vec{H}_0)}_{\text{precession}} + \underbrace{\gamma(\vec{M} \times \vec{H}_1)}_{\substack{\text{perturbation}\\ \text{(enegry}\\ \text{absorption)}}} - \underbrace{\frac{(\vec{i}M_x + \vec{j}M_y)}{T_2} - \frac{\vec{k}(M_z - M_0)}{T_1}}_{\text{relaxation}} \tag{2-10}$$

where $\vec{i}$, $\vec{j}$, and $\vec{k}$ are the unit vectors in the laboratory (x, y, z) coordinate system.

2.1.1. Bloch Equations in the Rotating Frame

For greater understanding of the implications of the Bloch equations, it is necessary to separate the equation into parts for each of its components M_x, M_y, and M_z. Each of the resulting three equations of motion are considerably simplified if the laboratory (x, y, z) coordinates are transformed into a rotating coordinate system (x', y', z) (see Fig. 2-3). In this rotating coordinate system, the x' and y' axes are rotating about the z axis at a

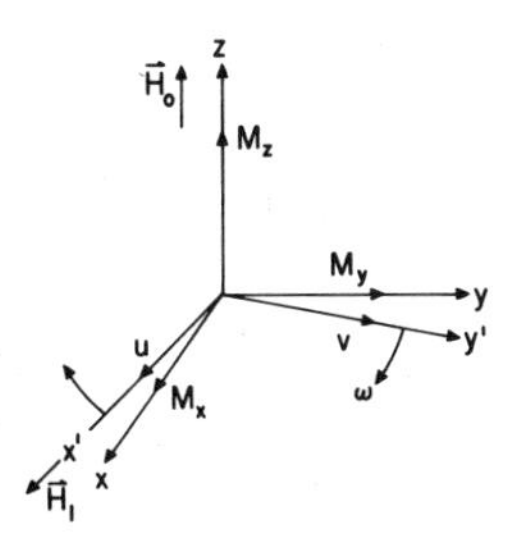

Fig. 2-3. Axes and components of the magnetization in the laboratory coordinate system (x, y, z) and the rotating coordinate system (x', y', z) rotating about the z axis with frequency ω.

frequency $\omega = -\gamma H_1$, where H_1 is the magnitude of the applied rf field $\vec{H}_1$. For NMR interest, ω will be close to ω_0, the Larmor frequency. In the rotating frame, u and v will be the transverse magnetization components along the x' and y' axes, respectively, and M_z will remain as the longitudinal component.

The equations of motion for the components in the rotating frame may be written as

$$\frac{dM_z}{dt} = \gamma H_1 v - \frac{(M_z - M_0)}{T_1} \tag{2-11a}$$

$$\frac{dv}{dt} = (\omega_0 - \omega)u - \gamma H_1 M_z - \frac{v}{T_2} \tag{2-11b}$$

$$\frac{du}{dt} = -(\omega_0 - \omega)v - \frac{u}{T_2} \tag{2-11c}$$

2.1.2. Steady-State Solution of the Bloch Equations

In general there are two methods of doing NMR experiments. One is to slowly sweep through the NMR resonance with a small H_1 field. The other is to apply prescribed sequences of pulses of a strong H_1 field at the resonance frequency. We will deal first with the sweeping or continuous wave (CW) method. In this method we make the approximation that we are sweeping sufficiently slowly so the magnetization is never very far from its steady state. Making the steady state approximation that

$$\frac{dM_z}{dt} = \frac{du}{dt} = \frac{dv}{dt} = 0$$

we solve for the components in Eq. 2-11 and obtain, for the rotating coordinate system

$$M_z = M_0 \frac{1 + T_2^2(\omega_0 - \omega)^2}{1 + T_2^2(\omega_0 - \omega)^2 + \gamma^2 H_1^2 T_1 T_2} \tag{2-12a}$$

$$v = M_0 \frac{\gamma H_1 T_2}{1 + T_2^2(\omega_0 - \omega)^2 + \gamma^2 H_1^2 T_1 T_2} \tag{2-12b}$$

$$u = M_0 \frac{\gamma H_1 T_2^2(\omega_0 - \omega)}{1 + T_2^2(\omega_0 - \omega)^2 + \gamma^2 H_1^2 T_1 T_2} \tag{2-12c}$$

Because $\vec{H}_1$ is applied along the x' axis, u is the in-phase component and v is the out-of-phase component of the transverse magnetization.

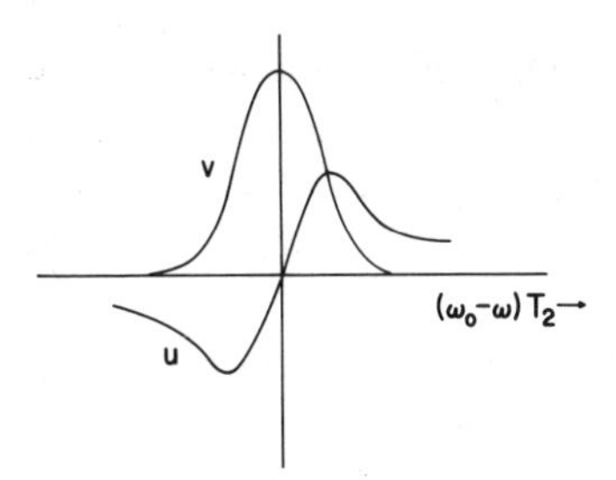

FIG. 2-4. Absorption or out-of-phase (v) and dispersion or in-phase (u) lineshapes from the Bloch equations.

The voltage induced in a coil the axis of which is in the fixed y direction will vary sinusoidally, giving an absorption signal proportional to v and a dispersion signal proportional to u. These lineshapes in the absence of saturation effects are illustrated in Fig. 2-4. This process for generating a voltage in a coil is termed "nuclear induction." In most NMR instruments, either the absorption or dispersion signal, or a combination of the two, may be observed.

A. *Lineshape and Saturation*

For a small $\vec{H}_1$ field, i.e., $\gamma^2 H_1^2 T_1 T_2 \ll 1$, the absorption v may be written as

$$v = (\gamma H_1 M_0) \frac{T_2}{1 + T_2^2(\omega_0 - \omega)^2} \tag{2-13a}$$

or

$$v = \tfrac{1}{2}(\gamma H_1 M_0) g(\nu) \tag{2-13b}$$

where

$$g(\nu) = \frac{2T_2}{1 + 4\pi^2 T_2^2(\nu_0 - \nu)^2} \tag{2-14}$$

is the lineshape function describing the Lorentzian lineshape of the absorption signal in Fig. 2-4. The fact that the linewidth at half-maximal amplitude is related to T_2 by Eq. 1-9

$$T_2 = 1/\pi W_{1/2}$$

is a direct consequence of the Lorentzian lineshape. In most liquids and solutions, the Lorentzian lineshape provides a satisfactory description of the experimentally observed resonance peak. However, NMR peaks from nuclei in very viscous solutions or in membranes may not be accurately described by the Lorentzian lineshape. In solids, a Gaussian curve is often observed for the lineshape:

$$g(\nu)_{\text{Gaussian}} = \frac{T_2}{(2\pi)^{1/2}} \exp[-2\pi^2 T_2^2(\nu_0 - \nu)^2] \tag{2-15}$$

In the case of a larger H_1 field, saturation effects may come into play. Increase in the saturation factor

$$1 + \gamma^2 H_1^2 T_1 T_2$$

will result in broader lines. Following the onset of saturation, the peak amplitude will initially continue to increase with increasing H_1 power, but it will eventually pass through a maximum and will then decrease with further increases in H_1. With continued increase in H_1 power, the peak will ultimately become saturated and disappear. Until disappearance, however, the peak will maintain a Lorentzian lineshape but the measured "T_2" from the linewidth will be too small according to

$$T_{2\,\text{meas}} = T_2/(1 + \gamma^2 H_1^2 T_1 T_2)^{1/2} \tag{2-16}$$

Saturation is obviously to be avoided for accurate T_2 measurements from the linewidth.

It is interesting to note that as the spin–lattice relaxation time decreases, the greater H_1 must be to achieve saturation. As the relaxation processes become more efficient, a greater H_1 power is required to saturate.

It will be observed from Eqs. 2-12b and 2-12c that the dispersion signal will maintain a finite signal in a large H_1 field even when the absorption signal has become saturated. As will be seen, the sensitivity of detection is improved by using as large an H_1 field as possible without saturating. It is therefore often advantageous to use large H_1 fields and to observe the dispersion signal when searching for the resonance of dilute samples having unknown relaxation times.

B. *Ringing*

If a spectrum is swept rapidly, i.e., if the H_0 field is scanned quickly, some distortion of the resonance peak may occur, and the slow-passage results presented in the previous section are not adequate. The phenomenon observed is called "ringing" (or "wiggles") and is evident after the magnetic field has passed through the Larmor frequency. An example of ringing is shown in the spectrum in Fig. 2-5. The ringing decays exponentially with time. The reason for ringing can be explained in the following way. At resonance, a nuclear spin system will have a transverse component of the magnetic moment. With slow passage of the H_0 sweep, this transverse

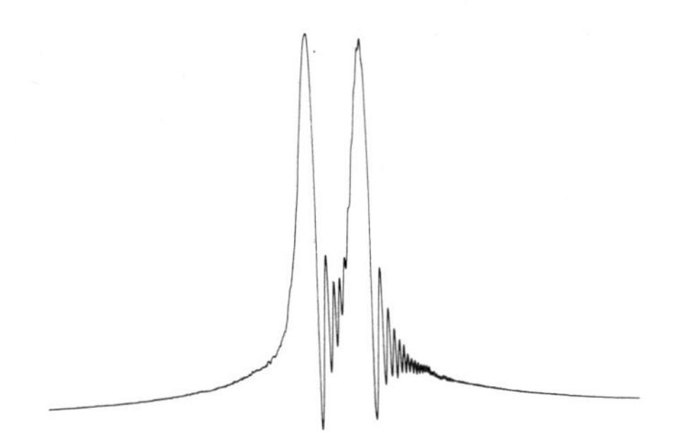

FIG. 2-5. Ringing (wiggles) on the acetaldehyde doublet.

component will return adiabatically to its equilibrium value (i.e., zero) after H_0 has passed through the resonance condition ($\nu_0 = \gamma H_0/2\pi$). However, if H_0 is swept quickly, the transverse component will still be finite even after H_0 has gone completely past the resonance condition. The remaining transverse component rotating at the Larmor frequency will beat against the $\vec{H}_1$ field, alternately going in and out of phase with $\vec{H}_1$ as $\vec{H}_1$ changes its rate of rotation. The beat pattern, or ringing, results.

Ringing will persist as long as the transverse magnetization is present. The transverse magnetization decays with a time constant T_2, the transverse relaxation time. So, in a very homogeneous field, the exponential decay of the ringing is a measure of T_2. With resonance lines having a short T_2 (broad lines) ringing will not be seen. Conversely, a strong symmetrical ringing pattern on a sharp resonance line with long T_2 is a sign of good magnetic field homogeneity. Inhomogeneity in the magnetic field will shorten or eliminate the ringing.

C. *Amplitude and Intensity*

The voltage V induced in a receiver coil the axis of which is in the fixed y direction is proportional to dM_y/dt, viz.,

$$V = K(dM_y/dt) \tag{2-17}$$

where the constant K is dependent on the geometry and sample filling factor of the coil.

For fixed T_1, T_2, M_0, and H_1, the peak value for the v-mode signal at $\omega = \omega_0$ under conditions of slow passage is

$$V_p \propto \frac{\gamma H_1 M_0 T_2}{1 + \gamma^2 H_1{}^2 T_1 T_2} \tag{2-18}$$

Under nonsaturating conditions, i.e., small H_1, $V_p \propto H_1 M_0 T_2$. The peak height will then increase with H_1 until saturation effects begin to enter. The peak height is proportional to M_0 which, resulting from the sum of individual nuclear spins, is proportional to the number of nuclei in the sample. Comparison of peak heights for purposes of quantitative analysis may be made if each has the same value for T_2.

A better method for quantitative comparison of the number of nuclei contributing to various resonance peaks is to compare the integrated areas under the peaks. The area under the peak is proportional to the integral

$$A = \int V \, dH \propto H_1 M_0 \tag{2-19}$$

which is independent of T_2. The intensities obtained by peak integration are often used for quantitative analysis of a sample's composition. The peak integrals are used even more often for spectral interpretation. For example, the hydroxyl, methylene, and methyl peaks of ethanol will have integrals in the ratio 1:2:3, corresponding to the ratio of the number of protons contributing to each peak.

2.1.3. Pulsed NMR

In the previous sections, aspects of the steady-state solution to the Bloch equations were discussed. The steady-state approximation with a small H_1 field is appropriate for continuous wave (CW) NMR in which the field or frequency is swept through resonance. However, a second experimental method has proved to be quite useful. With this method, the H_1 field is applied in very short, intense pulses. The intense pulse will rotate the macroscopic magnetic moment through an angle dependent on the length of time the pulse is applied. When the pulse is finished, the spin system will be in a nonequilibrium state. Relaxation processes will then control the behavior of the spin system without the nuclear spins being perturbed by the H_1 field. A transient signal called a "free induction decay" (FID) may be observed following application of the pulse. The amplitude $g(t)$ of the free induction decay may be measured as a function of time. This time-domain function $g(t)$ inherently contains all the information the frequency-domain spectrum contains. The time-domain NMR function and frequency-domain spectrum are related by the Fourier transform. The FID may therefore be Fourier transformed for presentation of the more familiar frequency spectrum. However, use of a given sequence of pulses can provide more information than is obtainable from the frequency spectrum. The most prevalent use of pulsed NMR, aside from Fourier transforming, is to measure the spin–lattice and spin–spin relaxation times. Pulsed NMR provides the most accurate methods for measuring T_1 and T_2. Greater detail about pulsed and Fourier transform NMR than that presented here may be found in the monograph by Farrar and Becker (3). Recent developments have been reviewed by Farrar *et al.* (4).

A. *The Pulsed NMR Experiment*

A collection of nuclei initially at equilibrium will have a net magnetic moment $\vec{M}$, with magnitude M_0, aligned along the z axis in the direction of the $\vec{H}_0$ field. The magnetization will be nutated away from the z axis through an angle θ (in radians) by application of an rf field pulse at the Larmor frequency ω_0. The nutation angle θ is determined by the gyromagnetic ratio of the nucleus, the amplitude H_1 of the rf pulse, and the length of time t_w the rf field is applied:

$$\theta = \gamma H_1 t_w \tag{2-20}$$

Figure 2-6 illustrates the nutation of $\vec{M}$ on application of the rf pulse for a sufficient time t_w for $\vec{M}$ to rotate 90° ($\theta = \pi/2$). That pulse is referred to as a 90° or $\pi/2$ pulse. Application of the H_1 field for a time twice as long ($\theta = \pi$) will result in inversion of $\vec{M}$. Such a pulse is called a 180° or π pulse.

When the pulse width t_w is very short compared to T_1 or T_2, essentially no relaxation occurs during the pulse. Typically t_w is on the order of 1–200 μsec, depending on the nucleus and the power available from the NMR instrument. The induction signal (free induction decay) that remains after the rf pulse is completed decays exponentially—$\exp[t/T_2^*]$—where T_2^* may equal T_2 in a sufficiently homogeneous magnetic field or may just be a time constant characterizing the relaxation caused by magnetic field inhomogeneity. Figure 2-7 shows, as an example, the free induction decay for the ^{23}Na NMR signal for a NaCl solution.

Transient phenomena using sequences of rf pulses can be used for meas-

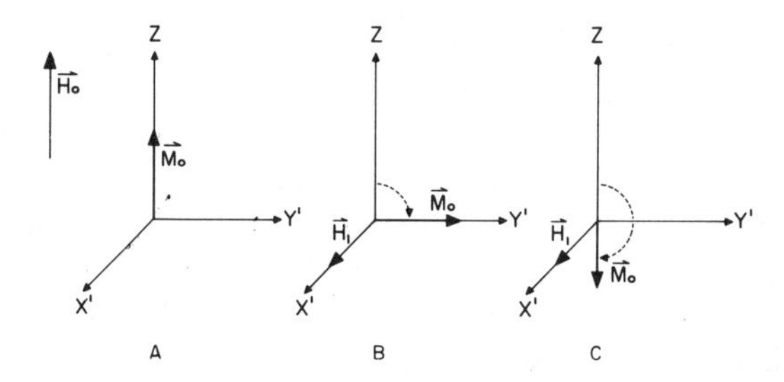

FIG. 2-6. Rotation of macroscopic magnetic moment $\vec{M}$ (with magnitude M_0) in the rotating coordinate system when $\omega = \omega_0$. (A) Spin system at equilibrium in $\vec{H}_0$ field. (B) Application of a 90° H_1 pulse ($\theta = \pi/2$). (C) Application of a 180° H_1 pulse ($\theta = \pi$).

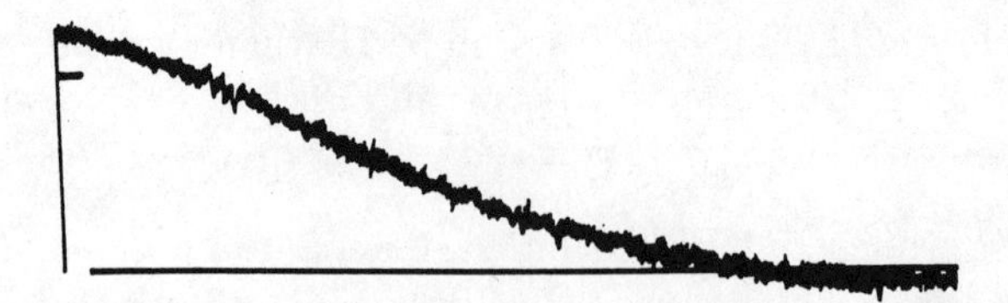

FIG. 2-7. ^{23}Na NMR free induction decay for a saturated aqueous NaCl solution. The position of the gate for monitoring the amplitude of the FID is illustrated.

uring T_1 or T_2. The pulses are used both as a method for perturbing the spin system from equilibrium and as a means for detecting the magnetization. Because components of $\vec{M}$ are detected along the y' axis, a 90° pulse ($\theta = \pi/2$) will nutate any z component M_z so it will be detected.

B. T_1 *Experiment*

The spin–lattice relaxation time is measured by following the return of the magnetization to its equilibrium value after it has been perturbed by an rf field at the resonance frequency. The pulsed NMR methods generally use a 90°–τ–90° or 180°–τ–90° two-pulse sequence to measure T_1 (5). Using the latter sequence for illustration, the processes for this sequence, considered in the rotating coordinate system, are the following. The magnetization in the $\vec{H}_0$ field is initially at its equilibrium value M_0 and oriented along the z axis. Application of an intense 180° rf pulse will nutate the magnetization into the negative z direction while maintaining a magnitude $-M_0$. As the magnetization relaxes back to its equilibrium value following the 180° pulse, the value of M_z at any time will be described by the Bloch Eq. 2-11a in the absence of an H_1 field,

$$\frac{dM_z}{dt} = \frac{-(M_z - M_0)}{T_1} \qquad (2\text{-}21)$$

Application of a 90° pulse at some time τ after the 180° pulse will cause M_z to be nutated through $\pi/2$ radians so the magnetization may be detected along the y' axis. By measuring the magnitude of the free induction decay following the 90° pulse as a function of τ, the time between pulses, the spin–lattice relaxation time may be determined.

A multiple exposure photograph of the 180°–τ–90° sequence for different values of τ is presented in Fig. 2-8a for the ^{23}Na NMR signal from a solution of NaCl. The rf pulses are too fast to be observed. The signal amplitude $M_z(\tau)$ following the second pulse describes the relaxation process by the equation

$$M_z(\tau) = M_0[1 - (1 - \cos\theta)e^{-\tau/T_1}] \qquad (2\text{-}22)$$

which may be developed from the Bloch equation. For the 180°–τ–90° pulse sequence where $\theta = 180°$, Eq. 2-22 becomes

$$M_z(\tau) = M_0(1 - 2e^{-\tau/T_1}) \tag{2-23}$$

Other pulse sequences have been used to measure T_1. Some of these will be considered briefly in Chapter 5.

C. T_2 *Experiment*

The spin–spin or transverse relaxation time is measured by studying the decay of the transverse component of the magnetization—the component in the xy plane:

$$\vec{M}_\perp = \vec{i}M_x + \vec{j}M_y$$

A commonly used pulse method for measuring T_2 utilizes a 90°–τ–180° sequence (6). The experimental procedure for this sequence, considered in the rotating coordinate system, is the following. Figure 2-9 illustrates the processes. Initially the magnetization is at equilibrium parallel to the direction of the stationary magnetic field, $\vec{H}_0$, which defines the z axis (A). The magnetic moment is then nutated by applying a 90° pulse so that, after the pulse, all of the individual magnetic spins are aligned along the y' axis, giving a nonzero value to the transverse component of the magnetization (B). This component commences to decay because of T_2 relaxation

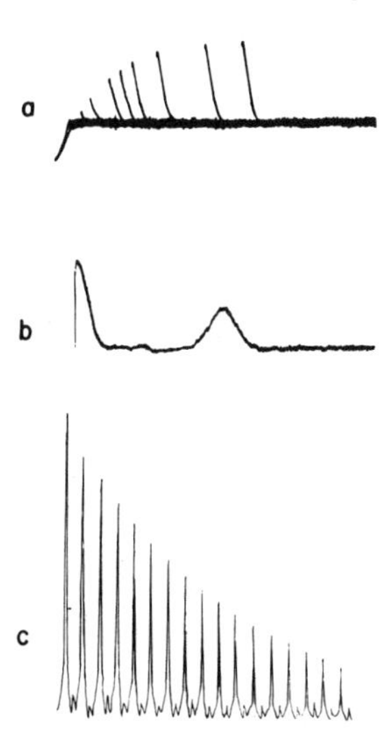

FIG. 2-8. (a) 180°–τ–90° sequence for various values of τ (multiple exposure photograph). (b) 90°–τ–180° sequence showing the spin echo. (c) Carr-Purcell sequence showing a train of echoes.

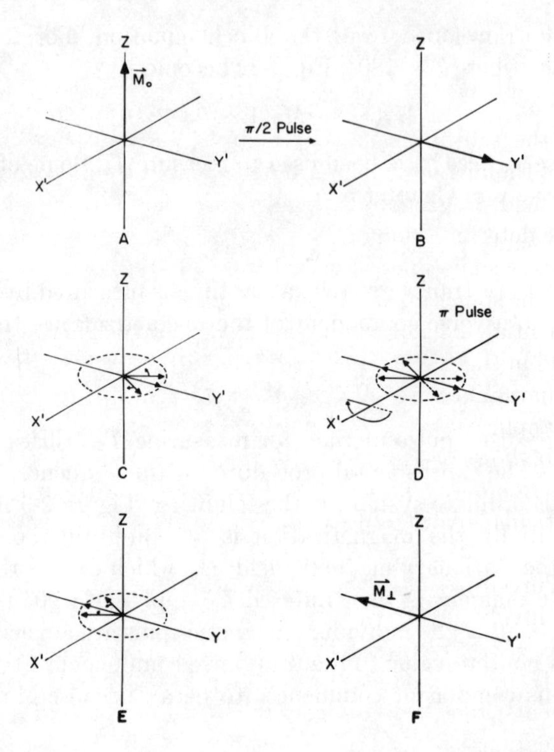

FIG. 2-9. The 90°–τ–180° T_2 experiment with formation of a spin echo.

processes and inhomogeneities in the magnetic field. It is preferred that the inhomogeneity effect dominate, so the magnetic field is intentionally made slightly inhomogeneous if it does not already dominate. This decay caused by magnetic field inhomogeneity is characterized by a decay constant T_2*. The field inhomogeneity causes the individual magnetic spins to have different precessional frequencies. This results in a loss of phase coherence of the individual spins and they will fan out in the $x'y'$ plane (C). If, after a time τ, a 180° pulse is applied, the spins will be flipped 180° (D). Because the individual spins are still precessing at their same individual frequencies as before (E), the spins will all be in phase again at time 2τ with the orientation in the negative y' direction (F). This will result in an induction signal being detected in the absence of an applied rf field; this signal is referred to as a "spin echo." Following the spin echo, the individual spins will lose phase and fan out again in the $x'y'$ plane. The amplitude of the spin echo is a measure of the magnitude of $\overrightarrow{M}_\perp$. The echo amplitude may be measured for different values of τ, the time between the two pulses, as a means of determining T_2. The height of the echo is given by

$$M_\perp(2\tau) = M_0 \exp[(-2\tau/T_2) - \tfrac{2}{3}\gamma^2 G^2 D \tau^3] \qquad (2\text{-}24)$$

where D is the self-diffusion coefficient of the molecule in the sample tube and G is the magnetic field gradient caused by inhomogeneity in the H_0 magnetic field. It is obvious that the diffusion term must be negligible if an accurate determination of T_2 is desired. The accurate measurement of long T_2 values is sometimes obviated by the diffusion term, which can be seen to affect the spin echo amplitude at long τ so that the decay is no longer simply exponential. The diffusion term in Eq. 2-24 originates from inclusion of such a term in the Bloch equation (Eq. 2-10). It is usually omitted from the Bloch equation because it is generally insignificant for most continuous wave NMR applications.

Figure 2-8b shows a photograph of the 90°–τ–180° pulse sequence with the free induction decay following the 90° pulse and the spin echo at time 2τ. That particular photograph recorded the ^{23}Na NMR signal of an NaCl solution.

Carr and Purcell (5) developed a variation of Hahn's spin echo method. This method involves a 90° pulse at time $t = 0$ followed by a train of 180° pulses at times τ, 3τ, 5τ, . . . which refocus the individual spins to form echoes at times 2τ, 4τ, 6τ, ... The most common pulsed NMR method used for T_2 measurements, and the most accurate, utilizes the Carr-Purcell sequence with a modification described by Meiboom and Gill (7). Without the modification any error in the spin echo magnitude caused by any inaccuracy in the 180° pulse width will be cumulative, becoming more serious as the number of 180° pulses in the Carr-Purcell sequence increases. The Meiboom-Gill modification eliminates this cumulative effect giving, in fact, exactly the correct amplitude for all even-numbered echoes and only slightly attenuated amplitudes for the odd-numbered echoes. This modification entails a 90° phase shift after the 90° pulse so the precise setting of the 180° pulse width is not so critical. The phase shift also eliminates effects caused by inhomogeneities in the $\vec{H}_1$ field. The height of the nth echo in the Carr-Purcell sequence is given by

$$M_\perp(2n\tau) = M_0 \exp[(-2n\tau/T_2) - \tfrac{2}{3}\gamma^2 G^2 D \tau^3] \qquad (2\text{-}25)$$

Again we note the presence of a diffusion term. However, if the pulse spacing 2τ is made sufficiently small, very little diffusion will occur before the next pulse hits. Long T_2 values may then be measured without interference from diffusion by using short τ values and running the Carr-Purcell pulse train for sufficiently long times.

A Carr-Purcell sequence with the Meiboom-Gill modification for the ^{23}Na

resonance of an aqueous NaCl solution is shown in the photograph trace of Fig. 2-8c. The rf pulses are too fast to be observed. T_2 may easily be calculated from the exponential decay of such traces using Eq. 2-25.

In addition to measuring T_2, the $90°-\tau-180°$ and Carr-Purcell sequences have also been used to advantage in measuring rapid chemical exchange rates (8), spin–spin coupling (9), and self-diffusion constants (10, 11).

2.1.4. Fourier Transform NMR

As already mentioned, a free induction decay (FID) is observed following an intense rf pulse at the resonance frequency. The FID is a time-domain function $g(t)$ that inherently contains the same information as the frequency-domain spectrum with lineshape function $g(\nu)$. Lowe and Norberg (12) have proved that the free induction decay and the high-resolution NMR spectrum are Fourier transforms of one another:

$$g(t) = k \int_{-\infty}^{\infty} g(\nu) e^{(i2\pi\nu t)} \, d\nu \tag{2-26a}$$

$$= k \int_{-\infty}^{\infty} g(\nu) [\cos(2\pi\nu t) + i \sin(2\pi\nu t)] \, d\nu \tag{2-26b}$$

where k is a constant obtained by normalization of $g(\nu)$.

The Fourier transform of $g(t)$ contains both the absorption spectrum and the dispersion spectrum. The absorption component is just

$$g(\nu)_{\text{Abs}} = k \int_{0}^{\infty} g(t) \cos(2\pi\nu t) \, dt \tag{2-27}$$

and the dispersion spectrum is obtained by taking the sine transform in Eq. 2-26b. In laboratory experiments, the NMR spectrometer is usually interfaced with a small digital computer that takes the FID signal and Fourier transforms it.

As discussed earlier in this chapter (Section 2.1.2, A), the Lorentzian function (Eq. 2-14)

$$g(\nu) = \frac{2T_2}{1 + 4\pi^2 T_2^2 (\nu_0 - \nu)^2}$$

describes the lineshape for liquid samples or solution samples. The Fourier transform of the Lorentzian lineshape is quite simply an exponentially decaying time function with a single decay constant T_2:

$$g(t) = g(0) e^{-t/T_2} \tag{2-28}$$

where $g(0)$ is the amplitude at the instant the pulse is applied ($t = 0$).

Figure 2-10 shows transient signals and the Fourier transformed spectra for three examples. A spectrum with more than one resonance line will have "beats" in its transient signal. For a two-line spectrum, the beat pattern will be fairly simple (Fig. 2-10b). The beat frequency will just be the chemical shift (in Hz) between the two resonance lines. If the resonance lines are split by spin–spin coupling, further beating in the transient signal can be observed at the frequency of the coupling constant. As can be seen in Fig. 2-10c, the beat pattern in a multiple-line spectrum is essentially incomprehensible.

A. *Comparison of Fourier Transform NMR with CW NMR*

The booming Fourier transform NMR field received its primary impetus from Ernst and Anderson (13), who detailed several tremendous advantages of Fourier transform NMR over conventional frequency or field sweep NMR (CW NMR). Some of those advantages can be ennumerated.

1. A major advantage of the Fourier transform method is a tremendous savings in the time required to obtain the NMR signal. The time required for a single sweep is $\sim 1/R$ sec, where R is the desired resolution (in Hz). To just resolve lines separated by 0.5 Hz in a spectrum of 1000 Hz width, approximately 2 sec are necessary, compared with a 2000 sec CW scan necessary to achieve the same resolution. In the case of CW NMR, only a very small amount of the spectrum is being observed at any one time as the frequency (or field) is swept across the spectrum. This is very inefficient use of time. In contrast, with the Fourier transform experiment, the entire spectrum may be observed at any instant. The effect is the same as if many different rf transmitters and receivers, all tuned for slightly different frequencies (say, 2500 Hz sweep range), were operating simultaneously. For a given signal-to-noise ratio the time savings of Fourier transform over CW

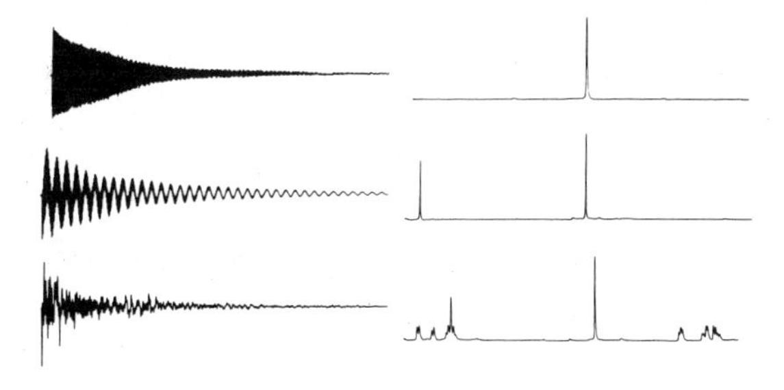

FIG. 2-10. The free induction decay signal and its Fourier transform (frequency spectrum) for (top) a single-line spectrum, (middle) a two-line spectrum, and (bottom) a complex spectrum.

NMR will be approximately given by the factor

$$SW/R + SW/W_{1/2} \tag{2-29}$$

where SW is the spectral width and $W_{1/2}$ is the resonance peak linewidth. In experiments involving NMR of biopolymers, it is necessary to use high magnetic fields (e.g., with superconducting magnets) to "spread out" the lines. With the lines "spread out" over a greater spectral width, the time enhancement ratio becomes larger, giving a greater advantage to Fourier transform NMR. For nonproton NMR (e.g., ^{13}C), the chemical shifts at a given field strength will usually be greater than for proton NMR, so Fourier transform NMR will be even more helpful. The time savings for proton resonance at 100 MHz (23 kGauss) will be on the order of 200, and the time savings for ^{13}C NMR at 25 MHz (23 kGauss) will be on the order of 5000.

By virtue of its rapid time basis, the Fourier transform method has distinct advantages for studying such phenomena as unstable species, chemical exchange rates, molecular dynamics, and T_1 and T_2 relaxation times of each of the resonance lines in a spectrum.

2. A second advantage of the Fourier transform technique comes as a result of the time savings. This advantage is enhanced sensitivity. With the use of well-known signal averaging techniques, the signal-to-noise ratio may be improved by repetitively scanning over the spectrum adding each spectrum to the previous ones as it is recorded. The rationale for this sensitivity gain is that the noise (essentially thermal noise from the receiver coil) is random, whereas the resonance peaks are coherent. When a number N of spectra are added, the random noise will have an amplitude $N^{1/2}$ times as large as the noise in an individual spectrum. The resonance peak is coherent so the signal amplitude is N times larger than in a single scan. Therefore, the net improvement in signal-to-noise improvement ratio for N scans is

$$SNIR \propto N^{1/2} \tag{2-30}$$

With pulsed NMR, the signal-to-noise may also be improved by $N^{1/2}$ if the free induction decays following N pulses are added coherently. With the great time advantage of Fourier transform NMR, a much larger number of spectra can be averaged in a certain period of time. For a given amount of time, the sensitivity advantage of Fourier transform NMR over CW NMR will be approximately given by the factor

$$(SW/W_{1/2})^{1/2}$$

It can again be seen that the Fourier transform method increases its advantage when several narrow peaks with large chemical shift differences are observed. At 25 MHz this advantage will be greater than at 15 MHz for ^{13}C spectra. In general practice, the sensitivity advantage of Fourier transform NMR is about one order of magnitude. In Fig. 2-11, a field sweep spectrum is compared with a Fourier transform spectrum using the same sample and same spectrometer.

3. Another advantage of Fourier transform NMR is the absence of ringing. In practice this effectively allows greater resolution to be achieved. This may be particularly important in looking at weak signals adjacent to fairly strong peaks.

4. Accurate time measurements in Fourier transform NMR replace the less accurate frequency calibration of conventional high-resolution NMR. This can be quite important for ultrahigh-resolution studies (13).

5. Possible signal distortion caused by saturation in CW NMR is eliminated.

To balance this list of advantages, some limitations will be briefly mentioned but not explained in detail here. The origin of some of these limitations is explained in Chapter 5 and elsewhere (3). The intensities and amplitudes of the peaks in the Fourier transform spectrum cannot be meaningfully compared unless a delay is introduced between pulses $\gtrsim 5T_1$. This delay, of course, cuts down on the time and sensitivity advantage of the Fourier transform technique.

Observation of weak signals in the presence of a strong peak is limited by the finite dynamic range of the computer used to accumulate and Fourier transform the signal. It is quite difficult to detect weak proton signals in

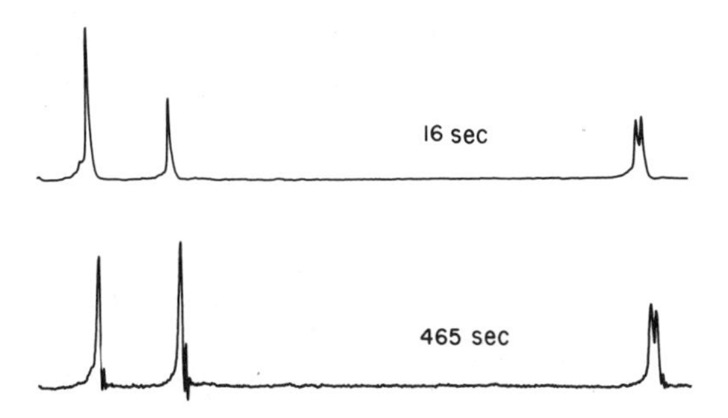

FIG. 2-11. 220 MHz proton NMR spectra of the H-8, H-2, and H-1′ peaks of 0.1 M ADP in D_2O. The lower spectrum was obtained by frequency sweep (CW) NMR. The upper spectrum was obtained by Fourier transform NMR. Note that the peak intensities in ordinary Fourier transform NMR are not valid representations of the number of nuclei contributing to the peak.

the presence of H_2O. To overcome the problem, D_2O is used. However, even in 99.9+% D_2O, the residual HDO peak will be limiting.

Obtaining the absorption phase of all peaks in a spectrum simultaneously is sometimes a practical problem. Occasionally an absorption peak is observed at the low-field end of the spectrum and a dispersion peak is observed at the high-field end of the spectrum. One other annoyance is the occasional appearance of spurious signals in the spectra. However, there are ways of checking whether or not these are spurious signals. The last disadvantage is one of financial consideration. The Fourier transform NMR spectrometer will be more expensive than the conventional spectrometer, primarily because of the high-power amplifier, digital computer, analog-to-digital converter, digital-to-analog converter, and associated software.

Instrumental requirements and other system considerations will be presented in Chapter 5.

B. *T_1 Measurement via Fourier Transform NMR*

Fourier transform NMR provides a means to measure the spin–lattice relaxation times of all lines in a spectrum simultaneously with good accuracy (14). The 180°–τ–90° two-pulse sequence can be used just as described in Section 2.1.3,B. for ordinary pulsed NMR. In this case, the 180° pulse is nonselective, being sufficiently strong to invert the magnetic moments of all spins in the sample regardless of chemical shift. The 90° pulse will likewise tilt the magnetic moments of all spins by the same amount, $\pi/2$ radians. The minimum amount of H_1 power necessary to affect all spins equally is given by

$$\gamma H_1/2\pi \geq SW \tag{2-31}$$

The free induction decay following the 90° pulse is monitored and the Fourier transform computed and recorded. The spectrum of the partially relaxed Fourier transform (PRFT) is recorded as it looks at time $t = \tau$. The procedure is repeated for various values of τ and a set of PRFT spectra are obtained as illustrated in Fig. 2-12 (15). Using Eq. 2-23, T_1 for each of the resonances giving a peak in the spectrum may be determined by measuring the peak amplitudes as a function of τ.

Signal averaging can, of course, be used with this relaxation method as it is with ordinary Fourier transform NMR. The only stipulation is that a pulse delay be inserted between successive 180°–τ–90° sequences to make the repetition time about five times as long as the longest T_1. The pulse delay is necessary for allowing the spin system to relax back to its equilibrium state before another pulse sequence is applied. Improved methods for measuring T_1 with Fourier transform NMR have been published (16, 17). These techniques will be discussed in Chapter 5.

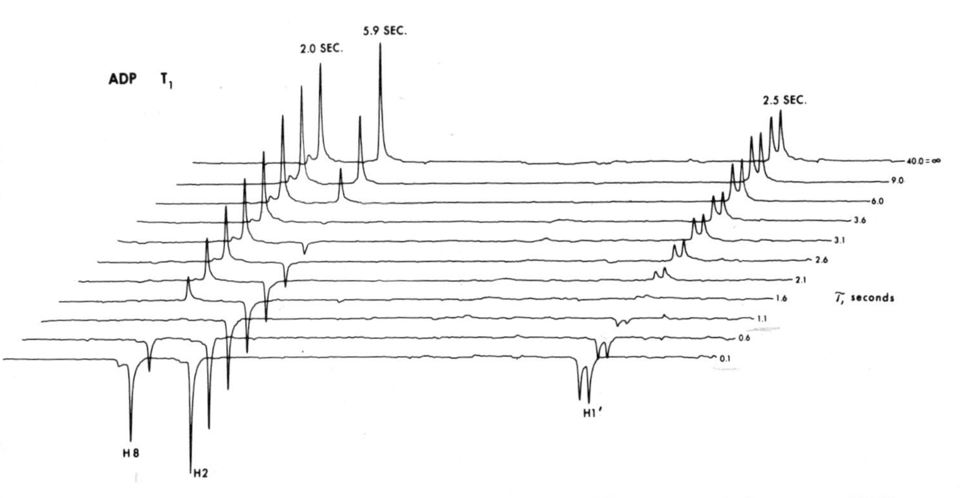

FIG. 2-12. Partially relaxed Fourier transform (PRFT) spectra of the proton NMR signals from a solution of 0.1 *M* ADP in D_2O. Each spectrum is the Fourier transform of the free induction decay following the second pulse of a 180°–τ–90° pulse sequence. The various spectra were obtained for different values of τ.

2.2. Quantum Mechanical Description of NMR

Quantum mechanics can be used to describe the behavior of an ensemble of nuclear spins (2, 18). In fact, the equations developed by Bloch from simple phenomenological considerations can be derived from fundamental quantum mechanical considerations. Quantum mechanical treatments, particularly the density matrix technique, are especially useful for describing relaxation phenomena. However, because most NMR experiments may be described quite adequately in classical (or certainly semiclassical) terms, the reader is referred to the texts of Slichter (2) and Abragam (18) for a quantum mechanical treatment.

2.3. Correlation Function and Correlation Time

NMR relaxation time and linewidth measurements provide information about molecular motions. These molecular motions are conveniently described in terms of correlation functions and correlation times, concepts that will be introduced in this section and used in later sections.

Following absorption of rf energy, a spin system will be in a state of nonequilibrium. The time required for relaxation of the spin system to thermal equilibrium will be related to the probability for transition from a

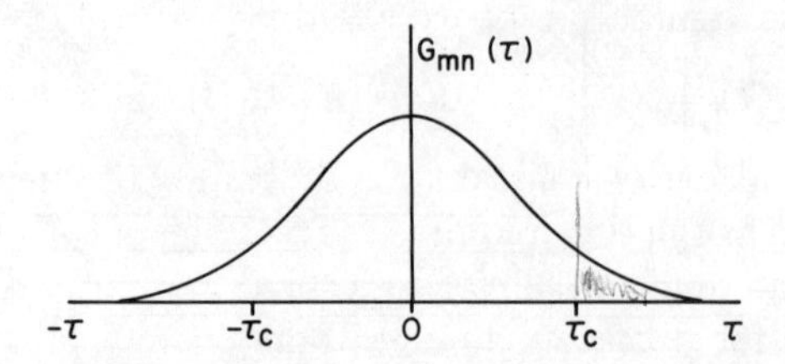

FIG. 2-13. Correlation function $G_{mn}(\tau)$ of $\mathcal{H}_1(t)$.

high-energy nuclear spin level to a low-energy level. Using time-dependent second-order perturbation theory, it may be shown that the probability per second of transition from state n with energy E_n to state m with energy E_m is (2)

$$\frac{1}{\hbar^2}\int_{-\infty}^{\infty} G_{mn}(\tau)e^{-i(E_m-E_n)\tau/\hbar}\,d\tau \tag{2-32}$$

where $G_{mn}(\tau)$ is the correlation function (or sometimes called the "autocorrelation function") of $\mathcal{H}_1(t)$ which is the Hamiltonian representing the interaction between the nuclear spin system and the magnetic (or electric) field causing relaxation. It is assumed that $\mathcal{H}_1(t)$ varies randomly with time but persists in a coherent manner for short periods of time. This last point leads to the definition of the correlation function

$$G_{mn}(\tau) = \overline{(m \mid \mathcal{H}_1(t) \mid n)(n \mid \mathcal{H}_1(t+\tau) \mid m)} \tag{2-33}$$

which is called the "correlation function of $\mathcal{H}_1(t)$" because it describes how the value of $\mathcal{H}_1(t)$ at time t is correlated to its value at some later time $t + \tau$. A plot of $G_{mn}(\tau)$ is shown in Fig. 2-13. It is obvious from the figure that there is a symmetry between the past and the future, i.e.,

$$G_{mn}(\tau) = G_{mn}(-\tau) \tag{2-34}$$

$\mathcal{H}_1(t)$ varies randomly with time because of thermal motion in the sample. For times less than some critical time τ_c, which is called the correlation time, the motion of $\mathcal{H}_1(t)$ is sufficiently small that $\mathcal{H}_1(t) \approx \mathcal{H}_1(t+\tau)$. For times much longer than τ_c, the correlation function becomes negligibly small. The correlation time may then be used as a measure of how long the nuclear magnetization is maintained in a given orientation with respect to the field causing relaxation.

For most cases of interest in biochemical systems (membranes being a possible exception), the motion leading to relaxation is random rotational and translational diffusion (Brownian motion), which means that the correlation function is assumed to be exponential

$$G_{mn}(\tau) = \overline{(m \mid \mathcal{H}_1(t) \mid n)(n \mid \mathcal{H}_1(t) \mid m)}e^{-|\tau|/\tau_c} \tag{2-35}$$

with an exponential decay constant τ_c. To gain a physical feeling for τ_c, it can be estimated as the time required for rotation of the nuclear magnetic moment through an angle of 33° for rotational motion of a molecule, or as the time required for diffusion through a distance equal to the length of the molecule for translational motion. Obviously, for a very small molecule, e.g., glucose, in solution the correlation time for rotational or translational motion will be very short, $\sim 10^{-12}$–10^{-10} sec. Within limits, which will soon be discussed, it may be seen from Eq. 2-32 that the slower the motion, i.e., the longer τ_c is, the greater will be the transition probability and, consequently, the faster a perturbed spin system will relax. That means a nucleus on a large macromolecule will generally relax faster than a nucleus on a small molecule.

The reader may note from the previous discussion of Fourier transform NMR that Eq. 2-32 describes a Fourier transform. The spectral density shall therefore be defined as the Fourier transform of the correlation function:

$$J_{mn}(\omega) = \int_{-\infty}^{\infty} G_{mn}(\tau)e^{-i\omega\tau}\,d\tau \tag{2-36}$$

A plot of the spectral density as a function of ω is given in Fig. 2-14. With the spectral density defined, the transition probability of Eq. 2-32 becomes

$$\frac{J_{mn}(E_m - E_n)}{\hbar^2} \tag{2-37}$$

It follows from Fig. 2-14 that only those motions having a frequency $\lesssim 1/\tau_c$ will effectively induce transitions leading to relaxation. The area under the curve in Fig. 2-14 is constant so that a motion with a longer τ_c

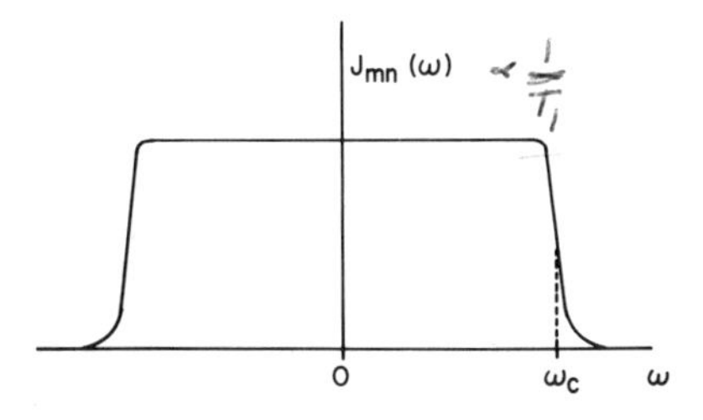

FIG. 2-14. The spectral density of the correlation function $G_{mn}(\tau)$ as a function of ω. $\omega_c \approx 1/\tau_c$.

will have a smaller ω_c and, consequently, a larger spectral density. Going back to the assumption that the sample has Brownian motion and an exponential correlation function $G_{mn}(\tau)$, then

$$J_{mn}(E_m - E_n) = J_{mn}(\omega) = \overline{(m \mid \mathcal{H}_1(t) \mid n)(n \mid \mathcal{H}_1(t) \mid m)} \frac{2\tau_c}{1 + \omega^2\tau_c^2} \tag{2-38}$$

Figure 2-15 shows the spectral density of Eq. 2-38 as a function of τ_c. For small molecules we will be to the left in the plot of Fig. 2-15, so a larger correlation time will result in a larger transition probability. However, for macromolecules the rates of motion are sufficiently slow that the nuclear spin may be affected by a correlation time $\tau_c \gtrsim \omega^{-1}$. The form of the spectral density in Eq. 2-38 will often be seen as the various relaxation mechanisms are encountered and will be of practical interest when the NMR of macromolecules, membranes (or membrane models), and small molecules bound to macromolecules are considered.

2.3.1. Rotational Correlation Times

The rotational correlation time (τ_c or τ_r) provides a one-parameter description of the fluctuations leading to intramolecular dipole–dipole relaxation and quadrupole relaxation. It is often assumed that the rotational diffusion limit describes the random rotational motion of the molecules. For liquids in the rotational diffusion limit, the motion of a molecule is Brownian, having many molecular collisions before it turns 1 radian. The rotational correlation time is given in terms of the rotational diffusion coefficient D_r by

$$\tau_r = 1/6D_r \tag{2-39}$$

In one correlation time, the molecule will rotate $3^{-1/2}$ radians (~33°).

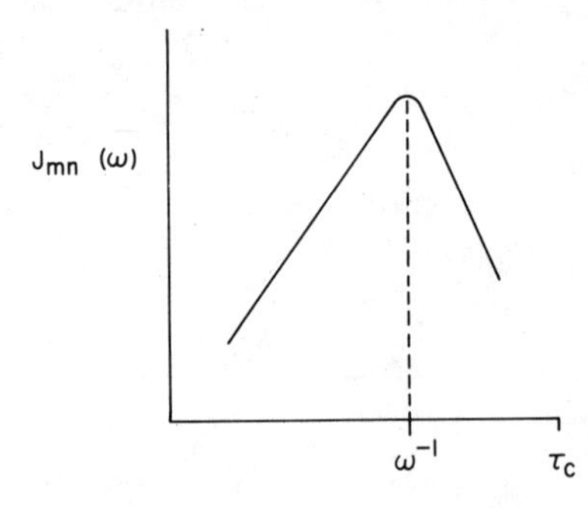

Fig. 2-15. Spectral density as a function of the correlation time.

Various models have been postulated for calculating the rotational diffusion coefficient, and therefore the rotational correlation time, from molecular properties. If it is assumed the molecule is a sphere of radius a in a continuous medium of viscosity η, the Debye-Stokes theory will give the rotational correlation time (18)

$$\tau_r = 4\pi a^3 \eta / 3kT \tag{2-40}$$

where k is the Boltzmann constant and T is the absolute temperature. This model qualitatively predicts the usually observed features that the correlation time increases with larger molecules, viscous solutions, and low temperatures. For a small molecule (e.g., H_2O), $\tau_r \sim 10^{-12}$ sec but for a large molecule (e.g., creatine kinase, molecular weight 81,000), $\tau_r \sim 10^{-7}$ to 10^{-8} sec. Equation 2-40 is simple and readily applicable because the sample parameters a and η, or reasonable estimates, are usually known. It is also noteworthy, on a qualitative basis, that Eq. 2-40 predicts a linear dependence of T_1 with T/η. This dependence has been demonstrated for some liquids, but exceptions are not uncommon. Several references have appeared in the literature indicating that the correlation times calculated from the Debye-Stokes theory are too large by an order of magnitude for relatively small molecules (19–22). In using Eq. 2-40, one must realize that only the upper limit on τ_r is obtained. One problem with application of the Debye-Stokes theory is that the viscosity η which is used is a macroscopic translational parameter; in fact, a rotational microviscosity would be more meaningful (if it were experimentally available).

The rotational correlation time calculated from the microviscosity model (23) has been demonstrated to agree more closely with experimental values (20–22). The microviscosity model attempts to correct the Debye-Stokes assumption that the solute molecule is dissolved in a continuous medium. The model may be expressed by

$$\tau_r(\text{microvisc}) = f_R \tau_r(\text{Stokes}) \tag{2-41}$$

The microviscosity factor f_R is

$$f_R = \left[6\frac{a_s}{a} + \frac{1}{\left(1 + \frac{a_s}{a}\right)^3} \right]^{-1} \tag{2-42}$$

where a_s is the effective radius of the solvent molecule and a is the effective radius of the solute molecule. The microviscosity model will give smaller correlation times for the tumbling of solute molecules than the Debye-Stokes model but, in the limit of very large solute molecules (macromolecultes), τ_r(microvisc) will approach τ_r(Stokes). For a small protein, such as myoglobin (molecular weight 18,000), we calculate f_R to be 0.77, which means, for most practical purposes, that the calculated rotational correlation time for macromolecules may be estimated adequately with the Debye-Stokes theory. The microviscosity model is useful for small molecules. The Hill theory (24) is also applicable to smaller molecules. Rodlike (eccentric) molecules will have a larger rotational correlation time. Correction factors for eccentricity have been discussed by Shimizu (25).

Wallach (26) has derived an expression especially useful for calculating the rotational correlation time of a nuclear spin on a flexible chain of a protein or on the fatty acid moiety of lipids when all the internal rotations of the chain are rapid compared to the overall motion of the protein or lipid vesicle, etc. Each internal rotation in the chain is considered between the point of attachment to the macromolecule and the end of the chain. A simplified treatment suffices if each succeeding internal rotation (progressing toward the end of the chain) is much faster than the preceeding one. If τ_0 is the rotational correlation time of the macromolecule, the correlation time of a nuclear spin on the attached flexible chain in the simplified treatment is (26)

$$\tau_r = \tau_0[\tfrac{1}{4}(3\cos^2\theta - 1)^2][\tfrac{1}{4}(3\cos^2\beta_{12} - 1)^2][\tfrac{1}{4}(3\cos^2\beta_{23} - 1)^2]\ldots \quad (2\text{-}43)$$

where θ is the angle between the nuclear spin and the first internal rotation axis (at the end of the chain), β_{12} is the angle between the first and second internal rotation axes, β_{23} is the angle between the second and third, etc. If Eq. 2-43 is used, it should also be experimentally verified that extreme narrowing conditions (i.e., $\omega^2\tau_r^2 \ll 1$) hold. Wallach's treatment has subsequently been extended such that the extreme narrowing limit is no longer a necessity (26a).

It is often found that over a limited temperature range the correlation time has an Arrhenius temperature dependence:

$$\tau_c = \tau_c^0 e^{-E_{\text{act}}/RT} \quad (2\text{-}44)$$

where E_{act} is an activation energy. Quite commonly, T_1 has a temperature dependence described by Eq. 2-44.

2.4. Relaxation Mechanisms

Nuclear magnetic resonance relaxation has already been discussed to some extent. An understanding of the possible mechanisms leading to relaxation requires knowledge of the magnitude and nature of the interaction of the nuclear spin system with the various degrees of freedom, such as rotational, translational, or (rarely) vibrational, present in the lattice. The term "lattice" refers to the environment surrounding the nucleus. It entails all ions or molecules in the sample, including the molecule containing the nucleus of interest. The random motions of the molecules result in interaction of the nuclear spin system with random, fluctuating magnetic and electric fields having Fourier transform components at the requisite frequency to induce transitions between the nuclear spin states. Interaction with the random, fluctuating fields leads to spin–lattice relaxation. The transitions caused by the fluctuating fields will continue to transfer energy from the spin system to the lattice until thermal equilibrium is achieved. Interaction with the random static fields leads to spin–spin relaxation.

The random magnetic or electric fields may originate from any of the following sources: (*1*) magnetic moments of other nuclei; (*2*) magnetic moments of unpaired electrons; (*3*) molecular magnetic moments; (*4*) angular variations in the electronic shielding of the stationary H_0 magnetic field; and (*5*) electric quadruple moment, if any, of the nucleus being irradiated. The interaction with local magnetic fields produced by other nuclei may be of two types: (*1*) direct dipole–dipole coupling and (*2*) indirect or scalar coupling (electron transmitted). The magnetic moment of an electron is much larger than that of a nucleus and provides a very efficient relaxation mechanism. The presence of small amounts of paramagnetic species, including O_2, can control the relaxation.

We can distinguish between the longitudinal (spin–lattice) and the transverse (spin–spin) relaxation times by considering the effects of the x, y, and z components (laboratory coordinate system) of the local fields. The H_z component will affect the spread in the individual precession frequencies. It should be recalled that a field will produce a torque on a magnetic moment oriented perpendicular to the direction of the magnetic field. The H_z component will not contribute to longitudinal relaxation because that requires change in the M_z component of the magnetization. Instead, H_z will contribute to relaxation of the transverse magnetization. The same argument holds for the effect of the H_x and H_y components of the local fields on longitudinal relaxation. The longitudinal relaxation time is given by

$$1/T_1 = K\overline{H^2}J(\omega) \tag{2-45}$$

where K is constant, $\overline{H^2}$ is the mean-square average of the local fields, and $J(\omega)$ is the spectral density. Actually, there may be a sum of spectral densities in Eq. 2-45. It is interesting that $\overline{H^2}$ possesses a finite value although the time average of the local fields average to zero. As usual, a simple exponential correlation function, with correlation time τ_c, is assumed. In that case, T_1 and T_2 are given simply by (2)

$$\frac{1}{T_1} = \gamma^2(\overline{H_x^2} + \overline{H_y^2}) \frac{\tau_c}{1 + \omega^2\tau_c^2} \tag{2-46}$$

and

$$\frac{1}{T_2} = \gamma^2\left(\overline{H_z^2}\tau_c + \overline{H_y^2}\frac{\tau_c}{1 + \omega^2\tau_c^2}\right) \tag{2-47}$$

In the extreme narrowing limit, i.e.,

$$\omega^2\tau_c^2 \ll 1$$

and with an isotropic fluctuating field, i.e.,

$$\overline{H_x^2} = \overline{H_y^2} = \overline{H_z^2}$$

T_1 and T_2 are equal. This is usually the case for small molecules, but the extreme narrowing conditions are often not met in biological systems; the large molecular or particle sizes lead to relatively long values for τ_c. Furthermore, in the case of membranes or lipid vesicles, the local field is often not isotropically fluctuating.

Log–log plots of T_1^{-1} and T_2^{-1} as a function of τ_c are shown in Fig. 2-16. It is also useful to remember the general shape of the curves in Fig. 2-16. It is also important to note that always

$$T_2 \leq T_1 \tag{2-48}$$

All causes of T_1 relaxation will contribute to T_2 relaxation. In addition, certain other low-frequency transitions, most notably chemical exchange, will contribute to T_2.

In this section the most prominent relaxation mechanisms will be covered, each one of which will have the same general features as Eqs. 2-46 and 2-47. The observed T_1 or T_2 relaxation time may have contributions from each

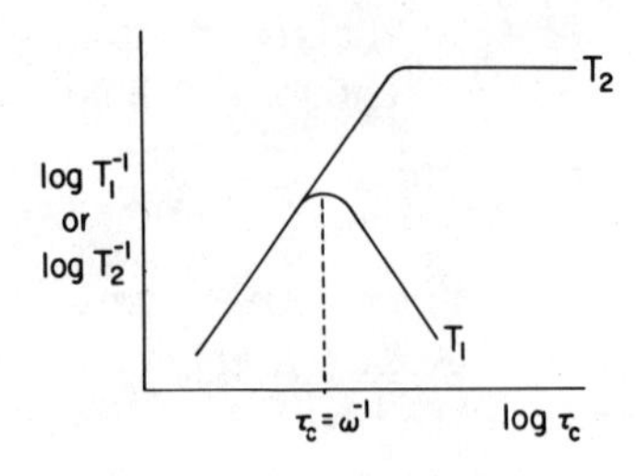

FIG. 2-16. Dependence of T_1 and T_2 on correlation time.

relaxation mechanism, e.g.,

$$1/T_1 = \sum_j (1/T_1)_j \tag{2-49}$$

where the sum is over all relaxation mechanisms, although quite often one mechanism will dominate.

2.4.1. Nuclear Dipole–Dipole Relaxation

For spin $\frac{1}{2}$ nuclei in liquids (with the exception of ^{19}F), the dipole–dipole relaxation mechanism is usually dominant. With this relaxation mechanism, the relaxation of nucleus I_1 will be facilitated by direct interaction with the magnetic dipole of nucleus I_2. The two nuclear spins separated by a distance r will couple with energy

$$E_{dd} \propto \vec{\mu}_1 \cdot \vec{\mu}_2 / r^3 \tag{2-50}$$

where $\vec{\mu}_1$ and $\vec{\mu}_2$ are the dipole moments of the two nuclei. Dipolar relaxation has been discussed by Abragam (18, p. 289ff.).

The contribution of dipolar relaxation may be separated into those from the same molecule (intramolecular) and those from nearby molecules (intermolecular),

$$1/T_1 = (1/T_1)_{\text{intra}} + (1/T_1)_{\text{inter}} \tag{2-51}$$

We will deal with the intramolecular contribution first.

For a two-spin system with equal spins ($I_1 = I_2 = I$), the intramolecular contribution to relaxation of nucleus I_1 is caused solely by dissipation of the nuclear spin energy to rotational motions and is given by

$$\left(\frac{1}{T_1}\right)_{\text{intra}} = \frac{2\gamma^4\hbar^2 I(I+1)}{5r^6}\left(\frac{\tau_r}{1+\omega^2\tau_r^2} + \frac{4\tau_r}{1+4\omega^2\tau_r^2}\right) \tag{2-52}$$

and

$$\left(\frac{1}{T_2}\right)_{\text{intra}} = \frac{\gamma^4\hbar^2 I(I+1)}{5r^6}\left(3\tau_r + \frac{5\tau_r}{1+\omega^2\tau_r^2} + \frac{2\tau_r}{1+4\omega^2\tau_r^2}\right) \tag{2-53}$$

The correlation time appropriate for use with these expressions is the rotational correlation time τ_r discussed in Section 2.3.1. In the limit of extreme narrowing ($\omega^2\tau_c^2 \ll 1$),

$$\left(\frac{1}{T_1}\right)_{\text{intra}} = \left(\frac{1}{T_2}\right)_{\text{intra}} = \frac{2\gamma^4\hbar^2 I(I+1)}{r^6}\tau_r \tag{2-54}$$

For unlike nuclei (I_1 having gyromagnetic ratio γ_1, and I_2 having gyromagnetic ratio γ_2), the dipolar contribution to I_1 will be

$$\left(\frac{1}{T_1}\right)_{\text{intra}}^{\text{unlike}} = \frac{3\gamma_1^2\gamma_2^2\hbar^2 I_2(I_2+1)}{r^6}\tau_r \qquad (2\text{-}55)$$

The difference between Eqs. 2-54 and 2-55 involves the "3/2 effect" discussed by Abragam (18, p. 297).

The relaxation of nuclear spin I_1 may also be hastened by a contribution from intermolecular dipole–dipole interactions, in which fluctuations arise from the random translational motion with respect to nuclei on neighboring molecules. For like nuclei, this contribution is given by Eqs. 2-52 and 2-53, replacing the rotational correlation time with a translational correlation time, τ_{tr}, which can be thought of as the time required for nucleus I_1 to move a distance $2a$, where a is the radius of the molecule. The theory of Brownian motion gives the translational correlation time

$$\tau_{tr} = \frac{a^2}{3D} = \frac{2\pi\eta a^3}{kT} \qquad (2\text{-}56)$$

As with the rotational correlation time, τ_{tr} can also be modified by a translational microviscosity factor (23).

Gutowsky and Woessner (27) have shown that the equations presented here for the inter- and intramolecular dipolar contributions to the relaxation rate $1/T_1$ can be easily extended to systems containing many nuclei.

Some observations concerning the dipole–dipole relaxation mechanism are in order. The first point concerns the relative importance of various magnetic nuclei in promoting relaxation. Table 2-1 is illustrative. Protons,

TABLE 2-1

THE RELATIVE MAGNITUDE OF DIPOLAR RELAXATION TERMS FOR PROTONS RELAXED BY VARIOUS NUCLEI

PROTON RELAXED BY	RELATIVE MAGNITUDE OF CONTRIBUTION[a]
H	100.0
D	4.0
^{31}P	10.4
^{14}N	0.8
^{39}K	0.8
^{23}Na	23.2

[a] Comparison made for an equal number of nuclei at the same distance from the proton.

TABLE 2-2

^{13}C RELAXATION RATES AS A FUNCTION OF THE NUMBER OF PROTONS BONDED TO THE CARBON ATOM FOR CHOLESTERYL CHLORIDE[a]

	OBSERVED RELAXATION RATE (SEC^{-1})	OBSERVED RATIO OF RELAXATION RATES	EXPECTED RATIO OF RELAXATION RATES (SAME τ_c)
CH	1.9	1	1
CH_2	3.8	2	2
CH_3	0.67	0.35	3

[a] From Allerhand *et al.* (28).

because of a greater gyromagnetic ratio, are usually the most important nuclei leading to dipolar relaxation.

The r^{-6} dependence of Eqs. 2-52 and 2-53 practically limits contributions to those from the closest neighbors. In the case of nondecoupled ^{13}C NMR, the only significant contribution to relaxation comes from protons directly bonded to the ^{13}C. The central atom of such molecules as neopentane are an obvious exception having no directly bonded hydrogen. Allerhand *et al.* (28) found the ^{13}C relaxation rates (T_1^{-1}) for the protonated carbon atoms on the ring backbone of cholesteryl chloride to be proportional to the number of attached hydrogens, as shown in Table 2-2. The CH and CH_2 carbons are in the ring backbone, which has the rotational correlation time 9×10^{-11} sec. If the CH_3 protons were limited to the same motion, we would expect a relaxation rate of 5.7 sec^{-1}. The observed rate of 0.67 sec^{-1} is a manifestation of the additional rotational freedom of the rotating CH_3 group, which has a correlation time $\gtrsim 5 \times 10^{-12}$ sec.

2.4.2. Scalar Relaxation

If there is spin–spin (scalar) coupling between nuclei I_1 and I_2, their NMR resonance lines will be split into multiplets with a coupling constant J (see Section 1.5). The term "scalar coupling" is used to distinguish this electron-mediated coupling from the direct interaction of magnetic moments, i.e., dipole–dipole coupling. Scalar coupling can be a cause of relaxation of I_1 either if J is time dependent (scalar relaxation of the first kind) or if spin I_2 relaxes rapidly (scalar relaxation of the second kind).

If I_2 has another relaxation mechanism (e.g., quadrupole relaxation) promoting rapid relaxation of I_2, the multiplet splitting of I_1 will collapse and I_1 will also be relaxed by virtue of its coupling with I_2. This was termed "scalar relaxation of the second kind" by Abragam (18). If I_2 has a relaxation time much less than $1/(2\pi J)$, the scalar relaxation contributions to I_1 will be (18, p. 311)

$$\frac{1}{T_1} = \frac{8\pi^2 J^2 I_2(I_2+1)}{3} \frac{(T_2)_2}{1+(\omega_1-\omega_2)^2 (T_2)_2^{\,2}} \tag{2-57}$$

and

$$\frac{1}{T_2} = \frac{4\pi^2 J^2 I_2(I_2+1)}{3} \left((T_1)_2 + \frac{(T_2)_2}{1+(\omega_1-\omega_2)^2 (T_2)_2^{\,2}} \right) \tag{2-58}$$

where $(T_1)_2$ and $(T_2)_2$ are the spin–lattice and spin–spin relaxation times, respectively, of nucleus I_2, and ω_1 and ω_2 are the angular precession frequencies of nuclei I_1 and I_2, respectively. It will be noticed from Eqs. 2-57 and 2-58 that $1/T_1$ and $1/T_2$ follow curves similar to those in Fig. 2-16. To have a scalar contribution to $1/T_1$ requires $(T_2)_2$ to be short and of the same order as the difference in precession frequencies. In practice this is usually not encountered. However, scalar relaxation of the second kind will contribute to $1/T_2$ ($\propto$ linewidth) more frequently via the frequency-independent term of Eq. 2-58. Broadening of ^1H resonances for protons bonded to ^{14}N are common. Ogg and Ray (29) illustrated this point nicely in the case of anhydrous ammonia. The ^1H spectrum of $^{14}NH_3$ shows three broad peaks but the ^1H spectrum of $^{15}NH_3$ (^{15}N has spin $\frac{1}{2}$) is a sharp doublet. The ^1H lines of other nonexchangable (or slowly exchanging) protons bonded to nitrogen will be similar; the indole N–H protons of the tryptophans in lysozyme are an example (30).

When J becomes a function of time, scalar relaxation of the first kind may occur. J is a function of time when chemical exchange takes place, being, in fact, J when I_1 and I_2 are in a covalent bond in the same molecule and being zero when I_1 and I_2 are not covalently bonded. Chemical exchange is explained in more detail in Section 2.6; it involves the rapid making and breaking of the bond joining I_1 and I_2. If the rate of exchange $1/\tau_{ex}$ is greater than J or $1/T_1$ for either I_1 or I_2, then the multiplet splitting will collapse, leading to a singlet resonance peak. The formulas for scalar relaxation of the first kind have the same form as Eqs. 2-57 and 2-58 with $(T_1)_2$ and $(T_2)_2$ replaced by τ_{ex}.

2.4.3. Paramagnetic Relaxation

The magnetic moment of an unpaired electron is about 1000 times larger than the magnetic moments of nuclei. Just as the dipole moment of a nucleus gives rise to a dipolar field that can lead to relaxation and chemical shifts, the very strong electron dipole moment results in relaxation and chemical shifts of nuclei. The chemical shift effects will be discussed in the next chapter. The relaxation effects will be discussed here.

The magnetic moment of an electron spin is 657 times greater than the magnetic moment of a proton. This means an unpaired electron has a relaxation mechanism 500,000 ($\sim 657^2$) times more efficient than a proton. Therefore, the presence of a paramagnetic species in the sample can easily provide the dominant relaxation mechanism. NMR studies are usually restricted to dilute solutions of the paramagnetic species. Otherwise, NMR resonances are broadened beyond detection with continuous wave NMR or relax too fast for observation with pulsed NMR. Some molecules of biological interest naturally contain paramagnetic species. In other cases, paramagnetic probes may be added and NMR relaxation time measurements used to study structure and kinetics, as described in Chapter 6.

There are two types of electron–nucleus interactions that contribute to T_1 and T_2 NMR relaxation: (*1*) a dipole–dipole coupling following the form of Eq. 2-50 and, therefore, dependent on the electron–nucleus distance and (*2*) a scalar coupling dependent on the electron spin density at the nucleus. Solomon (31) derived equations for the pure proton–electron dipole–dipole interaction, and Bloembergen (32) added the terms entailing scalar coupling for nuclei in solutions containing paramagnetic ions. The paramagnetic contributions to the T_1 and T_2 relaxation times of a spin $\frac{1}{2}$ nucleus with gyromagnetic ratio γ perturbed by a paramagnetic ion are given by the Solomon-Bloembergen equations:

$$\frac{1}{T_{1M}} = \frac{2}{15}\frac{S(S+1)\gamma^2 g^2 \beta^2}{r^6}\left(\frac{3\tau_c}{1+\omega_I^2\tau_c^2} + \frac{7\tau_c}{1+\omega_s^2\tau_c^2}\right) + \frac{2}{3}\frac{S(S+1)A^2}{\hbar^2}\left(\frac{\tau_e}{1+\omega_s^2\tau_e^2}\right) \tag{2-59}$$

$$\frac{1}{T_{2M}} = \frac{1}{15}\frac{S(S+1)\gamma^2 g^2 \beta^2}{r^6}\left(4\tau_c + \frac{3\tau_c}{1+\omega_I^2\tau_c^2} + \frac{13\tau_c}{1+\omega_s^2\tau_c^2}\right) + \frac{1}{3}\frac{S(S+1)A^2}{\hbar^2}\left(\tau_e + \frac{\tau_e}{1+\omega_s^2\tau_e^2}\right) \tag{2-60}$$

where S is the electron spin quantum number; g is the electronic g factor; β is the Bohr magneton; ω_I and ω_s ($= 657\ \omega_I$) are the Larmor angular precession frequencies for the nuclear spins and electron spins, respectively; r is the ion–nucleus distance; A is the hyperfine coupling constant; and τ_c and τ_e are the correlation times for the dipolar and scalar interactions, respectively. In both Eqs. 2-59 and 2-60, the first term represents the dipolar contribution and the second term represents the scalar contribution to the relaxation rates. Scalar coupling requires a finite electron spin density at the nucleus, so nuclei on molecules not in the first coordination sphere of the paramagnetic ion will have no relaxation contribution from scalar coupling.

The correlation times are given by

$$1/\tau_c = 1/\tau_r + 1/\tau_s + 1/\tau_M \tag{2-61}$$

and

$$1/\tau_e = 1/\tau_s + 1/\tau_M \tag{2-62}$$

where τ_r is the rotational correlation time, τ_s is the electron spin relaxation time,* and τ_M is the mean lifetime of a nucleus in the sphere of influence of the paramagnetic ion. Equations 2-61 and 2-62 can be compared with the resistance of a circuit of parallel resistors; if one resistor has a much lower resistance, the current will flow primarily through that resistor. Likewise, if one correlation time has a much lower value, the fluctuation modulating the interaction will be the one for that correlation time.

Several publications (e.g., 35–39) have provided experimental verification of the Solomon-Bloembergen equations for the solvent nuclei in water and methanol solutions of several paramagnetic ions. The aquo complexes of the first-row transition metal ions have $\tau_r \approx 10^{-11}$ sec and τ_M several orders of magnitude larger; for most of the ions, $\tau_s \lesssim \tau_r$ but for Mn(II), Cu(II), and Cr(III), $\tau_s > \tau_r$.

Aside from scientific curiosity pertaining to the other modes of relaxation, Eqs. 2-61 and 2-62 are important for two prime reasons when paramagnetic ions are used as probes in macromolecules. First, knowledge of τ_M gives information about exchange kinetics. Second, once τ_c is known, it can be used in Eqs. 2-59 and 2-60 to calculate the ion–nucleus distance, giving information about structure. The details of these considerations will be discussed in Chapter 6.

2.4.4. Electric Quadrupole Relaxation

A nucleus with nuclear spin I greater than $\frac{1}{2}$ possesses an electric quadrupole moment eQ. For those nuclei with $I \gtrsim 1$, the electric quadrupole

* Reuben *et al.* (33) noted that, strictly speaking, one should distinguish between the longitudinal and transverse relaxation times of the unpaired electron for the various terms of the Solomon-Bloembergen equations. Koenig (34) subsequently pointed out that the distinction generally need not be made. Therefore, for notational simplicity, we will consider only one electron spin relaxation time, namely, the longitudinal relaxation time.

moment is given in the Table of Nuclear Properties (Appendix 2). Interactions of the electric quadrupole moment with an electric field gradient eq at the nucleus and the modulation of these interactions by rotational motion usually provides the dominant relaxation mechanism for quadrupolar nuclei. It should be noted that the same type of motion (rotational) leading to intramolecular dipole–dipole relaxation for spin $\frac{1}{2}$ nuclei will lead to quadrupolar relaxation for a nucleus with spin $> \frac{1}{2}$. In the limit of extreme narrowing, the contribution of nuclear quadrupole relaxation to the relaxation rate may be expressed as (18, p. 313ff.)

$$\frac{1}{T_1} = \frac{1}{T_2} = \frac{3}{40}\frac{(2I+3)}{I^2(2I-1)}\left(1+\frac{\eta^2}{3}\right)\left(\frac{e^2qQ}{\hbar}\right)^2 \tau_r \tag{2-63}$$

where $e^2qQ/\hbar$ is the quadrupole coupling constant and η is the asymmetry parameter. The field gradient eq is actually the principal component of the field gradient tensor and η is a measure of how much the electric field gradient deviates from axial symmetry. The term $(1 + \eta^2/3)$ is usually small and can generally be ignored.

Two nitrogen compounds can be used to illustrate the importance of the electric field symmetry. If a nucleus is in a field of cubic symmetry, the electric field gradients at the nucleus will be vanishingly small. Such is the case for NH_4^+ where ^{14}N $(I = 1)$ T_1 values are $\gtrsim 40$ sec; this can be compared to a ^{14}N T_1 value of 0.51 msec for unsymmetrical ^{14}N in CCl_3CN (40).

The quadrupole contribution to the relaxation rate, expressed by Eq. 2-63, clearly depends on the magnitude of the interaction $(e^2qQ/\hbar)$ and the modulation of that interaction (τ_r). The values of the rotational correlation time usually increase with decreasing temperature or increasing solution viscosity, as discussed in Section 2.3.1, but the more interesting case is an increase in τ_r observed when a small quadrupolar ion [e.g., $^{23}Na^+(I = 3/2)$] is bound by a large slow-moving biopolymer. The value of eQ is fixed for any nucleus, but eq may vary. The source of the electric field gradient eq at the nucleus may be: (*1*) valence electrons of the nucleus in question, (*2*) distortion of the closed shells of electrons around the nucleus, and (*3*) charge distributions associated with adjacent atoms or ions. Because the charge sources contribute to eq with an r^{-3} dependence (2, p. 174), the electrons around the nucleus will have the greatest influence. Bond formation will therefore result in larger values of eq, with covalent bond formation having a much larger influence than ionic bond formation (41). For example, the ^{35}Cl relaxation time of Cl^- in aqueous solution is about 30 msec but the ^{35}Cl relaxation time in a covalent C–Cl bond (e.g., CCl_4) is about 20 μsec (42). If a quadrupole ion is complexed nonsymmetrically, the electronic environment around the nucleus will be perturbed sufficiently to give an increase in *eq*. For example, the ^{23}Na T_1 in dilute aqueous NaCl solution is 58 msec but is reduced in 1:1 solutions of NaCl with citrate to 18 msec (43) and with EDTA to 2 msec (44) because of complex formation. Although complex formation occurs, the bonding is predominantly ionic so the relaxation time is not decreased to the extent of ^{35}Cl in covalent C–Cl bonds.

For almost all quadrupolar nuclei in noncubic electron environments, the electric quadrupole relaxation mechanism will be controlling. An interesting nucleus that provides an exception to this is deuterium, D, which has a small quadrupole moment *eQ*. In most molecules quadrupole relaxation prevails, but other mechanisms can contribute significantly to deuterium relaxation.

2.4.5. Spin–Rotation Relaxation

The interaction of the nuclear magnetic moment with the magnetic field generated at the nucleus by the rotation of a molecular magnetic moment arising from the electronic structure of a molecule is termed the "spin–rotational interaction" (45, 46). The rotational collisions modulate the interaction providing a means for energy transfer from the nuclear spin system to the molecular rotation.

The spin–rotation mechanism is often the only mechanism available for spin $\frac{1}{2}$ nuclei in the gas phase. It is also an important mechanism in some cases in the liquid phase, generally with small molecules at reasonably high temperatures. As such, it would hold little interest for biochemists. However, a variant, termed "nuclear-spin–internal-rotation relaxation" (47) can be pertinent if the nucleus is in a freely rotating chemical group attached to a larger molecule. Such a mechanism may need to be considered for ^{31}P relaxation in nucleotides.

advantage

Hubbard (48) calculated the contribution of spin–rotational interactions to the relaxation of spin $\frac{1}{2}$ nuclei in isotropically rotating molecules to be

$$\frac{1}{T_1} = \frac{2\pi I'kT}{3\hbar^2} C^2\tau_j \tag{2-64}$$

where I' is the moment of inertia of the molecule, C^2 is the square of the average of the spin–rotation tensor that couples the nuclear spin angular momentum vector with the angular momentums of the molecule, and τ_j is the spin–rotational (angular velocity) correlation time. τ_j may be thought of as the time between rotation collisions.

τ_j may be related to the previously discussed rotational correlation time by (48)

$$\tau_j\tau_r = I'/(6kT) \tag{2-65}$$

for the case $\tau_j \ll \tau_r$. The spin–rotation relaxation mechanism has been shown to be an important source for ^{19}F spin–lattice relaxation (49).

The internal rotation of a small group relative to the larger molecule to which it is attached may also be an important source of relaxation (47, 50). The case considered was that of the ^{19}F NMR relaxation of benzotrifluoride where the $-CF_3$ group rotates freely with respect to the phenyl ring. The contribution from spin–internal-rotation can be given in an expression essentially identical to Eq. 2-64, with the exception that the moment of inertia I' is for the rotating group rather than for the entire molecule. In benzotrifluoride at room temperature, this is the important mechanism. Unlike the dipole–dipole mechanism, the spin–rotation and spin–internal-rotation mechanisms become more important at higher temperatures.

The spin–internal-rotation relaxation mechanism may be important in the ^{1}H NMR of freely rotating $-CH_3$, ^{31}P NMR of freely rotating $-OPO_3$, and proton decoupled ^{13}C NMR of freely rotating $-CH_2X$ groups attached to larger molecules.

2.4.6. Relaxation via Anisotropic Electronic Shielding

This is the only relaxation mechanism requiring the presence of an applied magnetic field H_0. As discussed in Section 1.4, the magnetic field experienced by a nucleus depends on its electronic shielding (chemical shift.) If the shielding is not equal in all directions around a nucleus, i.e., if it is not isotropic, the secondary magnetic field caused by electronic currents need not be parallel to the direction of the applied field and will therefore have a perpendicular component. This perpendicular component will fluctuate as the molecule tumbles, leading to relaxation. The contribution to relaxation from chemical shift anisotropy is given, in the extreme narrowing limit, by (51)

$$1/T_1 = (2/15)\gamma^2 H_0^2 (\Delta\sigma)^2 \tau_c \qquad (2\text{-}66)$$

$$1/T_2 = (7/45)\gamma^2 H_0^2 (\Delta\sigma)^2 \tau_c \qquad (2\text{-}67)$$

where $\Delta\sigma$ is the difference between the shielding constants parallel and perpendicular to the molecule's axis of symmetry.

The anisotropic electronic shielding contribution to relaxation is generally rare but may be a factor in the relaxation of nonprotonated carbons in ^{13}C NMR with superconducting magnets because Eqs. 2-66 and 2-67 show a dependence on H_0^2. The ^{19}F relaxation time at $-142°C$ of $CHFCl_2$ at 56.4 MHz provides an example of the anisotropic chemical shift mechanism (52). This mechanism can be distinguished from the others by its dependence on field strength.

2.5. Double Resonance

The term "double resonance" refers to any technique in which a second rf field $\vec{H}_2$ is applied to the sample in addition to the observing $\vec{H}_1$ rf field discussed so far in this chapter. The second rf field can be used for several purposes, but the most interesting experiments involve a strong saturating $\vec{H}_2$ field. This saturating $\vec{H}_2$ field can be used for spin decoupling, saturation transfer, or nuclear Overhauser effect studies.

2.5.1. Spin Decoupling

To "unravel" the spin–spin couplings in a complicated spectrum, it is often useful to apply a powerful second rf field $\vec{H}_2$ at the resonance frequency of some particular nucleus (53). If the $\vec{H}_2$ field applied at the resonance frequency of nucleus A, which is coupled with nucleus B, is sufficiently strong, the multiplet of B will collapse to a singlet. It was noted in Fig. 1-4 that the splitting of resonance B was determined by the spin orientations of the A nuclei. Application of the strong $\vec{H}_2$ field at the resonance frequency for A, however, will cause rapid transitions between spin states of A. These rapid transitions will result in nucleus B "seeing" only a single averaged orientation of A. For collapse to occur, it is necessary that the decoupling field be sufficiently strong that $\gamma H_2 \gg \pi J_{AB}$. An example of spin decoupling is shown in Fig. 2-17. When either resonance of thiouracil is irradiated with a strong H_2 field, the other resonance collapses into a singlet.

Spin decoupling greatly simplifies a complex spectrum and will lead to

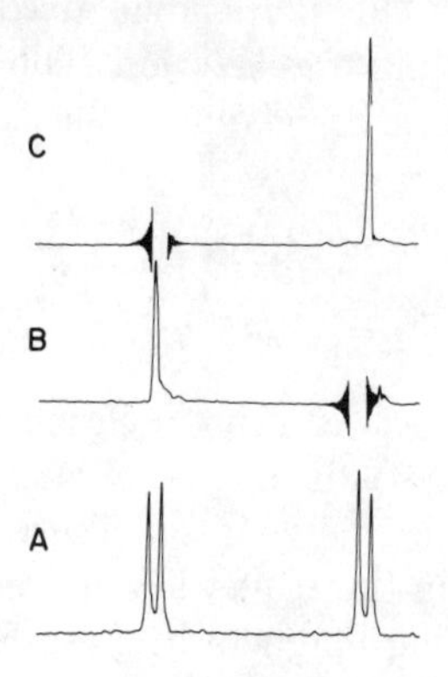

FIG. 2-17. 60 MHz proton NMR spectrum of thiouracil in D_2O at 35°C. (A) Normal spectrum. (B) Spectrum obtained in the presence of a strong irradiating rf field at the frequency of the upfield peak. (C) Spectrum obtained in the presence of a strong irradiating rf field at the frequency of the downfield peak.

an improvement in signal-to-noise ratio. For this purpose, as well as the resultant nuclear Overhauser effect, most ^{13}C spectra are proton decoupled. The ^{13}C spectra are simplified because there is no ^{13}C–^{1}H splitting. The multiplet collapse and nuclear Overhauser effect lead to higher signal-to-noise ratios in proton-decoupled ^{13}C spectra.

2.5.2. The Nuclear Overhauser Effect

The nuclear Overhauser effect (NOE) experiment is similar to that of the other double irradiation methods. One nuclear resonance is irradiated with a strong rf field $\vec{H}_2$, while another resonance is monitored with a nonsaturating $\vec{H}_1$ field. The intensity of the resonance monitored may change when the other resonance is saturated. That is the nuclear Overhauser effect. It is caused by changes in the populations of the nuclear energy levels. The population changes are a result of nuclear polarization of the monitored nuclei, coupled to the saturated nuclei, caused by spin–lattice relaxation mechanisms. A detailed account of the phenomenon and its chemical applications is found in the monograph by Noggle and Schirmer (54). Careful study of the NOE provides qualitative structural information and, in some favorable cases, estimated internuclear distances (55).

The process leading to the observed nuclear Overhauser effect may be viewed in the following manner. For simplicity discussion will be restricted to a two-spin system in which the spins $I_1 = I_2 = \frac{1}{2}$. I_1 and I_2 are not J coupled but are sufficiently close to one another that dipole–dipole coupling between the two will contribute to spin–lattice relaxation. A schematic for the energy levels in this two-spin system is given in Fig. 2-18. Transitions between the various energy levels will occur continuously.

The resonance of nucleus 1 is saturated so the I_1^- and I_1^+ levels are

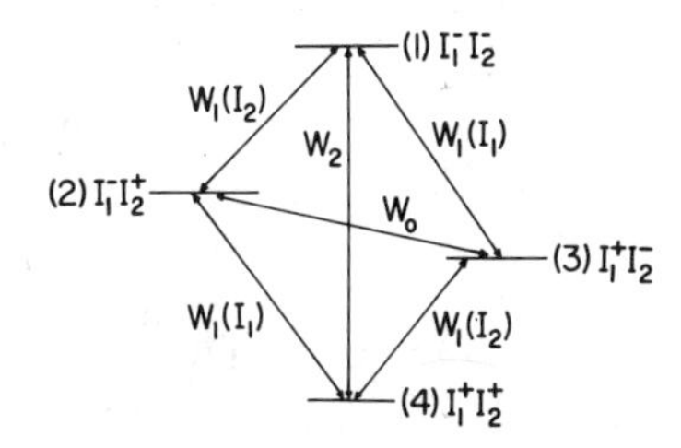

Fig. 2-18. Energy level diagram for a two-spin system ($I_1 = I_2 = \frac{1}{2}$).

Level (1). Spin I_1 is α; I_2 is α - - - $\alpha\alpha$ or I_1^-, I_2^-
Level (2). Spin I_1 is α; I_2 is β - - - $\alpha\beta$ or I_1^-, I_2^+
Level (3). Spin I_1 is β; I_2 is α - - - $\beta\alpha$ or I_1^+, I_2^-
Level (4). Spin I_1 is β; I_2 is β - - - $\beta\beta$ or I_1^+, I_2^+

equally populated. Spin–lattice relaxation processes will try to return nucleus 1 to thermal equilibrium so $I_1^- \rightarrow I_1^+$ transitions by relaxation predominate. Because I_1 and I_2 spins can be dipolar coupled, double transitions may occur, e.g., $I_1^-I_2^+ \rightarrow I_1^+I_2^-$. If the nature of the I_1–I_2 coupling is predominantly dipole–dipole, then the prevalent double transitions will be $I_1^-I_2^- \rightleftarrows I_1^+I_2^+$. With the resonance of nucleus 1 saturated and dipolar coupling predominant, the transition $I_1^-I_2^- \rightarrow I_1^+I_2^+$ will be prevalent. This transition leads to a higher population in the lower spin state of nucleus 2 and, consequently, to an NMR peak with enhanced absorption intensity. This process is often referred to as "dynamic polarization."

The fractional enhancement of the integrated intensity of spin I_2 when I_1 is saturated compared to its equilibrium intensity is given by

$$f_{I_2}(I_1) = \frac{W_2 - W_0}{2W_1(I_2) + W_0 + W_2} \cdot \frac{\gamma_{I_1}}{\gamma_{I_2}} \tag{2-68}$$

where γ_{I_2} and γ_{I_1} are the respective gyromagnetic ratios of spins I_2 and I_1. The double transition probabilities W_2 and W_0 and the single transition probabilities W_1 are indicated for the appropriate transitions in Fig. 2-18. Balaram *et al.* (56), using equations derived by Solomon (31), obtained the equation

$$f_{I_2}(I_1) = \frac{5 + \omega^2\tau_c^2 - 4\omega^4\tau_c^4}{10 + 23\omega^2\tau_c^2 + 4\omega^4\tau_c^4} \tag{2-69}$$

for two dipolar-coupled protons. For small molecules in which the dipole–dipole relaxation mechanism is the exclusive means of relaxation, the fractional enhancement will be

$$f_{I_2}(I_1) = \gamma_{I_1}/2\gamma_{I_2} \tag{2-70}$$

For protons the maximum enhancement will be 0.5 or 50%. Any other relaxation mechanisms will contribute to keeping the enhancement less than 50%. The additional reason for using proton decoupling with ^{13}C NMR spectra is implicit in Eq. 2-70. Carbon-13 nuclei are relaxed predominantly by dipole–dipole coupling with protons. Saturation of the proton resonances will lead to an increase of $\sim$2 in ^{13}C signal intensity from the nuclear Overhauser effect. The original report on the Overhauser effect involved electron–nuclear interactions (57). The gyromagnetic ratio of an electron is so much larger than that of a nucleus that saturation of an electron spin can give nuclear resonance enhancements on the order of 1000.

For large molecules, such as proteins and nucleic acids, it is possible to obtain a negative NOE. In such cases, $\omega^4\tau_c^4 > 1$ and, according to Eq. 2-69,

$f_{I_2}(I_1)$ can be negative. For proton–proton interactions, the limiting negative enhancement is -1, which means the signal can completely disappear. It is also apparent from Eq. 2-69 that the NOE will be frequency dependent for macromolecular systems.

Other possible causes for observation of negative nuclear Overhauser effects have been described by Noggle and Schirmer (54): (*a*) the presence of nuclei with gyromagnetic ratios of opposite sign, (*b*) the presence of scalar coupling modulated by chemical exchange or internal motion (time-dependent J), and (*c*) the effect of a third spin proximate to the spins.

Anet and Bourn (58) have shown that NOE measurements can be used to provide qualitative information about molecular geometry. However, an interesting facet of the NOE phenomenon is that, being caused by dipolar relaxation, the magnitude of the NOE enhancement can be correlated with the distance between the monitored nucleus and the saturated nucleus (59). Schirmer *et al.* (60) have used NOE observations to obtain quantitative information about molecular conformations.

The example of the dimethylformamide NOE in the pioneering study by Anet and Bourn (58) can be considered. The two methyl peaks are not clearly distinguished by chemical shifts or coupling constants. Saturation of the low-field methyl peak in an 8% solution of dimethylformamide gave an $18 \pm 1\%$ intensity enhancement of the formyl proton peak. Saturation of the high-field methyl peak gave a $2 \pm 1\%$ decrease in intensity of the formyl proton peak. On that basis, it was concluded that the low-field peak was caused by the methyl group closest to the formyl proton, i.e., the cis-methyl group.

2.5.3. INDOR

INDOR is an acronym derived from Internuclear Double Resonance. If there are two nuclei, P and Q, such that Q contributes to the dipolar relaxation of P, or P and Q are coupled by spin–spin interactions, it is possible to obtain the INDOR spectrum of P by saturating Q. The INDOR spectrum is obtained by observing P at its resonance frequency with a weak $\vec{H}_1$ field while a frequency sweep is made using a strong $\vec{H}_2$ field. A signal will be observed in the resulting INDOR spectrum of P at the frequency where Q resonates. The saturation of Q at this frequency causes a perturbation in the P resonance, which is then recorded. Before the advent of Fourier transform NMR, the INDOR spectra of ^{13}C and ^{14}N interacting with protons were used to achieve improved signal-to-noise ratios for those nuclei (61). INDOR spectra are still useful for determining which nuclei are coupled (either scalar or dipolar) to a nucleus of interest.

2.6. Chemical Exchange

Chemical reactions that proceed slowly can obviously be studied using nuclear magnetic resonance intensity measurements to monitor the concentration of a particular species. Such applications will not be of interest here. Instead, the unique features of magnetic resonance that permit the study of fast reaction kinetics on systems at chemical equilibrium will be considered. One important characteristic of NMR distinguishing it from other spectroscopies is the slow time scale of the phenomenon. It has already been noted that NMR relaxation times can be quite long. If the kinetics of chemical exchange reactions are of the same order of magnitude as the relaxation rates, the exchange kinetics can affect the spectrum. For diamagnetic systems, the kinetic processes that can be studied by NMR are approximately in the range of 10^{-1}–10^{5} sec^{-1}. Even faster processes (up to $\sim 10^{8}$ sec^{-1}) have been measured in studies of ligand exchange kinetics with paramagnetic ions (36). It should also be pointed out that electron paramagnetic resonance (EPR) spectra may be used to examine rate processes of 10^{5}–10^{12} sec^{-1} (e.g., 62).

This section will be concerned with how NMR spectral changes can be used for quantitative studies of intramolecular and intermolecular chemical exchange kinetics. The reader may consult the excellent article by Johnson (63) for a detailed discussion of the theory.

2.6.1. Bloch Equations and Rate Processes

For simplicity, how the exchange of a group of nuclei between two different chemical environments (or sites) affects the NMR will be considered. The approach is easily generalized to more than two sites. The nuclei will have different chemical shifts in the two different sites and, for the moment, effects of spin–spin coupling will be considered negligible. For example, in an aqueous ethanol solution, the water protons quickly exchange with the hydroxyl protons on ethanol. When exchange is slow, two individual NMR resonances for the exchanging protons are observed. When exchange is rapid, only one peak is observed with an average chemical shift. Although the solution is at equilibrium, it is a dynamic equilibrium with exchange of protons between ethanol and water molecules. NMR is a unique spectroscopic technique in that it can tell how fast that exchange is.

The effects of exchange are most easily understood in terms of modification of the Bloch equations (Section 2.1). Gutowsky *et al.* (64) were the first to use this approach. McConnell (65) used a more direct approach in modifying the Bloch equations. McConnell's modifications will be presented here.

A magnetic nucleus is reversibly exchanging between sites A and B, having chemical shifts $\omega_A/2\pi$ and $\omega_B/2\pi$. It is assumed that the transition state exists for a negligibly short period of time and that no spin dephasing occurs during this time. The lifetimes of the nucleus in sites A and B are τ_A and τ_B. The probability of the nucleus jumping from A to B is $1/\tau_A$ and the probability from B to A is $1/\tau_B$. A nucleus jumping from one A site to another A site will have no effect on magnetization.

Just as any rate expression may be modified when another rate process is included, the Bloch equations in the rotating frame (see Section 2.1.1), modified by exchange, are

$$\frac{dM_z^A}{dt} = \gamma H_1 v_A - \frac{M_z^A - M_0^A}{T_{1A}} - \frac{M_z^A}{\tau_A} + \frac{M_z^B}{\tau_B} \tag{2-71a}$$

$$\frac{dM_z^B}{dt} = \gamma H_1 v_B - \frac{M_z^B - M_0^B}{T_{1B}} - \frac{M_z^B}{\tau_B} + \frac{M_z^A}{\tau_A} \tag{2-71b}$$

$$\frac{dv_A}{dt} = \Delta\omega_A u_A - \gamma H_1 M_z^A - \frac{v_A}{T_{2A}} - \frac{v_A}{\tau_A} + \frac{v_B}{\tau_B} \tag{2-71c}$$

$$\frac{dv_B}{dt} = \Delta\omega_B u_B - \gamma H_1 M_z^B - \frac{v_B}{T_{2B}} - \frac{v_B}{\tau_B} + \frac{v_A}{\tau_A} \tag{2-71d}$$

$$\frac{du_A}{dt} = -\Delta\omega_A v_A - \frac{u_A}{T_{2A}} - \frac{u_A}{\tau_A} + \frac{u_B}{\tau_B} \tag{2-71e}$$

$$\frac{du_B}{dt} = -\Delta\omega_B v_B - \frac{u_B}{T_{2B}} - \frac{u_B}{\tau_B} + \frac{u_A}{\tau_A} \tag{2-71f}$$

where the superscripts and subscripts A and B refer to the nucleus at sites A and B; M_z is the z component of the magnetization, M_0 is the z component of the magnetization at equilibrium; u and v are the respective components in phase and out of phase with the rotating rf field; $\Delta\omega_A = \omega_A - \omega$; and $\Delta\omega_B = \omega_B - \omega$ (ω is the frequency of the rf field).

Equations 2-71 can be solved in the slow-passage case characteristic of frequency- or field-swept high-resolution NMR by setting

$$\frac{dM_z^A}{dt} = \frac{dM_z^B}{dt} = \frac{dv_A}{dt} = \frac{dv_B}{dt} = \frac{du_A}{dt} = \frac{du_B}{dt} = 0$$

Combining the equations for u and v components into one expression for the total complex moment (66),

$$G = G_A + G_B$$

$$= -i\gamma H_1 M_0 \frac{\tau_A + \tau_B + \tau_A\tau_B(\alpha_A P_A + \alpha_B P_B)}{(1 + \alpha_A\tau_A)(1 + \alpha_B\tau_B) - 1} \quad (2\text{-}72)$$

where $i = \sqrt{-1}$,

$$\alpha_A = 1/T_{2A} - i\,\Delta\omega_A \quad (2\text{-}73a)$$

$$\alpha_B = 1/T_{2B} - i\,\Delta\omega_B \quad (2\text{-}73b)$$

and the fraction of nuclei in sites A and B, P_A and P_B $(= 1 - P_A)$, are related to τ_A and τ_B by

$$P_A = \frac{\tau_A}{\tau_A + \tau_B} \quad (2\text{-}74a)$$

and

$$P_B = \frac{\tau_B}{\tau_A + \tau_B} \quad (2\text{-}74b)$$

The complex moments for sites A and B are

$$G_A = u_A + iv_A \quad (2\text{-}75a)$$

and

$$G_B = u_B + iv_B \quad (2\text{-}75b)$$

Equation 2-72 also uses the following relations:

$$M_z^A \approx M_0^A = P_A M_0 \quad (2\text{-}76a)$$

$$M_z^B \approx M_0^B = P_B M_0 \quad (2\text{-}76b)$$

where M_0 is the total equilibrium magnetization, and it is assumed that the z components are not perceptibly different from their equilibrium magnetizations in the presence of a weak H_1 field. It will now be seen how the rate of exchange can affect the spectrum of two lines that are not split into multiplets. Reference to Fig. 2-19 will aid understanding.

2.6.2. Exchange between Two Different Sites

A. *Slow Exchange*

In the limit of very slow exchange, two separate lines are observed at $\nu_A = \omega_A/2\pi$ and $\nu_B = \omega_B/2\pi$. The limit of slow exchange is expressed as:

$$\tau_A, \tau_B \gg 1/(\omega_A - \omega_B)$$

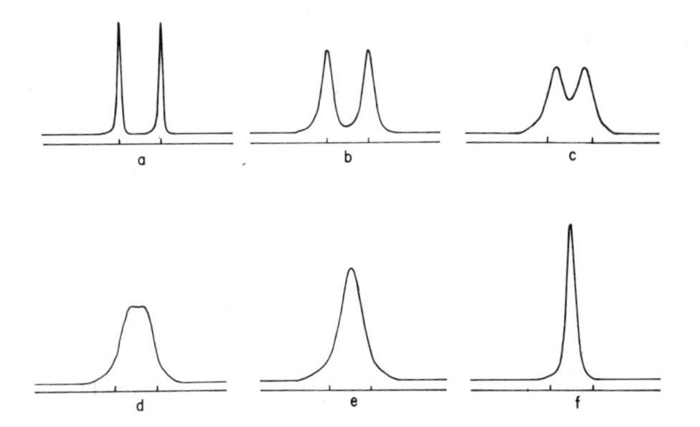

FIG. 2-19. Effect of exchange rate on the spectrum for a system of two sites with equal populations. The exchange rate increases from spectrum (a) through spectrum (f): (a) $2\pi\tau\Delta\nu = 14.7$; (b) 3.7; (c) 1.8; (d) $\sqrt{2}$; (e) 1; and (f) 0.5.

In this case each line can be considered separately. The lineshapes will be Lorentzian, as described by Eq. 2-14, and the intensity of each line will be weighted by the fraction of nuclei at each site. The out-of-phase component at site A, giving the absorption signal at ν_A, will be

$$v_A = -\gamma H_1 P_A M_0 \frac{T'_{2A}}{1 + (T'_{2A})^2(\Delta\omega_A)^2} \tag{2-77}$$

where the effective transverse relaxation time T'_{2A}, determined from the linewidth, is related to the relaxation time in the absence of exchange T_{2A} by

$$1/T'_{2A} = 1/T_{2A} + 1/\tau_A \tag{2-78}$$

The rate constant k_{ex} for exchange is simply $1/\tau_A$, which is obtained from the increased broadening caused by exchange. A similar consideration is made for the nucleus at site B having a peak at ν_B. A drawback for determining rate constants from the line broadening is that T_{2A} must be accurately known.

B. *Rapid Exchange*

In the limit of rapid exchange, τ_A and τ_B are small. In that case, the imaginary part of Eq. 2-72 gives the absorption signal (66):

$$v = -\gamma H_1 M_0 \frac{T_2'}{1 + T_2'^2(P_A\omega_A + P_B\omega_B - \omega)^2} \tag{2-79}$$

which is a single resonance peak centered at frequency

$$\nu = P_A\nu_A + P_B\nu_B \tag{2-80}$$

with a linewidth $W_{1/2}$ [$= 1(\pi T_2)$] such that

$$\frac{1}{T_2} = \frac{P_A}{T_{2A}} + \frac{P_B}{T_{2B}} \tag{2-81}$$

(see Fig. 2-19).

In the case of an exchange rate just a bit slower than the rapid exchange limit, the exchange rate will contribute to the linewidth and Eq. 2-81 is replaced by

$$\frac{1}{T_2} = \frac{P_A}{T_{2A}} + \frac{P_B}{T_{2B}} + P_A{}^2P_B{}^2(\omega_A - \omega_B)^2(\tau_A + \tau_B) \tag{2-82}$$

The experimental consequence of this is that exchange may lead to a T_2 much smaller that T_1. With the conventional sweep experiment, measurement of the linewidth will ultimately be limited by field inhomogeneity. Spin echo measurements (see Section 2.1.3,C) of T_2 will not be limited by magnetic field inhomogeneity and permit faster exchange rates to be measured (9).

C. *Intermediate Rates of Exchange*

When τ_A and τ_B are on the order of $1/(\omega_A - \omega_B)$, the complete expression for the imaginary part of G must be considered. Gutowsky and Holm (67) gave the complete expression as well as the following simplified equation for the absorption signal:

$$v = -\tfrac{1}{4}\gamma H_1 M_0 \frac{\tau(\omega_A - \omega_B)^2}{\{\tfrac{1}{2}(\omega_A + \omega_B) - \omega\}^2 + \tau^2(\omega_A - \omega)^2(\omega_B - \omega)^2} \tag{2-83}$$

which holds for the conditions of (*a*) equal lifetimes and populations such that

$$P_A = P_B = \tfrac{1}{2}$$

and

$$\tau = \frac{\tau_A\tau_B}{\tau_A + \tau_B} = \frac{\tau_A}{2} = \frac{\tau_B}{2}$$

and (*b*) the transverse relaxation times are sufficiently large that

$$1/T_{2A} = 1/T_{2B} \approx 0$$

The coalescing peaks in Fig. 2-19 are represented by Eq. 2-83. Under the exact conditions of coalescence,

$$\tau = \frac{\sqrt{2}}{\omega_A - \omega_B} \tag{2-84}$$

Sykes and Parravano (68) have used line broadening measurements in the fast-exchange limit to measure the rate constants k_1 and k_{-1} for exchange of the inhibitors *N*-acetyl-D-glucosamine and di-*N*-acetyl-D-glucosamine with lysozyme. To use equations in the fast-exchange limit, it is necessary to ascertain that the rate of exchange is larger than both the linewidth of the acetyl proton resonance for the bound inhibitor and the difference in acetyl resonance frequency between free and bound inhibitor. The rate constants reported by Sykes and Parravano (68) are given in Table 2-3. The "on" rate constant $k_1 = 1/\tau_A$ where τ_A is determined from the line broadening. The "off" rate constant k_{-1} is obtained using the dissociation constant K_d and the relationship $K_d = k_1/k_{-1}$. This is a valid procedure for obtaining the "off" rate constant because the "on" rate constant is measured with the system at equilibrium.

2.6.3. Techniques for Determination of Exchange Rates

For study of chemical exchange kinetics, it is desirable to measure τ (or τ_A or τ_B). Obtaining the point of coalescence by variation of temperature or concentration is one method. Linewidth and peak separation measurements have also been used to obtain the exchange rate with the fast- and slow-exchange equations presented above. The intensity ratio method (69) utilizes the ratio of the peak intensities to the intensity midway between the peaks in order to obtain the exchange rate. More complicated systems may be analyzed by computer calculation of the entire lineshape (70) as a function of exchange rate and comparison of the theoretical curves generated with observed spectra. Calculations incorporating the effects of chemical exchange and spin–spin splitting (which have so far been ignored) have also been reported (9, 69, 71). Spin echo techniques have also proved useful

TABLE 2-3

Rate Constants for Exchange of Inhibitors with Lysozyme[a]

Inhibitor	k_{-1} (sec^{-1})	k_1 (molal^{-1} sec^{-1})	Solvent	pH
Methyl-α-GlcNAc	$(5.5 \pm 2.1) \times 10^3$	$(1.4 \pm 0.7) \times 10^5$	D_2O	5.3
Methyl-β-GlcNAc	$(4.5 \pm 1.7) \times 10^3$	$(1.6 \pm 0.8) \times 10^5$	D_2O	5.3
α-GlcNAc	$(8.5 \pm 2.5) \times 10^3$	$(3.5 \pm 1.2) \times 10^5$	H_2O	5.2
β-GlcNAc	$(3.3 \pm 1.2) \times 10^3$	$(1.4 \pm 0.5) \times 10^5$	H_2O	5.2
DiGlcNAc	720 ± 120	3.6×10^6	H_2O, 0.1 *M* NaCl	6.0

[a] From Sykes and Parravano (68).

for determining exchange rates (9, 72). Critiques of the various methods have appeared (73, 74). The spin echo techniques seem to have the advantages over the high-resolution CW techniques in theory, but in practice the usual spin echo instruments lack the stability of high-resolution instruments. Inglefield *et al.* (74), in particular, have criticized the spin echo technique for giving activation energies 10–15% too low. Use of Fourier transform NMR spectrometers in conjunction with the spin echo methods may overcome the instrumental stability problems.

2.6.4. Multiplet Collapse Caused by Exchange

In the previous section, spin–spin splitting was neglected and concern was for the collapse of signals with different chemical shifts. Exchange kinetics could be examined by observing the collapse of the chemically shifted resonances. Exchange kinetics can also be studied by observing the collapse of multiplets. As discussed in Section 2.4.2, this is termed "scalar relaxation of the first kind" by Abragam (18), and the rate of exchange $1/\tau_e$ may be determined using Eqs. 2-57 and 2-58, replacing $(T_1)_2$ and $(T_2)_2$ with τ_{ex}. Multiplet structure exists when τ_{ex} is much greater than the coupling constant J. Any rapid exchange process that causes a fluctuation in the value of J can lead to collapse of the multiplets.

Takeda and Stejskal (75) have systematically examined the effects of exchange on doublets. In their study on the fast proton exchange kinetics of methylammonium ion in water, Grunwald *et al.* (69) examined the effect of exchange on the collapse of the quartet (caused by CH_3 coupling with NH_3^+), using the ratio of the maximum intensity to the minimum intensity between multiplet peaks and comparing observed ratios with theoretical ratios from calculated lineshapes.

The reader is referred to Johnson's article (63) for a discussion of theoretical approaches to exchange other than modified Bloch equations. In particular, the complete density matrix treatment and the relaxation matrix method for exchange are covered. A review of the applications of NMR to study of exchange processes in systems of biochemical interest has recently appeared (76).

References

1. F. Bloch, *Phys. Rev.* **70,** 460 (1946).
2. C. P. Slichter, "Principles of Magnetic Resonance." Harper, New York, 1963.
3. T. C. Farrar and E. D. Becker, "Pulse and Fourier Transform NMR." Academic Press, New York, 1971.
4. T. C. Farrar, A. A. Maryott, and M. S. Malmberg, *Annu. Rev. Phys. Chem.* **23,** 193 (1972).

5. H. Y. Carr and E. M. Purcell, *Phys. Rev.* **94,** 630 (1954).
6. E. L. Hahn, *Phys. Rev.* **80,** 580 (1950).
7. S. Meiboom and D. Gill, *Rev. Sci. Instrum.* **29,** 688 (1958).
8. A. Allerhand and H. S. Gutowsky, *J. Chem. Phys.* **41,** 2115 (1964).
9. H. S. Gutowsky, R. L. Vold, and E. J. Wells, *J. Chem. Phys.* **43,** 4107 (1965).
10. D. C. Douglass and D. W. McCall, *J. Phys. Chem.* **62,** 1102 (1958).
11. D. C. Douglass and D. W. McCall, *J. Phys. Chem.* **71,** 987 (1967).
12. I. J. Lowe and R. E. Norberg, *Phys. Rev.* **107,** 46 (1957).
13. R. R. Ernst and W. A. Anderson, *Rev. Sci. Instrum.* **37,** 93 (1966).
14. R. L. Vold, J. S. Waugh, M. P. Klein, and D. E. Phelps, *J. Chem. Phys.* **48,** 3831 (1968).
15. R. Freeman and R. C. Jones, *J. Chem. Phys.* **52,** 465 (1970).
16. J. L. Markley, W. J. Horsley, and M. P. Klein, *J. Chem. Phys.* **55,** 3604 (1971).
17. G. G. McDonald and J. S. Leigh, *J. Magn. Resonance* **9,** 358 (1973).
18. A. Abragam, "The Principles of Nuclear Magnetism." Oxford Univ. Press, London and New York, 1961.
19. J. G. Hindman, A. Svirmickas, and M. Wood, *J. Phys. Chem.* **72,** 4188 (1968).
20. D. E. O'Reilly, *J. Chem. Phys.* **49,** 5416 (1968).
21. L. Petrakis, *J. Phys. Chem.* **72,** 4182 (1968).
22. A. M. Pritchard and R. E. Richards, *Trans. Faraday Soc.* **62,** 1388 (1966).
23. A. Gierer and K. Wirtz, *Z. Naturforsch. A* **8,** 532 (1953).
24. N. E. Hill, *Proc. Phys. Soc., London, Sect. B* **67,** 149 (1954).
25. H. Shimizu, *J. Chem. Phys.* **40,** 754 (1964).
26. D. Wallach, *J. Chem. Phys.* **47,** 5258 (1967).
26a. A. G. Marshall, P. G. Schmidt, and B. D. Sykes, *Biochemistry,* **11,** 3875 (1972).
27. H. S. Gutowsky and D. E. Woessner, *Phys. Rev.* **104,** 843 (1956).
28. A. Allerhand, D. Doddrell, and R. Komoroski, *J. Chem. Phys.* **55,** 189 (1971).
29. R. A. Ogg and J. D. Ray, *J. Chem. Phys.* **26,** 1515 (1957).
30. J. D. Glickson, W. D. Phillips, and J. A. Rupley, *J. Amer. Chem. Soc.* **93,** 4031 (1971).
31. I. Solomon, *Phys. Rev.* **99,** 559 (1955).
32. N. Bloembergen, *J. Chem. Phys.* **27,** 572 and 595 (1957).
33. J. Reuben, G. H. Reed, and M. Cohn, *J. Chem. Phys.* **52,** 1617 (1970).
34. S. H. Koenig, *J. Chem. Phys.* **56,** 3188 (1972).
35. N. Bloembergen and L. O. Morgan, *J. Chem. Phys.* **34,** 842 (1961).
36. T. J. Swift and R. E. Connick, *J. Chem. Phys.* **37,** 307 (1962).
37. Z. Luz and S. Meiboom, *J. Chem. Phys.* **40,** 2686 (1964).
38. J. Eisinger, R. G. Shulman, and B. M. Szymanski, *J. Chem. Phys.* **36,** 1721 (1962).
39. H. G. Hertz, *Progr. NMR Spectrosc.* **3,** 159 (1967).
40. K. T. Gillen and J. H. Noggle, *J. Chem. Phys.* **53,** 801 (1970).
41. W. Gordy, *Discuss. Faraday Soc.* **19,** 14 (1955).
42. T. L. Stengle and J. D. Baldeschwieler, *Proc. Nat. Acad. Sci. U. S.* **55,** 1020 (1966).
43. T. L. James and J. H. Noggle, *Bioinorg. Chem.* **1,** 425 (1972).
44. T. L. James and J. H. Noggle, *J. Amer. Chem. Soc.* **91,** 3424 (1969).
45. H. S. Gutowsky, I. J. Lawrenson, and K. Shimomura, *Phys. Rev. Lett.* **6,** 349 (1961).
46. C. S. Johnson, Jr., J. S. Waugh, and J. N. Pinkerton, *J. Chem. Phys.* **35,** 1128 (1961).
47. A. S. Dubin and S. I. Chan, *J. Chem. Phys.* **46,** 4533 (1967).
48. P. S. Hubbard, *Phys. Rev.* **131,** 1155 (1963).
49. D. K. Green and J. G. Powles, *Proc. Phys. Soc., London* **85,** 87 (1965).
50. T. E. Burke and S. I. Chan, *J. Magn. Resonance* **2,** 120 (1970).

51. H. M. McConnell and C. H. Holm, *J. Chem. Phys.* **25,** 1289 (1956).
52. E. L. Mackor and C. MacLean, *Progr. NMR Spectrosc.* **3,** 152 (1967).
53. W. McFarlane, *Annu. Rev. NMR* (*Nucl. Magn. Resonance*) *Spectrosc.* **1,** 135 (1968).
54. J. H. Noggle and R. E. Schirmer, "The Nuclear Overhauser Effect: Chemical Applications." Academic Press, New York, 1971.
55. R. A. Hoffman and S. Forsén, *Progr. NMR Spectrosc.* **1,** 35 (1966).
56. P. Balaram, A. A. Bothner-By, and J. Dadok, *J. Amer. Chem. Soc.* **94,** 4015 (1972).
57. A. W. Overhauser, *Phys. Rev.* **91,** 476 (1953).
58. F. A. L. Anet and A. J. R. Bourn, *J. Amer. Chem. Soc.* **87,** 5250 (1965).
59. R. A. Bell and J. K. Saunders, *Can. J. Chem.* **48,** 1114 (1970).
60. R. E. Schirmer, J. H. Noggle, J. P. Davis, and P. A. Hart, *J. Amer. Chem. Soc.* **92,** 3266 (1970); erratum: **92,** 7239 (1970).
61. E. B. Baker, *J. Chem. Phys.* **37,** 911 (1962).
62. R. G. Pearson and T. Buch, *J. Chem. Phys.* **36,** 1277 (1962).
63. C. S. Johnson, Jr., *Advan. Magn. Resonance* **1,** 33 (1965).
64. H. S. Gutowsky, D. W. McCall, and C. P. Slichter, *J. Chem. Phys.* **21,** 279 (1953).
65. H. M. McConnell, *J. Chem. Phys.* **28,** 430 (1958).
66. H. S. Gutowsky and A. Saika, *J. Chem. Phys.* **21,** 1688 (1953).
67. H. S. Gutowsky and C. H. Holm, *J. Chem. Phys.* **25,** 1228 (1956).
68. B. D. Sykes and C. Parravano, *J. Biol. Chem.* **244,** 3900 (1969).
69. E. Grunwald, A. Loewenstein, and S. Meiboom, *J. Chem. Phys.* **27,** 630, 646 (1957).
70. R. J. Kurland, M. B. Rubin, and W. B. Wise, *J. Chem. Phys.* **40,** 2426 (1964).
71. A. Paterson, Jr. and R. Ettinger, *Z. Electrochem.* **64,** 98 (1960).
72. A. Allerhand and H. S. Gutowsky, *J. Chem. Phys.* **41,** 2115 (1964).
73. A. Allerhand, H. S. Gutowsky, J. Jonas, and R. A. Meinzer, *J. Amer. Chem. Soc.* **88,** 3185 (1966).
74. P. T. Inglefield, E. Krakower, L. W. Reeves, and R. Stewart, *Mol. Phys.* **15,** 65 (1968).
75. M. Takeda and E. O. Stejskal, *J. Amer. Chem. Soc.* **82,** 25 (1960).
76. B. D. Sykes and M. D. Scott, *Annu. Rev. Biophys. Bioeng.* **1,** 27 (1972).

CHAPTER 3

CHEMICAL SHIFTS AND STRUCTURE

The electronic shielding of a nucleus in a magnetic field and the resultant chemical shift in the nuclear magnetic resonance frequency were concepts introduced in Section 1.4. In general, the NMR resonance frequencies of nuclei from different elements will occur in vastly different regions of the spectrum for a given magnetic field. However, the topic of interest here is more subtle differences in resonance frequencies for nonequivalent nuclei of the same isotope, e.g., Why do the two types of carbon-13 in CH_3COOH have different NMR resonance frequencies? The different resonance frequencies result from the different electronic environments around the two carbon-13 nuclei, leading to variations in electronic shielding. A measure of the different resonance frequencies is the chemical shift that denotes the resonance position of the NMR signal with respect to a reference. In this section we will cover primarily those aspects of shielding and chemical shifts that have proved pertinent to biochemical studies.

3.1. Shielding

The presence of electrons around a nucleus causes the nucleus to be "shielded" to a certain extent from the applied magnetic field. The applied magnetic field induces electronic currents, setting up local magnetic fields around the nucleus that vary depending on the strength of the applied magnetic field, the type of bonds containing the electrons, the electronegativity of other elements involved in the bonds, and interactions of other molecules with the molecule containing the nucleus of interest. Such an array of contributions to the chemical shift presently makes an "exact" *ab initio* calculation of the chemical shift of any nucleus improbable. Nevertheless, calculations (usually approximate) have provided insight into the origin of chemical shifts.

The early theory of chemical shifts was developed by Lamb (1), who calculated the electronic shielding of a spherically symmetric atom. Ramsey (2, 3) used second-order perturbation theory to calculate the electronic currents produced in a molecule by an external field and the magnetic field produced at the nucleus by these electronic currents. Numerous additional theoretical calculations have been made. The articles by Lipscomb (4) and Musher (5) and the monograph by Memory (6) may be consulted for discussions of various theoretical approaches to the calculation of chemical shifts. This section will be concerned only with a qualitative picture of the shielding mechanism.

3.1.1. Contributions to Shielding

The shielding that a nucleus experiences is derived from at least three sources of electronic circulations: (*a*) local diamagnetic effects; (*b*) diamagnetic and paramagnetic effects from neighboring atoms; and (*c*) effects from interatomic currents. For any particular nucleus, these three sources will have different relative contributions. We may express the shielding constant as the sum of its parts from sources (*a*), (*b*), and (*c*) listed above:

$$\sigma = \sigma_a + \sigma_b + \sigma_c \tag{3-1}$$

where σ_a is positive, but σ_b and σ_c may be either positive or negative depending on the electronic environment of the nucleus.

A. *Diamagnetic Effects*

A fairly crude physical interpretation of σ_a, the shielding contribution from atomic diamagnetism, is depicted in Fig. 3-1 for an atom with spherical electronic symmetry. When a magnetic field H_0 is applied to an atom, electronic currents are induced in the plane perpendicular to the applied magnetic field. These diamagnetic currents, in turn, produce a magnetic field that acts to partially cancel the applied field and thus to shield the nucleus. This action means a larger magnetic field must be applied to

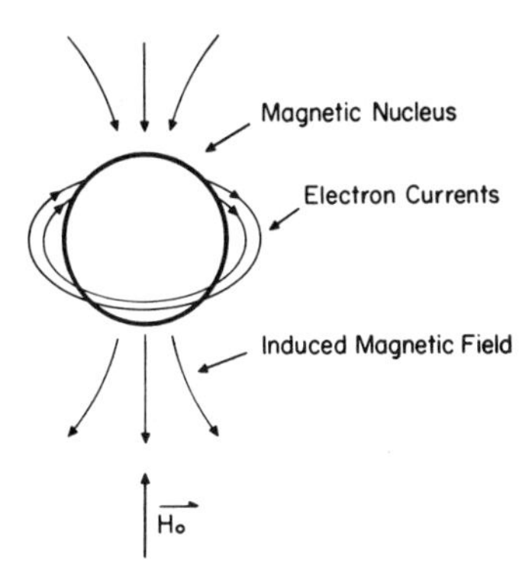

FIG. 3-1. Diamagnetic atomic currents around a nucleus and the resultant magnetic field opposed to the applied magnetic field H_0.

achieve the resonance condition ($\nu_0 = \gamma H_0/2\pi$) than would be necessary for a bare nucleus. Lamb (1) theoretically calculated the atomic diamagnetism and found the contribution from σ_a to be greatest for spherical electron distributions.

The contributions of neighboring atoms to the chemical shift is embodied in σ_b. These effects may be either diamagnetic or paramagnetic in nature. A diamagnetic source will be considered first. The diamagnetic shielding is roughly proportional to the electron density around the nucleus. For simple saturated molecules, the electron density will be increased if substituent groups contain electropositive atoms and decreased if substituent groups contain electronegative atoms. This dependence on the chemical shift of the proton resonance is illustrated in Fig. 3-2. A remarkably good correlation of the proton chemical shift in methyl halides with electronegativity of the halide atom is apparent. As a first approximation, the electronegativity effect may be used to estimate relative chemical shifts. For example, CH_3S— protons are more shielded than the protons in CH_3O—. If simple diamagnetic effects made the only contributions to shielding, the chemical shift could easily be predicted by summation of the contributions from the various substituent groups. However, the situation is frequently not that simple.

B. *Paramagnetic Effects*

If diamagnetic effects were the sole source of shielding, a shift of the ^{14}N resonance to lower fields would be expected as the number of more electronegative oxygen atoms bonded to nitrogen increased. Instead the

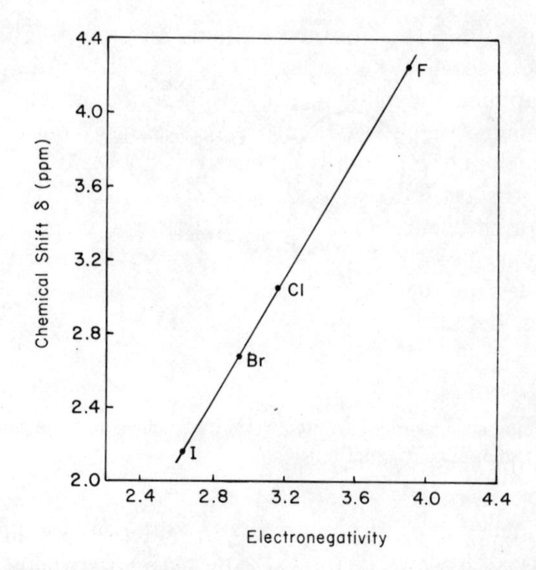

FIG. 3-2. The chemical shift of the methyl protons in the methyl halides as a function of the electronegativity (Pauling's) of the halide atom.

observed ^{14}N chemical shifts are as shown in the following tabulation:

COMPOUND	CHEMICAL SHIFT (ppm)[a]
N_2	14
NO_2^-	−241
NO_3^-	0

[a] Relative to NO_3^-.

As would be expected from purely diamagnetic considerations, nitrite has a -255 ppm shift downfield from molecular nitrogen. However, nitrate has nearly the same chemical shift as molecular nitrogen. The cause of this discrepancy has been ascribed to paramagnetic contributions from the nitrogen lone pair electrons mixing the low-lying excited electronic states, $n \rightarrow \pi^*$ transitions in the case of ^{14}N (7). Pople (8) has shown that the paramagnetic term arises from mixing of electronic ground states and excited states by the applied magnetic field. The importance of paramagnetic terms has also been noted in the case of hydrocarbons (9).

Paramagnetic shielding originates in the electronic currents induced by the applied magnetic field in bonding electrons, nonbonding electrons, or electrons completely localized on other atoms or groups. These induced currents produce a secondary magnetic field parallel to (and thereby reinforcing) the applied magnetic field, as illustrated in Fig. 3-3, which provides a paramagnetic contribution to the shielding. Whether or not this paramagnetic shielding contributes a negative or a positive term to Eq. 3-1 depends on the orientation of the magnetic nucleus relative to the induced field. For the illustration in Fig. 3-3, the paramagnetic contribution σ_b would be negative. In the case of proton NMR, paramagnetic contributions from electron currents around neighboring atoms, such as carbon, nitrogen, oxygen, or sulfur, frequently come into play.

C. *Shielding Anisostropy*

On the basis of present knowledge of diamagnetic effects, a linear decrease in the proton resonance frequencies of ethane, ethylene, and acetylene, respectively, might be expected in accordance with the electron-withdrawing tendency of the attached group. However, the proton resonances are observed to be 0.96δ (ethane), 5.84δ (ethylene), and 2.88δ (acetylene). Obviously, acetylene is out of line if only diamagnetic effects are considered; acetylene is more shielded than ethylene. The high degree of shielding in acetylene can be satisfactorily explained on the basis of anisotropy in the shielding from neighboring groups.

As was just discussed, the shielding of a nucleus is affected by electronic

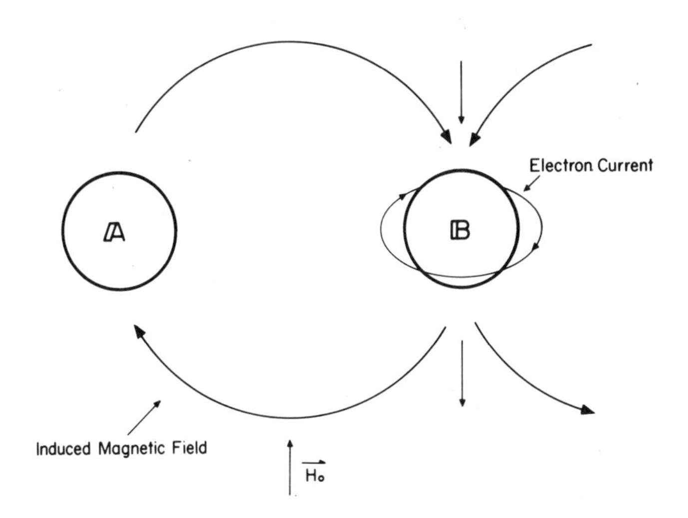

FIG. 3-3. Paramagnetic effects at nucleus A caused by the secondary magnetic field arising from induced electronic currents at nucleus B.

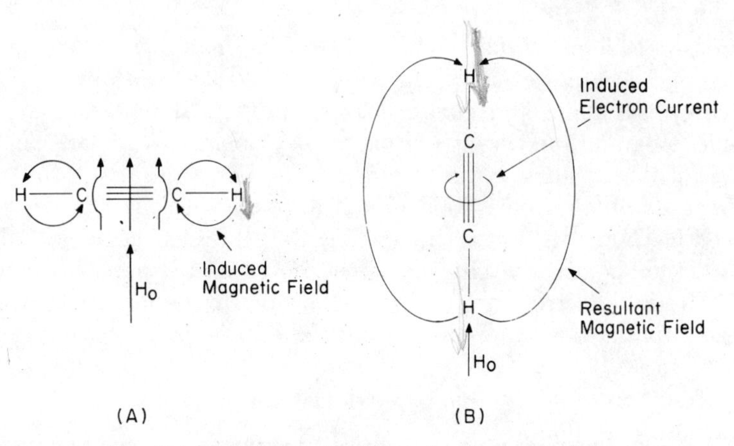

FIG. 3-4. Anisotropic shielding of acetylene protons.

currents in other parts of the same molecule. The fields caused by such currents will be averaged to zero by rapid rotation about single bonds, with no net shielding effect. However, with molecules that do not have all moieties freely rotating, these electronic currents can produce shielding effects. Whether the resultant effect is shielding or deshielding depends on the orientation of the nucleus with respect to the magnetic field produced by the electronic currents. This aspect of the dependence on orientation is embodied in the term "anisotropy."

The anomalous chemical shift of the acetylene proton resonance has been ascribed to diamagnetic shielding caused by electronic currents at the carbon atoms when the molecule is perpendicular to the direction of the magnetic field (10), as shown in Fig. 3-4a. There is an additional anisotropic effect when the linear acetylene is aligned parallel to the direction of the applied magnetic field, as shown in Fig. 3-4b. In the parallel orientation, electron currents will be induced perpendicular to the molecular axis in the cylindrical π electron cloud, resulting in a diamagnetic shielding field at the proton.

Deshielding of an aldehydic proton provides another example of anisotropic shielding. The observed field position ($\sim 10\delta$) for aldehydic proton resonances is anomalous. The anisotropic deshielding is illustrated in Fig. 3-5. Electronic currents are induced in the π electron cloud of the C=O bond when the planar aldehyde group is oriented perpendicular to the direction of the applied field. The resultant magnetic field reinforces the applied field at the proton culminating in an anisotropic deshielding. This anisotropic deshielding undoubtedly plays a role in the chemical

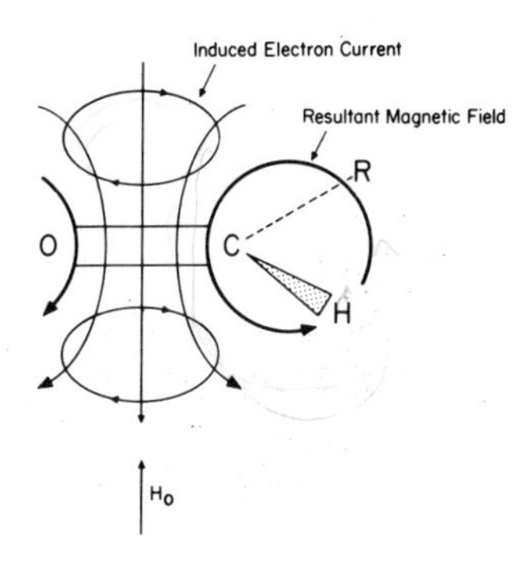

FIG. 3-5. Anisotropic deshielding of the aldehydic proton.

shifts of any $\begin{matrix} R \\ H \end{matrix}\!\!>\!\text{C}{=}\text{X}$ proton chemical shift, just as the acetylene explanation holds for R(H)—C≡X.

D. *Ring Currents*

Protons on aromatic rings resonate at anomalously low fields. The deshielding arises from the large electronic currents induced in the π electron orbital loops above and below the aromatic ring (11). The effect is illustrated in Fig. 3-6 for benzene and its six π electrons. The stationary magnetic field H_0 induces a large electron current to flow in the π orbital loops when the plane of the benzene ring is perpendicular to the direction of the H_0 field. The electron current sets up a resultant magnetic field, as

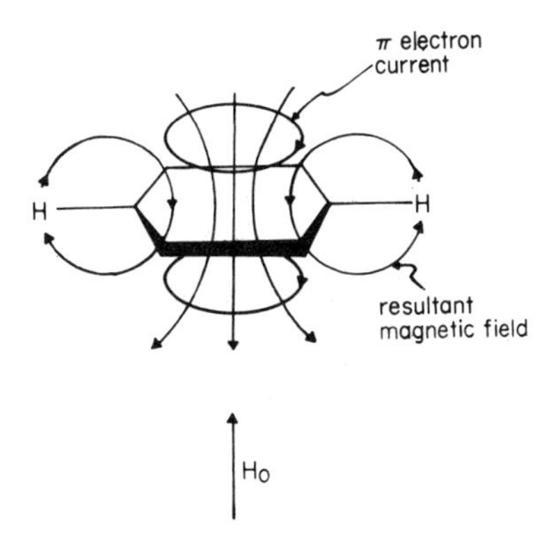

FIG. 3-6. Ring current effect causing deshielding of benzene protons.

shown in Fig. 3-6, which has the effect of reinforcing H_0 at the aromatic protons and leads to deshielding. Consequently, aromatic protons resonate at low fields. For example, Waugh and Fessenden (11) found the proton resonance of benzene at 1.48 ppm lower field than the proton resonance in the analogous olefinic compound 1,3-cyclohexadiene.

Figure 3-6 also indicates that protons which lie above or below the face of the ring will experience additional shielding. It can be shown that the anisotropic shifts caused by ring currents (or other electronic currents for that matter) are proportional to

$$(3 \cos^2 \theta - 1)/r^3$$

where the anisotropic group (i.e., the ring) is represented by a point dipole at the center of the ring where r is the distance from the point dipole to the monitored nucleus and θ is the angle between the radius vector (from the point dipole to the nucleus) and the axis normal to the plane of the ring (sixfold axis of symmetry for a benzene ring).

Johnson and Bovey (12) used Pauling's free electron model (13) to calculate the chemical shift caused by the ring current effect for a proton placed at any point in space. The calculations were made using benzene but were found to give fairly good chemical shift estimates for other aromatic compounds. Protons located directly above or below the plane of the ring will be shielded; those in the plane of the ring will be deshielded. The semiempirical theory of Johnson and Bovey overestimates the deshielding effects in aromatic hydrocarbons (14). Therefore, Haigh and Mallion (15) have developed a new theory and have published numerical tables of the predicted chemical shift for a proton placed at any distance or angle from the plane of the benzene ring.

One of the most complicated "ring current problems" likely to be encountered in the biochemical trade is the prediction of chemical shifts in heme proteins caused by the extensively conjugated porphyrin ring (cf. Fig. 3-7).

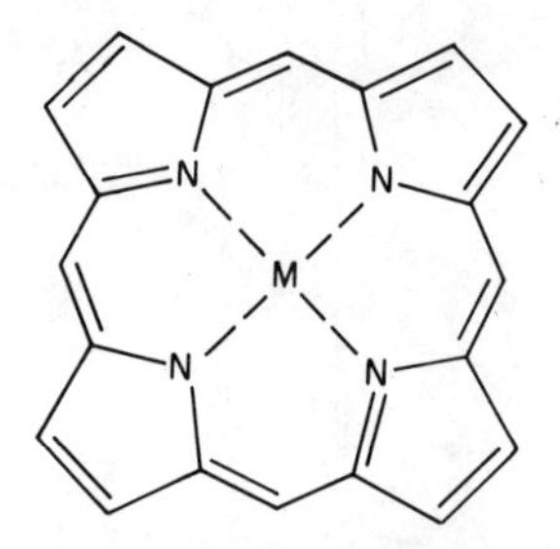

FIG. 3-7. Porphyrin ring system.

The porphyrin ring system has 18 electrons that can circulate around four pyrrole rings and a large "inner" ring. Abraham (16) tackled the problem using essentially the same theoretical approach as Johnson and Bovey for each of the individual rings. However, the calculated shifts for protons on the prophyrin ring were about 50% larger than the observed shifts. The more recent calculations by Shulman *et al.* (17), neglecting the spin–orbital coupling, have been quantitatively more accurate.

Many aromatic rings of biochemical interest (e.g., pyrimidine) are heterocyclic. In such cases, the inductive effect of the heteroatom or atoms, as well as the ring size and number of rings, should be taken into account if accurate chemical shift predictions are desired (17–20a).

3.1.2. Effect of Molecular Interactions on Shielding

Solvent influences, as well as such specific interactions as molecular association, complex formation, and hydrogen bonding, will affect chemical shifts.

A. *Solvent Effects*

The NMR chemical shifts for nuclei in a molecule are also influenced by surrounding molecules. In the absence of specific molecular interactions, solvent molecules will have the most profound effect. For example, the proton resonance of methane gas shifts downfield 1.13 ppm when it is dissolved in benzene and 1.99 ppm when it is dissolved in carbon disulfide. NMR studies on the solvent effect have been reviewed by Pople *et al.* (21) and by Foreman (21a). The factors pertaining to solvent molecules are also applicable to other solute molecules in the absence of actual bond formation between molecules. Buckingham *et al.* (22, 23) considered the "long-range" and "short-range" forces affecting the chemical shift in solution and proposed four shielding contributions from the solvent (23):

$$\sigma_{\text{solvent}} = \sigma_b + \sigma_a + \sigma_w + \sigma_E \tag{3-2}$$

where σ_b is a "long-range" term arising from the bulk magnetic susceptibility of the medium; σ_a arises from the anisotropy in the molecular susceptibility of the solvent molecules; σ_w arises from van der Waals or dispersion forces between solute and solvent; and σ_E is the "polar effect" caused by the charge distribution in the neighboring solvent molecules leading to permanent electric dipole or quadrupole fields acting on the solute. Each of the effects was experimentally demonstrated for proton resonances by an appropriate choice of solute–solvent systems (23). It was found that σ_b and σ_w act to deshield the proton; σ_a shields in such disk-shaped solvents as benzene but deshields in such rod-shaped solvents as carbon disulfide; and σ_E can either shield or deshield, depending on the orientation of the proton nucleus relative to the polar groups on the solute molecule. The effect from σ_E will become more important as the dielectric constant of the solvent is increased. For example, the proton resonance of CH_3CN as the solute in a 5 mole % solution with solvents of differing dielectric constants (ϵ) was found to be linearly dependent on $(\epsilon - 1)/(2\epsilon + 2.5)$, as predicted by theory (23). The data are given in Table 3-1. Solvent shifts that cannot be explained by Eq. 3-2 may be considered as evidence for a more specific interaction, such as molecular association or hydrogen bonding.

B. *Hydrogen Bonding*

Hydrogen bonding is a very common solution phenomenon. Both intramolecular and intermolecular hydrogen bonding will have an effect on the chemical shift. Very early in the history of NMR, Arnold and Packard (24) noted that the hydroxyl peak of ethanol shifted downfield about 1.5 ppm, relative to the methylene peak, on cooling from 78°C to −117°C. The deshielding of the hydroxyl proton was ascribed to hydrogen bond formation involving the hydroxyl proton, with more ethanol dimers being produced at low temperature. This increased deshielding has also been observed in concentration dependence studies. For example, Huggins *et al.* (25) measured the chemical shift of the hydroxyl proton for several phenols as a function of concentration in carbon tetrachloride and found the hydroxyl proton resonances to shift upfield as the phenol concentration decreased. They interpreted the results in terms of an equilibrium between the phenol monomer and a hydrogen bonded phenol dimer.

TABLE 3-1

PROTON RESONANCE SHIFT OF THE POLAR SOLUTE CH_3CN IN A 5 MOLE % SOLUTION IN VARIOUS SOLVENTS[a]

SOLVENT	ϵ (DIELECTRIC CONSTANT)	$(\epsilon - 1)/(2\epsilon + 2.5)$	CHEMICAL SHIFT[b] (Hz)
Cyclohexene	2.22	0.176	6.2
Ethyl ether	4.34	0.299	10.4
Ethyl formate	7.1	0.365	14.5
Cyclohexanone	18.3	0.443	16.8
Acetone	20.7	0.449	17.5

[a] From Buckingham *et al.* (23).
[b] Relative to 5 mole % CH_3CN in *n*-hexane at 60 MHz.

Further developments in the treatment of data in terms of hydrogen bonded molecular association have been made (26, 27).

The deshielding of the proton involved in hydrogen bonding poses some theoretical problems. Pople *et al.* (28) have pointed out that the decreased shielding arises from two general effects in the case of intermolecular hydrogen bonding between molecules XH and Y to form X—H···Y:

a. The stationary magnetic field will induce currents in Y, with the resulting magnetic field caused by those currents tending to shield the hydrogen bonding proton.

b. The electron donor molecule Y will distort the electronic structure in the X—H bond, leading to different diamagnetic currents. The result is a deshielding of the proton.

The observed downfield chemical shifts with hydrogen bonding indicate that effect (b) predominates in most cases; the exceptions involve a few aromatic systems.

Hydrogen bonding plays a very important role in many biological situations. One of these is the formation of Watson-Crick base pairs via bonding of complementary bases (adenine–thymine, adenine–uracil, and guanine–cytosine) in nucleic acids. Nuclear magnetic resonance studies have shown that sharp resonances exist at lower fields for protons involved in the Watson-Crick base pairing (29, 30). As an example, NMR studies of mixtures of the various purine and pyrimidine components of nucleic acids in deuterated dimethyl sulfoxide revealed a downfield shift of the proton resonances of the hydrogen bonding protons for mixtures containing the Watson-Crick base pairs but no downfield shift for mixtures containing any other combination of bases (29).

The shifts are summarized in Table 3-2. It is apparent from the results that both amino groups and the guanine–NH are involved in specific hydrogen bonding in the guanine–cytosine pairing. This also supports the idea of the specificity of the Watson-Crick pairing requiring three hydrogen bonds for the guanine–cytosine pair. Scheit (31) extended the studies by Shoup *et al.* (29) to show that inosine is capable of forming hydrogen bonded base pairs with adenine as well as uracil. As will be seen in Section 7.2, the downfield shift produced by hydrogen bond formation has enabled the proton resonances of the hydrogens involved in base pairing in transfer RNA's to be studied.

C. *Protolysis*

When a proton is added to $RCOO^-$, RNH_2, RS^-, etc., to form RCOOH, RNH_3^+, and RSH, respectively, the chemical shifts of nonexchanging

TABLE 3-2

INTERACTION SHIFTS ON BASE PAIRING[a,b,c]

BASES[d]	CONC. (M)	T(°C)	INTERACTION SHIFT (PPM)				
			G–NH	G–NH_2	C–NH_2	T–NH	A–NH_2
G+C	0.1 + 0.1	40	0.5	0.24	0.29		
		20	0.71	0.34	0.36		
	0.2 + 0.2	40	0.7	0.34	0.46		
		20	0.94	0.41	0.43		
A+T	0.1 + 0.1	20				0.05	0.02
	0.2 + 0.2	20				0.08	0.02
G+T	0.2 + 0.2	20	0.00	0.00		0.00	
G+A	0.2 + 0.2	20	0.02	0.00			0.00
C+T	0.2 + 0.2	20			0.03	0.00	
G+C	0.2 + 0.2	40[e]	1.1	0.52	0.5		
		20[e]	1.29	0.59	0.5		
		2[e]	1.46	0.66	{0.43, 0.96}		
		−10[e]	1.57	0.72	{0.44, 1.02}		
	0.1 + 0.2	−10[e]	1.73	0.82	{0.24, 0.51}		

[a] All measurements made in DMSO-d_6 solution except where indicated.

[b] No interaction shift was observed for G–8–H; a small downfield shift ($\leq$0.1 ppm) was found for C–5–H and C–6–H.

[c] From Shoup *et al.* (29).

[d] A, 9-ethyladenine; C, 1-methylcytosine; G, 9-ethylguanine; T, 1-methylthymine.

[e] In DMSO-d_6–DMF-d_7 (1 : 1 by volume).

proton resonances on the R moiety are shifted downfield with the deshielding effect attenuating with increasing distance between the nonexchangeable proton and the site of protonation. For this reason, the particular chemical shift of nuclei on many compounds of biological interest will be pH dependent. The glycine CH_2 proton resonance moves 1.05 ppm downfield, going from a basic to an acidic solution. The H8 and H1′ proton signals of guanosine triphosphate shift downfield 0.90 and 0.16 ppm, respectively, going from a basic to an acidic solution.

An example of the pH dependence of chemical shifts is illustrated in Fig. 3-8 for ethoxyaceteic acid, the spectrum at high pH of which was shown in Fig. 1-3. When a proton associates with a basic site on the ligand, a deshielding effect is produced on the ligand protons that results in a downfield shift of their resonances. Because the deshielding effect attenuates with distance from the site of perturbation, the magnitudes of the downfield shifts will depend on which donor atom on the ligand associates with the acidic proton. The resonances for those protons that are further from the site of acidic proton association will be shifted less than the resonances for the ligand protons that are closer to the protonation site. In the case of ethoxyacetic acid (cf. Fig. 3-8), protonation of the carboxylate group results in downfield shifts between pH 6 and 2 with the magnitude of the shifts being in the order "a" > "b" > "c" as would be expected. In the very acidic solutions (pH 1 to −1), the "a" resonance of ethoxyacetic acid shifts 16.4 Hz downfield, the "b" resonance shifts 15.3 Hz downfield, and the "c" resonance shifts 7.7 Hz downfield. These downfield shifts suggest that the site of further protonation is the ether oxygen because the "b" resonance shifts about the same amount as the "a" resonance. This

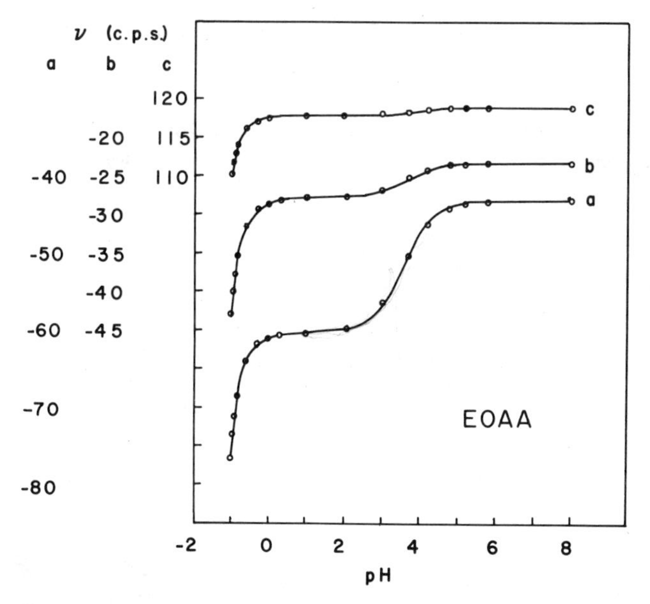

FIG. 3-8. Chemical shifts (60 MHz) of ethoxyacetic acid (EOAA) proton resonances as a function of pH using tetramethylammonium ion as an internal standard (32).

is in distinct contrast to the results for ethylthioacetic acid, in which the site of protonation in acidic solutions is not the thioether sulfur but the carbonyl oxygen (32).

The acid dissociation constant may be easily calculated using a titration plot, such as one in Fig. 3-8.

$$K_a = \frac{[H^+][L]}{[HL]} = [H^+]\left(\frac{\delta - \delta_{HL}}{\delta_L - \delta_{HL}}\right) \tag{3-3}$$

where δ, δ_{HL}, and δ_L are, respectively, the observed chemical shift at a given pH, the chemical shift for the protonated species, and the chemical shift for the unprotonated species. Such expressions as Eq. 3-3 may be used for other equilibria manifesting chemical shifts, including more complicated equilibria.

It will be seen in Chapter 7 that the pH dependence of proton resonances of enzymes can be used as a monitor of denaturation as well as a means of assigning histidine peaks. It has been observed that the histidine proton resonance titration curves for some proteins exhibit inflection points as well as shifts to high or low pH with respect to monomeric histidine. As a basis for understanding the titration curves, Sachs *et al.* (33) examined the chemical shifts of the C2 and C4 proton resonances of imidazole derivatives as a function of pH. As shown in Table 3-3, the presence of neighboring carboxyl and amino groups influences the pK_a values. If the neighboring charged group retains its charge through the pH range for titration of the imidazole ring, the C2 and C4 titration curves will be symmetrical, the entire titration curve being shifted to lower or higher pH values according to the charge on the neighboring group. If a neighboring charged group

TABLE 3-3

pK_a VALUES OBTAINED FROM THE TITRATION CURVES (CHEMICAL SHIFT VS. pH) FOR IMIDAZOLE DERIVATIVES[a]

COMPOUND	NEIGHBORING CHARGE	IMIDAZOLE pK_a (POTENTIOMETRIC)	IMIDAZOLE pK_a (C–2 NMR)	IMIDAZOLE pK_a (C–4 NMR)
Imidazole	0	7.08	7.20	7.15
L-Histidine	±	6.00	6.12	6.20
N-Acetylhistidine	−	7.08	7.19	7.05
Histidine methyl ester	+	5.3	5.52	5.52

[a] From Sachs *et al.* (33).

has a pK_a within 2 pH units of the imidazole pK_a, the C2 and C4 titration curves will exhibit assymmetry. The position of the inflection point depends on whether the neighboring group titrates at a lower or higher pH than pK_a of imidazole.

As illustrated with Fig. 3-8, information about the order of protonation as well as pK_a values may be obtained. Blumenstein and Raftery (34) have used this approach in a study of the ^{31}P and ^{13}C chemical shifts of nicotinamide adenine dinucleotide (NAD), nicotinamide adenine dinucleotide phosphate (NADP), and related compounds. Differences between the oxidized and reduced species were attributed to an interaction of the positively charged nitrogen atom of the pyridine ring with the negatively charged diphosphate backbone in the case of the oxidized species.

D. *Metal Complex Formation*

Just as protolysis leads to movement of nuclear resonance peaks downfield, complex formation with diamagnetic metal ions will give chemical shifts at lower fields for ligand nuclear resonances. The shifts for ligands coordinated to paramagnetic metal ions will be discussed in Section 3.3.

We can consider the effect of formation of a 2:1 complex with Pb(II) on the proton resonance chemical shifts of cysteamine-*N*-acetic acid,

$$HSCH_2\underset{b}{CH_2}N\underset{a}{CH_2}COOH$$

Relative to the chemical shifts of the uncomplexed ligand at pH 12, the chemical shifts for the 2:1 Pb(II) complex are downfield 0.22 ppm for the "a" resonance and 0.64 ppm for the "b" resonance. The downfield shifts, in fact, are evidence for the existence of a complex.

Nitrogen-15 NMR chemical shifts have provided information about metal ion coordination to ATP enriched to 70% ^{15}N at all nitrogen positions (35). No significant ^{15}N chemical shift, within experimental error, was detected on addition of Mg(II) to ATP at pH 9.5. However, it was found on addition of Zn(II) to an ATP solution that the N7 and N9 ring resonances and the N6 amino resonance were shifted. Those observations were taken as evidence for complexation of Zn(II) by the nitrogen atoms of ATP and lack of complexation of Mg(II).

E. *Association of Aromatic Molecules*

Aromatic compounds have been found to associate extensively in solution. This association has considerable implications for nucleic acid structure and interactions. The nature of the interaction has been studied by proton magnetic resonance and has been shown to involve vertical ring stacking in aqueous solution (36–40). In studies of stacking in solutions of monomers (either bases or nucleosides), it was found that the resonances of the ring protons of the bases, the methyl protons of the bases, and the H1′ proton were shifted upfield progressively as the concentration of bases or nucleosides was increased (37, 38). The upfield shift is attributed to the magnetic anisotropy of the ring current effect, as discussed in Section 3.1.1, D. The resonances shifted upfield are caused by protons either above or below the plane of the aromatic rings. The protons will be found in this configuration if the aromatic rings of the bases or nucleosides tend to stack like a deck of cards. It was mentioned in Section 3.1.2, B that NMR evidence also exists for hydrogen bonding between complementary base pairs. The relative importance of stacking and hydrogen bonding depends on solvent conditions. Hydrogen bonded complexes are predominant in nonpolar solvents (29–31, 41), and aromatic stacking complexes are predominant in polar solvents, especially at high salt concentrations (37, 42).

As an example of the molal concentration dependence of the proton chemical shifts, an abbreviated listing of the upfield change in chemical shifts is given in Table 3–4 for some purine nucleosides. On the basis of the chemical shift dependence, the order of the self-association tendency is

TABLE 3-4

MOLAL CONCENTRATION DEPENDENCE (0.0–0.2 MOLAL) OF THE PROTON CHEMICAL SHIFTS OF SOME PURINE NUCLEOSIDES IN D_2O [a]

COMPOUND	TEMPERATURE (°C)	Δσ (Hz)		
		H–2	H–8	H–1′
Inosine	32	6.4	5.3	7.1
1-Methylinosine	33	8.9	6.4	6.8
Ribosylpurine	30	10.7	6.4	13.1
2′-*O*-Methyladenosine	31	13.7	7.5	8.8
2′-Deoxyadenosine	30	19.8	13.0	13.6
N-6-Methyl-2′-deoxyadenosine	32	26.0	15.8	14.0
N-6-Dimethyladenosine	28	27.2	14.5	14.4
N-6-Methyladenosine	26	32.6	17.5	12.6
2′-Deoxyadenosine[b]	30	14.8	10.0	9.8
Adenosine	32	14.8	8.3	6.9

[a] From Broom *et al.* (38).

[b] Differences measured over concentration range 0.0–0.1 *m* because of solubility limitations.

N-6-methylated adenosines > *O*-methyladenosine, deoxyadenosine, and adenosine > 1-methylinosine, ribosylpurine > inosine. This order correlates reasonably well with the calculated polarizability but not with the dipole moments of the nucleoside bases (38).

The precise geometry of the aromatic stacking complexes is difficult to ascertain unambiguously because there is a distinct possibility that more than one geometric form of the complex exists, and there is rapid exchange among the different geometric forms and the uncomplexed nucleoside. Nevertheless, some conclusions have been proffered (36, 39). Scheit *et al.* (39) examined the proton NMR spectra of some deoxytrinucleotides containing thymine, adenine, and guanine in D_2O solution. In particular, their chemical shift measurements showed that thymine bases which follow purine bases in the (3′ → 5′) sequence of a polynucleotide strand exhibit an upfield shift of the thymine CH_3 resonance. A model was suggested in which the three rings were arranged in a partially overlapping "stair–step" configuration. That the rings are not stacked in a completely vertical manner was shown by the spectrum of 5′-*O*-alkylphosphoryldeoxythymidylyl-(3′ → 5′)-deoxyguanylyl-(3′ → 5′)-deoxythymidine, RpdTpdGpdT, which exhibits two different resonances for the thymine methyl protons. The "stair–step" model has a configuration in which only the 3′-terminal thymine methyl group can be influenced by ring currents in guanine.

The measurement of nuclear magnetic resonance chemical shifts and linewidths provides an unique method by which aromatic intermolecular and intramolecular stacking configurations may be investigated.

3.2. Chemical Shifts

The chemical shift for any particular nucleus is determined by the sum of all diamagnetic and paramagnetic effects accounting for solvent, temperature, association, and pH (in aqueous solutions). The usual positions of the proton NMR peaks for hydrogen in various structural environments are shown in Table 3-5. In some instances a peak will not fall in the range indicated in Table 3–5. The deviations can easily be understood in terms of the effects presented in Section 3.1. Extensive surveys of proton chemical shifts have been presented (43–46). Compilations of proton NMR chemical shifts for various types of compounds, such as amino acids (47, 48), have also appeared.

The shielding of proton resonances in methyl, methylene, and methine groups generally falls in the order methyl > methylene > methine.

Simple additivity rules have been observed for proton chemical shifts.

TABLE 3-5

PROTON CHEMICAL SHIFTS (IN PPM FROM TETRAMETHYLSILANE)

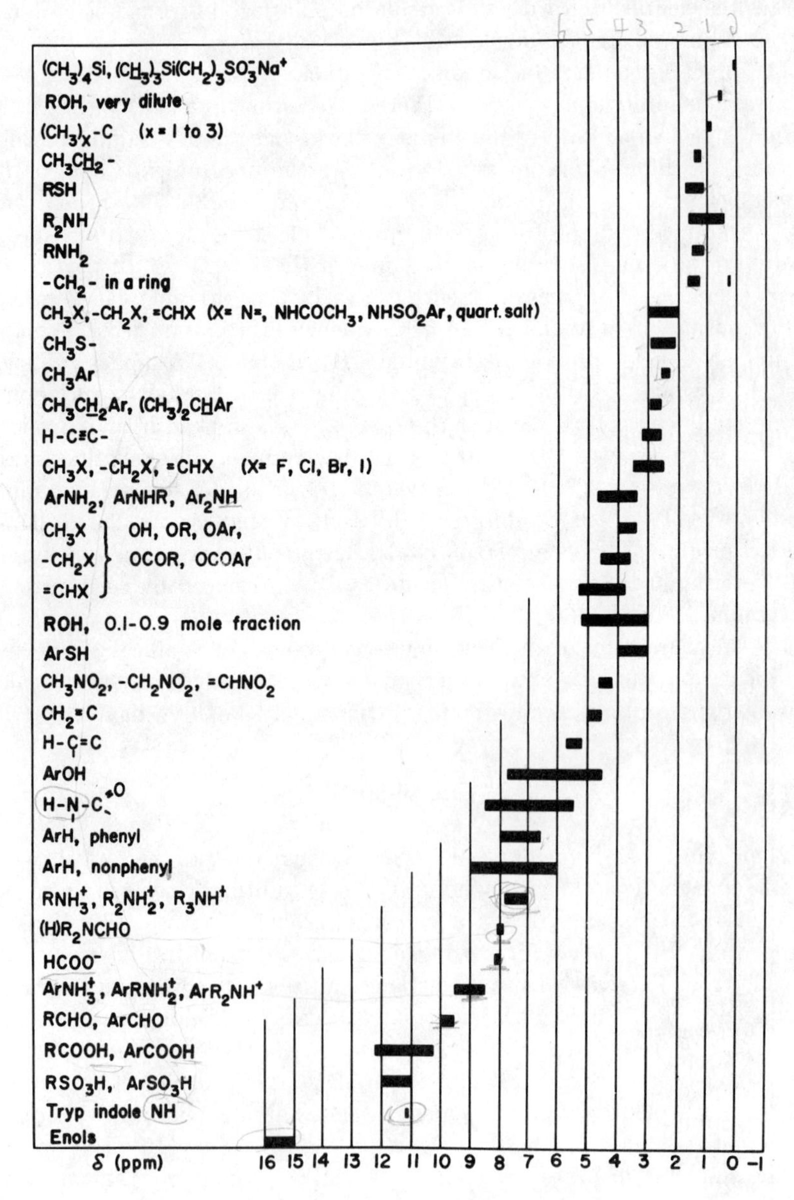

For example, the proton chemical shifts of disubstituted benzenes may be obtained by the sum of the shifts for the monosubstituted benzenes (49, 50). The α-proton chemical shift for a series of γ-substituted pyridines,

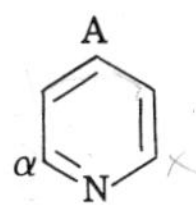

has been found to be proportional to the shift of the proton meta to A in substituted chlorobenzenes (51).

Carbon-13 NMR spectra have also been recorded for several classes of compounds (e.g., 52–54), including the amino acids (54). Table 3-6 provides a summary of the expected ^{13}C chemical shifts for carbon in various functional groups. It will be noticed that the range of chemical shifts is much greater (~250 ppm) for carbon-13 than for protons (~15 ppm). Recent monographs have described several aspects of ^{13}C NMR (55, 56) and a catalog of 500 ^{13}C NMR spectra has also been compiled (57).

TABLE 3-6

TYPICAL ^{13}C CHEMICAL SHIFTS (IN PPM FROM TETRAMETHYLSILANE)

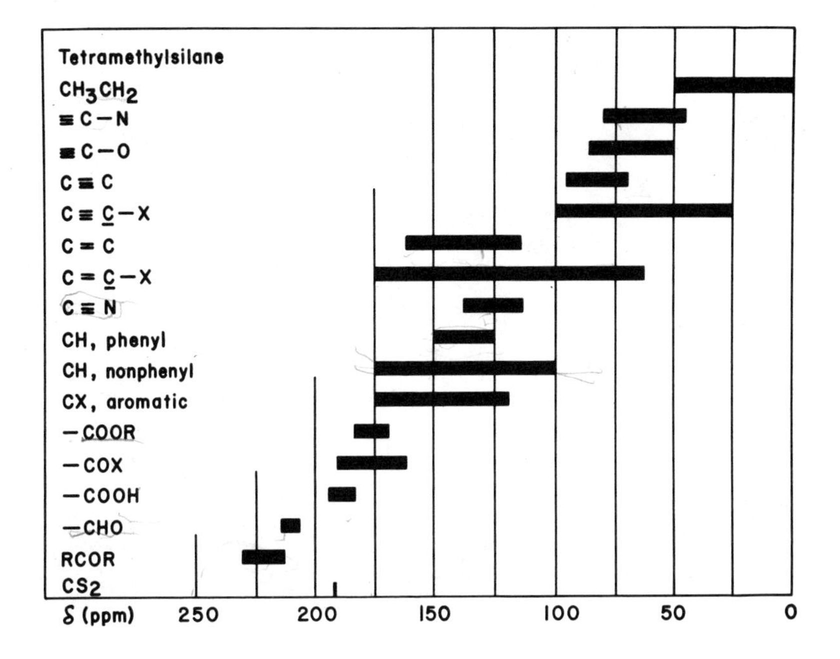

3.3. Paramagnetic Shifts

Chemical shifts in compounds containing paramagnetic sites can be unusually large. For example, as shown in Fig. 3-9, proton resonance peaks for an aqueous solution of sperm whale cyanometmyoglobin, which contains paramagnetic Fe(III), were observed from −27 ppm to +3 ppm (58). The shift arising from the presence of an unpaired electron can be expressed as the sum of two parts:

$$\delta_e = \delta_c + \delta_p \tag{3-4}$$

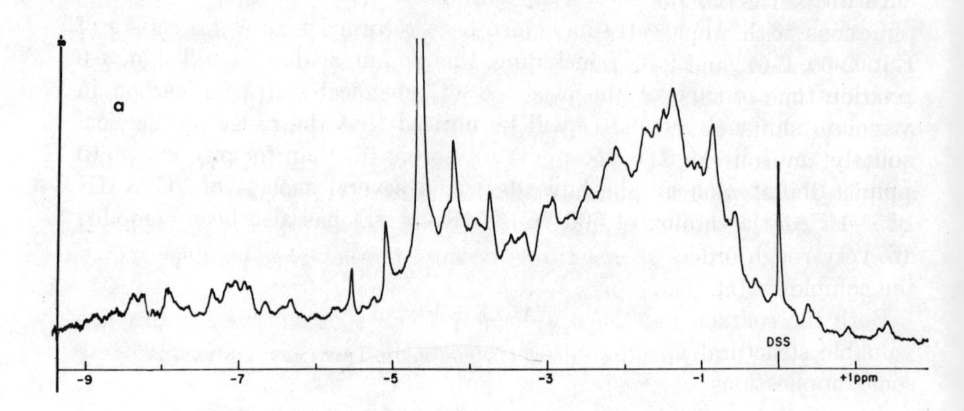

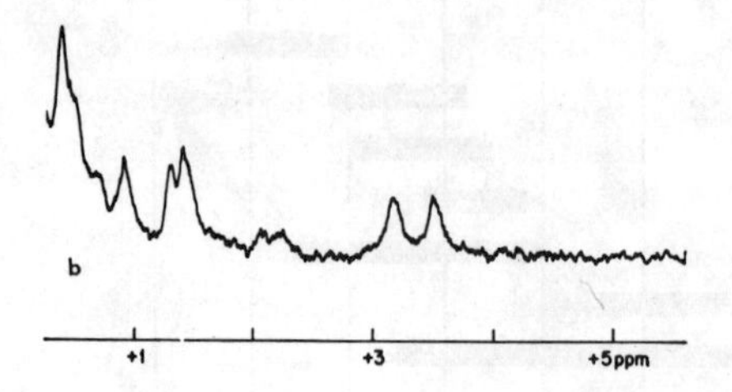

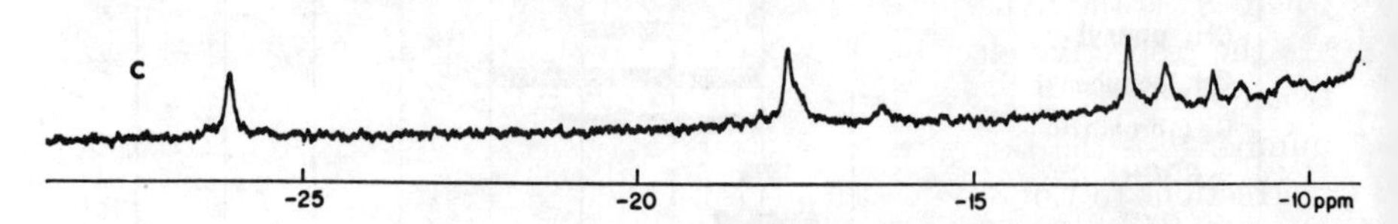

FIG. 3-9. 220 MHz proton NMR spectrum of sperm whale cyanometmyoglobin (5 m*M*) in deuterated phosphate buffer, pD 6.6, at 35°C. (a) Range +1.5 to −10 ppm from internal DSS standard. The five sharp lines between −3.5 and −6 ppm are caused by HDO and its first and second spinning side bands. (b) High-field range from 0 to +5 ppm. (c) Low-field range from −10 to −30 ppm (58).

where δ_c represents the contact shift and δ_p represents the pseudocontact shift. These effects are caused by hyperfine interactions of nuclei with the unpaired electron spin. The contact shift requires a finite unpaired electron density at the nucleus (hence the term "contact"), which occurs by delocalization of the unpaired electron, providing overlap with the s orbitals at the nucleus. The shift will be proportional to the unpaired electron density at the nucleus. These contact effects are transmitted through chemical bonds. Pseudocontact shifts are caused by a combination of spin–orbital and dipolar interactions. The pseudocontact effects are transmitted through space.

In Section 2.4.3, the relaxation mechanism resulting from nuclear interactions with unpaired electrons was discussed. It is apparent from Eqs. 2-60, 2-61, and 2-62 that it is necessary for τ_s, the electron spin relaxation time of the paramagnetic species, to be quite short so the paramagnetic shifts be not obscured by excessive line broadening. Consequently, aqueous Ni(II) and Co(II) complexes ($\tau_s \sim 10^{-13}$ sec) exhibit contact shifts, whereas aqueous Mn(II) and Gd(III) complexes ($\tau_s \sim 10^{-10}$–10^{-9} sec) exhibit line broadening. The τ_s values given in parentheses are very rough orders of magnitude because the actual value depends on the complexes (cf. Table 6-1).

Both the contact and the pseudocontact shifts can be used to provide valuable structural information. Eaton and Phillips (59) have reviewed some applications of contact and pseudocontact shifts.

3.3.1. Contact Shifts

The Fermi contact interaction occurs only if there is a finite probability of an unpaired electron spin being found at the nucleus. The Fermi contact shift for nucleus i is given by (59)

$$(\delta_c)_i = \left(\frac{\Delta H}{H_0}\right)_i = -A_i \frac{\gamma_e}{\gamma_N} \frac{g\beta S(S+1)}{3kT} \tag{3-5}$$

where A_i is the hyperfine coupling constant (frequency units), γ_e and γ_N are the respective electron and nuclear gyromagnetic ratios, g is the electronic g factor, β is the Bohr magneton, S is the electron spin quantum number, k is the Boltzmann constant, and T is the absolute temperature. Corrections to Eq. 3-5 due to anisotropic g factors have been presented by Jesson (60).

Equation 3-5 is valid under the conditions that the ground state of the system is paramagnetic, and

$$1/\tau_s \gg A_i \qquad 1/\tau_M \gg A_i$$

where τ_s is the electron spin relaxation time and τ_M is the characteristic exchange time for the paramagnetic species. If these conditions are not fulfilled, no contact shift will be manifested. If the conditions hold, different contact shifts may be observed for different nuclei in the same molecule because A_i may vary in sign and magnitude for the different nuclei.

Fermi (61) first described the direct electron–nucleus interaction, $\vec{\mathrm{I}} \cdot \vec{\mathrm{S}}$, which gives rise to hyperfine splitting in electron spin resonance (ESR) spectra when $1/\tau_s \ll A_i$ and to contact shifts in NMR spectra when $1/\tau_s \gg A_i$. The isotropic component of the interaction is given by Fermi's formula

$$A_i = (8\pi/3)\gamma_N \hbar g \beta \rho_i \tag{3-6}$$

where ρ_i is the unpaired electron spin density centered on the s orbital around the nucleus. The spin density can be related to the electronic wave function, permitting information concerning the electronic structure to be derived from contact shift measurements (62). The total spin density must equal the number of unpaired electrons in the paramagnetic species:

$$\sum_i \rho_i = 2S \tag{3-7}$$

where the sum is taken over the entire paramagnetic species. Therefore, the hyperfine coupling constant may be expressed as

$$A = \frac{\rho_i}{2S} A_s \tag{3-8}$$

where A_s is the hyperfine coupling constant for one unpaired electron in an s orbital. A value for A_s is characteristic for a particular atom. Values of A_s have been tabulated by Goodman and Rayner (63).

It has also been shown that the hyperfine coupling constants for aromatic protons (64) and for methyl groups attached to the aromatic ring (65) can be related to unpaired electron spin density on the neighboring ring carbon π orbital by

$$A_i = Q(\rho_c)_i \tag{3-9}$$

where Q is a proportionality constant that has been found to be approximately -6.3×10^7 Hz for all aromatic protons and approximately $+7.5 \times 10^7$ Hz for most methyl groups attached to aromatic rings. Relationship 3-9 arises from spin polarization, a mechanism by which spin, but not an electron, is transferred from the π orbital to the σ orbital between C and H.

3.3.2. Pseudocontact Shifts

In certain paramagnetic compounds or complexes, the combination of an anisotropic g tensor and a nucleus–electron dipolar interaction will lead to a pseudocontact shift which will be determined by the molecular geometry and the principal values of the electronic g tensor (66). The pseudocontact shift depends on the relationship of the rotational correlation time τ_r of the molecule with the electron spin relaxation time τ_s and the g tensor anisotropy, with primary interest being in the case when $\tau_r \gg \tau_s$.

With the assumption that the g tensor is axially symmetric ($g_{||} = g_z$, $g_\perp = g_x = g_y$) and possible effects of spin delocalization on the pseudocontact interactions are negligible, the pseudocontact shift for a paramagnetic complex in solution is given by

$$\delta_p = \frac{-\beta^2 \nu_0 S(S+1)}{27kT} \frac{(3\cos^2\theta - 1)}{r^3} (g_{||} - g_\perp)(g_{||} + 2g_\perp) \quad (3\text{-}10)$$

where r is the distance from the paramagnetic metal ion to the nucleus, ν_0 is the nuclear resonance frequency, and θ is the angle between the radius vector $\vec{r}$ and the principal axis of symmetry of the complex that defines the z direction. Jesson (60) and Kurland and McGarvey (67) have examined the case of nonaxial symmetry, i.e., $g_x \neq g_y$. Bleaney (68) has also considered less symmetric ligand fields, calculating the pseudocontact contribution to the chemical shift for lanthanide ion complexes in solution. It will be noted for Eq. 3-10 that the angular dependence of the pseudocontact shift means that the different nuclei in the same paramagnetic complex may have shifts differing in magnitude and in sign, because $(3\cos^2\theta - 1)$ has positive values for $\theta < 54°44'$ and negative values for $\theta > 54°44'$. The anisotropic g factors can be determined from the electron spin resonance spectrum in certain cases. It is apparent that the pseudocontact shift can provide valuable information concerning the orientation and the distance of the nucleus from the paramagnetic center once the $(3\cos^2\theta - 1)$ term is separated from the r^{-3} term. The r^{-3} dependence also implies that the pseudocontract shift will be important only for those nuclei in the vicinity of the paramagnetic center.

It is often difficult to clearly distinguish between a contact and a pseudocontact shift in the experimental NMR spectrum. The pseudocontact shift differs from solution to solid state, permitting distinction on this basis (66). In certain instances, the origin of the shift may be ascertained by comparing the NMR spectra of similar molecules or by considering the molecular geometry in terms of the "through bonds" (contact shift) and "through space" (pseudocontact shift) nature of the effects (69). It may also be possible to identify the shift as being Fermi contact in origin on the basis of the different dependence of T_1 and T_2 on the hyperfine coupling constant (cf. Eqs. 2-59 and 2-60) (70).

3.3.3. Shift Reagents

The use of particular lanthanide chelates as "shift reagents" has provided a powerful technique for the simplification of complicated NMR spectra. The shift reagents are usually β-diketone chelates of lanthanide ions, which are capable of increasing their coordination number to accommodate coordination with electronegative sites on the compound of interest. Large chemical shifts for the coordinated compound result as a consequence of the pseudocontact mechanism.

Hinckley (71) first recognized the ability of a lanthanide chelate to simplify an NMR spectrum. A dipivalomethane (DPM) complex of Eu(III), $Eu(DPM)_3 \cdot 2$ pyridine, was added to a solution of cholesterol in carbon tetrachloride. The spectrum of cholesterol in the absence and presence of the shift reagent is shown in Fig. 3–10. Hinckley made several interesting observations about the proton NMR spectrum of cholesterol in solutions containing Eu(DPM)·2 pyridine:

a. Very little broadening occurs.

b. The resonances are shifted substantially.

c. Only one sharp peak is observed for each cholesterol hydrogen, and the chemical shift depends on the concentration of $Eu(DPM)_3 \cdot 2$ pyridine. This indicates rapid metal complex–cholesterol exchange.

d. The chemical shifts were found to have an r^{-3} dependence, as expected for pseudocontact shifts (cf. Eq. 3–10), in accordance with the distance of the cholesterol hydrogens from Eu(III), which is coordinated to the cholesterol hydroxyl group.

It was subsequently reported that $Eu(DPM)_3$, which is free of the pyridine adduct, is a superior shift reagent (72).

In the previous section it was noted that those paramagnetic metal ions with the shortest electron spin relaxation times experienced much less line broadening so that paramagnetic shifts could be observed. Horrocks and Sipe (73) quantitatively studied the substrate ligand shifts and linewidths using DPM chelates with 11 lanthanide ions. The results for *n*-hexanol, 4-picoline *N*-oxide, and 4-vinylpyridine were found to be similar for each of the lanthanides with few differences being noted. It was found that the "shifting abilities" of the trivalent ions Tb, Dy, Ho, and Tm are greater than those of Eu and Pr, which are in more general use. The primary reason for the greater popularity of Eu and Pr is that they yield negligible line broadening, whereas Tb, Dy, Ho, and Tm cause some line broadening (although Tm did not cause extensive broadening).

Another reason for the use of both Eu and Pr is that they are often complementary. It has been observed that a substrate which gives a down-

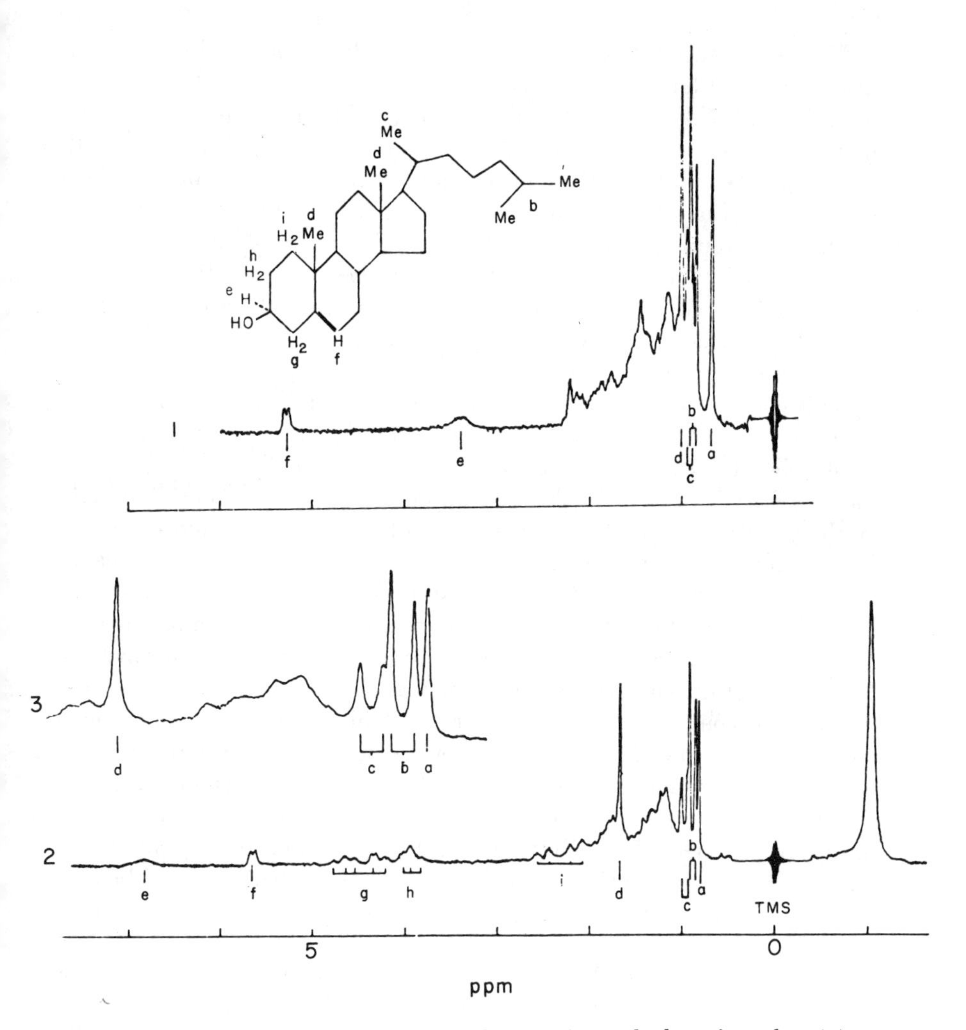

FIG. 3-10. (1) Proton NMR spectrum of cholesterol monohydrate in carbon tetrachloride. (2) Spectrum of carbon tetrachloride solution containing 0.05 *M* Eu(DPM)$_3$·2 pyridine and 0.1 *M* cholesterol monohydrate. (3) Expansion of that region of spectrum 2 which includes the methyl resonances. Peak assignments are designated by a letter corresponding to hydrogens in the molecular structure. The resonance 1 ppm upfield from tetramethylsilane (TMS) is caused by the metal complex (71).

field shift with Eu(DPM)$_3$ gives an upfield shift with Pr(DPM)$_3$ (74). The cause of this lies in differences in reagent–substrate geometry; the $(3 \cos^2 \theta - 1)$ term in Eq. 3-10 can assume either negative or positive values.

Most shift reagent work has made use of DPM chelates. However, it was discovered that lanthanide chelates using a fluorinated ligand 1,1,1,2,2,3,3,-heptafluoro-7,7-dimethyl-4,6-octanedione (FOD) had superior properties, namely greater solubility and ease of handling plus greater Lewis acidity (75). It was subsequently shown that for the same number of molar equivalents, $Eu(DPM)_3$ gives rise to larger shifts than $Eu(FOD)_3$ (76), so the real superiority of the FOD chelate lies in its solubility.

Shift reagents are capable of expanding their coordination number to accommodate coordination with several functional groups, including alcohols, amines, aldehydes, ketones, sulfoxides, and esters. The magnitude of the shifts varies with the functional group and in fact, with the specific compound. Several groups that do not complex, or complex weakly, with shift reagents are nitriles, nitro, imines, phosphines, halides, and double bonds (72). It has also been noted that acidic functions, such as carboxylic acids and phenols, decompose shift reagents (77). The nature of the particular solvent chosen also has a large effect. The most important is the effect of water. Water apparently complexes with the shift reagents more strongly than does the substrate of interest (78). Use of the shift reagents in aqueous solution is therefore obviated. Water-soluble lanthanide ion salts can be used to form complexes, inducing pseudocontact shifts in the ligand nuclear resonances. Barry *et al.* (79), for example, have developed a computer treatment for determining the conformations of flexible nucleotides in aqueous solution from the combined pseudocontact shift and relaxation time data derived from nucleotide complexes containing different lanthanide ions for promoting shifts [e.g., Eu(III)] or relaxation [e.g., Gd(III)]. In principle, it may be seen from a combination of Eqs. 3-10 and 2-59 that the geometry of the nucleotide could be established if the lanthanide ions complex the nucleotide in the same manner and the assumption of axial symmetry is valid.

Shift reagents have also been shown to be useful with ^{13}C spectra. For example, $Eu(DPM)_3$ has been employed as an aid for distinguishing between cis and trans isomers of 3-methylcyclopentanol and 1,3-dimethylcyclopentanol (80).

3.3.4. Chiral Solvents and Reagents

On occasion it might be desired to determine the enantiomeric purity of a compound. It has been observed that different enantiomers have nonequivalent proton NMR spectra in solutions in which the solvent is optically active (81, 82). The basis for nonequivalent spectra for enantiomers is the manner in which the optically active solvent solvates the optically active solute. Solvation results in formation of short-lived diastereomers that can be "resolved" in their NMR spectra. Consider, for example, a racemic mixture of methyl ethyl sulfoxide,

$$\mathrm{O}\blacktriangleleft\ddot{\mathrm{S}}\blacktriangleleft\overset{\alpha}{\mathrm{CH_2}}\ \overset{\beta}{\mathrm{CH_3}}$$
$$\mathrm{CH_3}\ (\text{on S})$$

in the optically active solvent, (−)-(*R*)-2,2,2-trifluorophenylethanol. It was found that there were two sets of nonequivalent proton resonances, one set from each diastereomer. Using a 100 MHz instrument, the chemical shift difference between the two diastereomers was 2.4 Hz for the α protons, 2.5 Hz for the β protons, and 2.5 Hz for the methyl protons (82). Such shifts are not great but do permit a determination of the enantiomeric composition of a mixture by measuring the peak intensities of the nonequivalent methyl resonances. The reason for the nonequivalence was postulated to be solute–solvent interactions producing short-lived, hydrogen bonded diastereomeric solvates.

Whitesides and Lewis (83) improved the NMR "resolution" of enantiomers by introducing an optically active lanthanide ion chelate, tris[3-*tert*-butylhydroxymethylene)-*d*-camphorato]europium(III), into racemic mixtures of primary amines or secondary alcohols. The advantage in using optically active shift reagents is, of course, the greater shifts and nonequivalence produced in the diastereomers. Subsequently, it was found that a fluorinated chelate gave better results. The addition of tris[3-heptafluoropropylhydroxymethylene-(+)-camphorato]europium(III) to racemates of alcohols, sulfoxides, an epoxide, and an aldehyde produced separate spectra for each of the enantiomers (84). For example, racemic α-methylbenzyl alcohol in the presence of the optically active shift reagent had its proton resonances at 100 MHz shifted downfield and separated for the two enantiomers (84) (see tabulation below).

PhCH(Me)OH	AVERAGE DOWNFIELD SHIFT (Hz)	SEPARATION OF ENANTIOMERIC PEAKS (Hz)
Me	265	5
CH	464	7
H_{ortho}	227	2

A chiral solvent and a chiral shift reagent have also been used to distinguish the meso, *d*, and *l* stereoisomers of the pesticide dieldrin (85).

References

1. W. Lamb, *Phys. Rev.* **60,** 817 (1941).
2. N. F. Ramsey, *Phys. Rev.* **78,** 699 (1950).
3. N. F. Ramsey, *Phys. Rev.* **86,** 243 (1952).
4. W. N. Lipscomb, *Advan. Magn. Resonance* **2,** 137 (1966).
5. J. I. Musher, *Advan. Magn. Resonance* **2,** 177 (1966).
6. J. D. Memory, "Quantum Theory of Magnetic Resonance Parameters." McGraw-Hill, New York, 1968.
7. J. B. Lambert and J. D. Roberts, *J. Amer. Chem. Soc.* **87,** 4087 (1965).
8. J. A. Pople, *J. Chem. Phys.* **37,** 53 and 60 (1962).
9. J. A. Pople, *Mol. Phys.* **7,** 301 (1964).
10. J. A. Pople, *Proc. Roy. Soc., Ser. A* **239,** 541 and 550 (1957).
11. J. S. Waugh and R. W. Fessenden, *J. Amer. Chem. Soc.* **79,** 846 (1957); **80,** 6697 (1958).
12. C. E. Johnson, Jr. and F. A. Bovey, *J. Chem. Phys.* **29,** 1012 (1958).
13. L. Pauling, *J. Chem. Phys.* **4,** 673 (1936).
14. R. B. Mallion, *J. Chem. Soc., B* p. 681 (1971) (and references therein).
15. C. W. Haigh and R. B. Mallion, *Org. Magn. Resonance* **4,** 203 (1972).
16. R. J. Abraham, *Mol. Phys.* **4,** 145 (1961).
17. R. G. Shulman, K. Wüthrich, T. Yamane, D. J. Patel, and W. E. Blumberg, *J. Mol. Biol.* **53,** 143 (1970).
18. M. S. Gil and J. N. Murrell, *Trans. Faraday Soc.* **60,** 248 (1964).
19. T. Schaefer and W. G. Schneider, *Can. J. Chem.* **41,** 966 (1963).
20. B. M. Lynch and H. J. M. Dou, *Tetrahedron Lett.* p. 2627 (1965).

20a. C. Giessner-Prettre and B. Pullman, *J. Theor. Biol.* **27,** 87 (1970).

21. J. A. Pople, W. G. Schneider, and H. J. Bernstein, "High Resolution Nuclear Magnetic Resonance," Chapter 16. McGraw-Hill, New York, 1959.

21a. M. I. Foreman, *Nucl. Magn. Resonance* **1,** 295 (1972).

22. A. D. Buckingham, *Can. J. Chem.* **38,** 300 (1960).
23. A. D. Buckingham, T. Schaefer, and W. G. Schneider, *J. Chem. Phys.* **32,** 1227 (1960).
24. J. T. Arnold and M. E. Packard, *J. Chem. Phys.* **19,** 1608 (1951).
25. C. M. Huggins, G. C. Pimentel, and J. N. Shoolery, *J. Phys. Chem.* **60,** 1311 (1956).
26. E. D. Becker, U. Liddel, and J. N. Shoolery, *J. Mol. Spectrosc.* **2,** 1 (1958).
27. M. Saunders and J. B. Hyne, *J. Chem. Phys.* **29,** 1319 (1958).
28. J. A. Pople, W. G. Schneider, and H. J. Bernstein, "High Resolution Nuclear Magnetic Resonance," Chapter 15. McGraw-Hill, New York, 1959.
29. R. R. Shoup, H. T. Miles, and E. D. Becker, *Biochem. Biophys. Res. Commun.* **23,** 194 (1966).
30. L. Katz and S. Penman, *J. Mol. Biol.* **15,** 220 (1966).
31. K. H. Scheit, *Angew. Chem., Int. Ed. Engl.* **6,** 180 (1967).
32. T. L. James and R. J. Kula, *J. Phys. Chem.* **73,** 634 (1969).
33. D. H. Sachs, A. N. Schecter, and J. S. Cohen, *J. Biol. Chem.* **246,** 6576 (1971).
34. M. Blumenstein and M. A. Raftery, *Biochemistry* **11,** 1643 (1972).
35. J. A. Happe and M. Morales, *J. Amer. Chem. Soc.* **88,** 2077 (1966).
36. P. O. P. Ts'o, M. P. Schweizer, and D. P. Hollis, *Ann. N.Y. Acad. Sci.* **158,** 256 (1969).
37. M. P. Schweizer, S. I. Chan, and P. O. P. Ts'o, *J. Amer. Chem. Soc.* **87,** 5241 (1965).

38. A. D. Broom, M. P. Schweizer, and P. O. P. Ts'o, *J. Amer. Chem. Soc.* **89,** 3612 (1967).
39. K. H. Scheit, F. Cramer, and A. Franke, *Biochim. Biophys. Acta* **145,** 21 (1967).
40. S. I. Chan and G. P. Kreishman, *J. Amer. Chem. Soc.* **92,** 1102 (1970).
41. L. Katz, *J. Mol. Biol.* **44,** 279 (1969).
42. P. O. P. Ts'o, N. S. Kondo, M. P. Schweizer, and D. P. Hollis, *Biochemistry* **8,** 997 (1969).
43. N. F. Chamberlain, *Anal. Chem.* **31,** 56 (1959).
44. G. V. D. Tiers, *J. Phys. Chem.* **62,** 1151 (1958).
45. E. Mohacsi, *J. Chem. Educ.* **41,** 38 (1964).
46. F. A. Bovey, "NMR Data Tables for Organic Compounds," Vol. 1. Wiley (Interscience), New York, 1967.
47. O. Jardetzky and C. D. Jardetzky, *J. Biol. Chem.* **233,** 383 (1958).
48. B. Bak, C. Dambmann, F. Nicolaisen, E. J. Pederson, and N. S. Bhacca, *J. Mol. Spectrosc.* **26,** 78 (1968).
49. J. S. Martin and B. P. Dailey, *J. Chem. Phys.* **39,** 1722 (1963).
50. B. Bak, J. B. Jensen, A. L. Larson, and J. Rastrup-Andersen, *Acta Chem. Scand.* **16,** 1031 (1962).
51. T. K. Wu and B. P. Dailey, *J. Chem. Phys.* **41,** 3307 (1964).
52. R. J. Pugmire and D. M. Grant, *J. Amer. Chem. Soc.* **90,** 4232 (1968) (and references cited therein).
53. D. E. Dorman and J. D. Roberts, *J. Amer. Chem. Soc.* **92,** 1355 (1970).
54. W. Horsley, H. Sternlicht, and J. S. Cohen, *J. Amer. Chem. Soc.* **92,** 680 (1970).
55. G. C. Levy and G. L. Nelson, "Carbon-13 Nuclear Magnetic Resonance for Organic Chemists." Wiley (Interscience), New York, 1972.
56. J. B. Stothers, "Carbon-13 NMR Spectroscopy." Academic Press, New York, 1972.
57. L. F. Johnson and W. C. Jankowski, "Carbon-13 NMR Spectra." Wiley (Interscience), New York, 1972.
58. K. Wüthrich, R. G. Shulman, and J. Peisach, *Proc. Nat. Acad. Sci. U.S.* **60,** 373 (1968).
59. D. R. Eaton and W. D. Phillips, *Advan. Magn. Resonance* **1,** 103 (1965).
60. J. P. Jesson, *J. Chem. Phys.* **47,** 579 and 582 (1967).
61. E. Fermi, *Z. Phys.* **60,** 320 (1930).
62. H. S. Jarrett, *Solid State Phys.* **14,** 215 (1963).
63. B. A. Goodman and J. B. Rayner, *Advan. Inorg. Chem. Radiochem.* **13,** 135 (1970).
64. H. M. McConnell, *J. Chem. Phys.* **24,** 764 (1956).
65. R. Bersohn, *J. Chem. Phys.* **24,** 1066 (1956).
66. H. M. McConnell and R. E. Robertson, *J. Chem. Phys.* **29,** 1361 (1958).
67. R. J. Kurland and B. R. McGarvey, *J. Magn. Resonance* **2,** 286 (1970).
68. B. Bleaney, *J. Magn. Resonance* **8,** 91 (1972).
69. A. Kowalsky, *Biochemistry* **4,** 2382 (1965).
70. K. Wüthrich, *Struct. Bonding (Berlin)* **8,** 53 (1970).
71. C. C. Hinckley, *J. Amer. Chem. Soc.* **91,** 5160 (1969).
72. J. K. M. Sanders and D. H. Williams, *J. Amer. Chem. Soc.* **93,** 641 (1971).
73. W. D. Horrocks, Jr. and J. P. Sipe, III, *J. Amer. Chem. Soc.* **93,** 6800 (1971).
74. J. Briggs, G. H. Frost, F. A. Hart, G. P. Moss, and M. L. Staniforth, *Chem. Commun.* p. 749 (1970).
75. R. E. Rondeau and R. E. Sievers, *J. Amer. Chem. Soc.* **93,** 1522 (1971).
76. B. L. Shaprio, M. D. Johnston, Jr., A. D. Godwin, T. W. Proulx and M. J. Shaprio, *Tetrahedron Lett.* p. 3233 (1972).

77. J. E. Maskasky and M. E. Kenny, *J. Amer. Chem. Soc.* **93,** 2060 (1971).
78. I. Armitage and L. D. Hall, *Can. J. Chem.* **49,** 2770 (1971).
79. C. D. Barry, A. C. T. North, J. A. Glasel, R. J. P. Williams, and A. V. Xavier, *Nature* (*London*) **232,** 236 (1971).
80. M. Christl, H. J. Reich, and J. D. Roberts, *J. Amer. Chem. Soc.* **93,** 3463 (1971).
81. W. H. Pirkle and S. D. Beare, *J. Amer. Chem. Soc.* **89,** 5485 (1967).
82. W. H. Pirkle and S. D. Beare, *J. Amer. Chem. Soc.* **90,** 6250 (1968).
83. G. M. Whitesides and D. W. Lewis, *J. Amer. Chem. Soc.* **92,** 6979 (1970).
84. R. R. Fraser, M. A. Petit, and J. H. Saunders, *Chem. Commun.* p. 1450 (1971).
85. M. Kainosho, E. Ajisaka, W. H. Pirkle, and S. D. Beare, *J. Amer. Chem. Soc.* **94,** 5924 (1972).

CHAPTER 4

SPIN–SPIN SPLITTING AND STRUCTURE

In Section 1.5 the origin of spin–spin coupling and the rules governing resonance peak multiplicities and intensities were qualitatively presented. Certain aspects of spin–spin coupling will be reviewed in more detail in this chapter. In Chapter 1 the discussion was limited to simple first-order coupling, i.e., the chemical shift difference $\Delta\delta$ between any two resonances is much greater than the coupling constant J. When the coupling constants and chemical shifts are of the same order and this first-order rule is violated, complex splitting patterns can result that bear little resemblance to the predicted first-order spectrum. Second-order splittings and drastic changes in line intensities and peak positions result. For example, Fig. 4-1 (1) shows schematically the proton NMR spectrum of a X—CH_2CH_2—Y group for various values of $J/\Delta\delta$. The spectra at high values of $J/\Delta\delta$ are not immediately recognizable as originating from the interaction of two methylene groups.

In this chapter the analysis of complex spectra and the use of certain techniques to aid spectral analysis will be briefly discussed. The concluding section of the chapter will cover the relationship of spin–spin coupling with molecular geometry, including some important biochemical implications of that relationship.

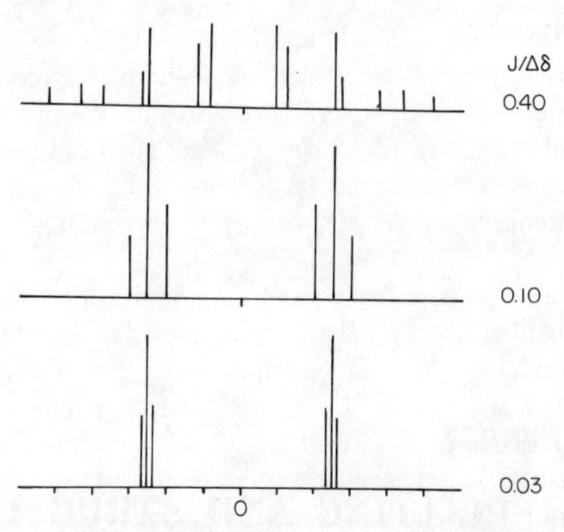

FIG. 4-1. A_2B_2 theoretical splitting patterns as a function of the ratio of the coupling constant J to the chemical shift $\Delta\delta$ between the two groups. Data taken from Corio (1).

4.1. Origin of Spin–Spin Splitting

The first successful theoretical description of spin–spin splitting was presented by Ramsey and Purcell (2), with later amplification by Ramsey (3). Barfield and Grant (4) and Murrell (5) have discussed various theoretical approaches to the problem, such as use of the perturbation method or the variational procedure and various molecular wave formulations used in the calculations. In-depth theoretical knowledge of the origin of electron–nuclear spin scalar interactions is usually not necessary in biochemical applications of this particular NMR parameter. For the purpose of this discussion the following version of the origin will suffice.

Consider the ^{13}C—^{1}H scalar coupling in a molecule such as ^{1}H—$^{13}COOD$. The ^{13}C nucleus will have its magnetic moment vector either parallel or antiparallel to the direction of the stationary magnetic field. Whichever the orientation, the magnetic moment vector of a bonding electron around the ^{13}C nucleus will tend to assume an orientation parallel to that of the nuclear magnetic moment vector (the spin angular momentum vectors will be antiparallel but the gyromagnetic ratio is positive for the ^{13}C nucleus and negative for an electron). However, the second electron in the ^{13}C—^{1}H bond will be antiparallel to the first with the proton magnetic moment vector tending to be parallel to the second electron magnetic moment vector. It can therefore be seen how one nuclear magnetic moment perturbs the orientation of a second nuclear magnetic moment via this electron mediation mechanism. The ^{13}C nuclear magnetic moment has two possible orientations that will affect the proton spin in different ways, giving rise to a doublet in the proton NMR spectrum. Additional orientations are possible with a greater number of nuclei in a chemically equivalent group, as illustrated in Fig. 1-5. Each of these orientations will perturb the nuclear spin orientation of other nearby nuclei, passing the information by way of intervening electrons. The result is a multiplet in the NMR spectrum with the number of components of the multiplet depending on the possible orientations.

Coupling constants have signs as well as magnitudes. With a first-order spectrum, only the magnitude of the coupling constants may be determined from the spectrum. The $^{13}C—^{1}H$ example just cited will give a positive coupling constant. However, a $^{13}C—C—^{1}H$ coupling may give a negative coupling constant because there will be additional intervening electrons. The coupling constant is negative if the coupled nuclear spins are parallel. The sign of the coupling constant will depend on the number of bonds and the bond hybridization (6).

4.2. Analysis of NMR Spectra

In a two-spin system, the two nuclei A and X may interact with a spin–spin coupling constant J_{AX}, and the characteristics of the resulting NMR spectrum will be given by the first-order splitting rules in Section 1.5. The system of nomenclature stipulates that an ethyl group obeying first-order splitting rules gives rise to an A_3X_2 pattern in its proton NMR spectrum. If the first-order conditions do not hold, the system is referred to as A_3B_2. For the two-spin system, if first-order rules are not observed, the spectrum is denoted AB. The same system of nomenclature would denote a three-spin system by ABX, AMX, or ABC. In general, the interacting groups of nuclei are labeled A, B, C, etc., if their chemical shift values are of the same order of magnitude as the coupling constants; they are labeled M, N, etc., or X, Y, etc., if the chemical shifts of the nuclei are much larger than the chemical shifts of the interacting nuclei (A, B, etc.).

In the following discussion, the procedure for calculating and analyzing spectra up to the AB case will be covered as a simple example. For more details and for more complicated systems, the reader is referred to the monographs by Roberts (7) and Corio (8), and the recent review by Jones (9).

4.2.1. The One-Spin System A

For simplicity the case of a single nucleus with spin $I = \frac{1}{2}$ in an external magnetic field will be considered first. This is essentially the system discussed in Sections 1.2 and 1.4. A nuclear magnetic moment in the presence of a magnetic field can be oriented in $2I + 1$ possible directions. The spin $\frac{1}{2}$ nucleus can be oriented either opposed to or parallel to the direction of the stationary magnetic field H_0. The nuclear spin with the parallel orientation is in a lower energy state, denoted by the single spin wave function α. The antiparallel orientation has the higher energy and is denoted by the wave function β. As in the energy level diagram of Fig. 1-1, α is associated with the magnetic quantum number $m_I = \frac{1}{2}$ and β is associated with $m_I = -\frac{1}{2}$. The spin wave functions α and β form an orthonormal set.

The values of the two energy levels E_1 and E_2 corresponding to the two states α and β are obtained as solutions of the time-independent Schrödinger equation (10)

$$\mathcal{H}\psi_1 = E_1\psi_1 \tag{4-1}$$

where $\mathcal{H}$ is the Hamiltonian operator and the eigenfunction $\psi_1 = \alpha$. The eigenvalues E_1 and E_2 are given by

$$E_1 = \int \psi_1\mathcal{H}\psi_1 \, dV = \int \alpha\mathcal{H}\alpha \, dV \tag{4-2}$$

and

$$E_2 = \int \psi_2\mathcal{H}\psi_2 \, dV = \int \beta\mathcal{H}\beta \, dV \tag{4-3}$$

where the integration is taken over all space and dV is an incremental volume of space. The form of the eigenfunction for the one-spin case is shown so that only the nuclear Zeeman Hamiltonian is necessary. It is given in frequency units by

$$\mathcal{H} = -\gamma_A(1 - \sigma_A)\vec{I}_A \cdot \vec{H} \tag{4-4}$$

where $\vec{I}_A$ is the nuclear spin of nucleus A, $\vec{H}$ is the magnetic field, and σ_A is the shielding constant of nucleus A.

The energies of the α and β states are, respectively,

$$E_1 = +(\tfrac{1}{2})(\gamma_A/2\pi)(1 - \sigma_A)H_0 \tag{4-5}$$

and

$$E_2 = -(\tfrac{1}{2})(\gamma_A/2\pi)(1 - \sigma_A)H_0 \tag{4-6}$$

where H_0 is the magnitude of the applied field and the energies are given in frequency units.

The single resonance observed in the NMR spectrum for the one-spin system arises from transitions between the two states when the nucleus is said to "flip" between the α and β states. The frequency of the NMR transition is

$$\Delta E = |E_2 - E_1| = (\gamma_A/2\pi)(1 - \sigma_A)H_0 \tag{4-7}$$

It is apparent from Eq. 4-7 that the chemical shift of nucleus A is a simple function of the shielding constant as given in Chapter 1.

4.2.2. The Two-Spin AX System with No Coupling

It is obvious in the AX case in which the coupling constant is zero that the NMR spectrum will be first order, consisting of two lines, one at frequency $(\gamma_A/2\pi)(1 - \sigma_A)H_0$ and the other at frequency $(\gamma_X/2\pi)(1 - \sigma_X)H_0$. Nevertheless, it will be instructive to consider this case before the coupling interaction is added to the Hamiltonian. It is certainly possible to treat each nucleus independently because there is no coupling. However, it will be necessary to treat both spins simultaneously when the spin–spin coupling is brought in, so the two spins will be treated simultaneously here. To simplify nomenclature, both nuclei will be assumed to be protons so the subscripts can be dropped from the gyromagnetic ratio.

For the general case of n spin $\frac{1}{2}$ nuclei, there will be 2^n possible stationary states, each with its characteristic spin function ψ_n. The different spin functions ψ_n are obtained as the product of the spin functions for the individual nuclei. For the AX case, there will be four states with one of these energy states having a product spin function

$$\psi_1 = \alpha(A)\alpha(X) = \alpha\alpha \tag{4-8}$$

The nomenclature is simplified and it is understood that the nth symbol refers to the nth nucleus. The other spin states are $\alpha\beta$, $\beta\alpha$, and $\beta\beta$. Each of these spin functions is normalized, i.e.,

$$\int (\alpha\alpha)^2 \, dV = 1 \tag{4-9}$$

The energy level diagram of Fig. 4-2A shows the energy levels labeled with the appropriate product spin function for the AX case in the absence of coupling.

The total Hamiltonian for the AX system is just the sum of the individual operators $\mathcal{H}_A$ and $\mathcal{H}_X$:

$$\mathcal{H} = \mathcal{H}_A + \mathcal{H}_X \tag{4-10}$$

and the energy eigenvalues are calculated with the use of equations such as Eq. 4-2. For example, for the $\alpha\alpha$ spin state, the energy is

$$E_{\alpha\alpha} = \int \psi_1 \mathcal{H} \psi_1 \, dV$$

$$= \int \alpha(A)\alpha(X)(\mathcal{H}_A + \mathcal{H}_X)\alpha(A)\alpha(X) \, dV \tag{4-11}$$

$\mathcal{H}_A$ operates only on $\alpha(A)$ and $\mathcal{H}_X$ operates only on $\alpha(H)$ so the energy may be given by

$$E_{\alpha\alpha} = \int \alpha(A)\mathcal{H}_A\alpha(A)\, dV \int \alpha(X)\alpha(X)\, dV + \int \alpha(X)\mathcal{H}_X\alpha(X)\, dV$$

$$\times \int \alpha(A)\alpha(A)\, dV$$

$$= \tfrac{1}{2}(\gamma/2\pi)(1 - \sigma_A)H_0 + \tfrac{1}{2}(\gamma/2\pi)(1 - \sigma_X)H_0$$

$$= \tfrac{1}{2}(\gamma/2\pi)(2 - \sigma_A - \sigma_X)H_0 \qquad (4\text{-}12)$$

where Eqs. 4-5 and 4-9 were utilized. The energy eigenvalues for the other eigenfunctions may be calculated in the same manner. The energies are given in Fig. 4-2A.

Transitions between all four energy levels of Fig. 4-2A are not allowed. It is useful to define a new term

$$F_z = \sum_{\substack{\text{all} \\ \text{nuclei}}} m_I \qquad (4\text{-}13)$$

The allowed transitions between states are limited by the selection rule

$$\Delta F_z = \pm 1 \qquad (4\text{-}14)$$

for small rf (i.e., nonsaturating) fields. The value of F_z associated with each state is shown in Fig. 4-2A. The A ($4 \rightarrow 2$ and $3 \rightarrow 1$) and the X ($4 \rightarrow 3$ and $2 \rightarrow 1$) transitions shown in Fig. 4-2A are clearly allowed and lead to the expected two-line spectrum of Fig. 4-2B. Two lines from the $4 \rightarrow 1$ and $3 \rightarrow 2$ transitions do not appear at the usual rf power levels because they are forbidden by the selection rule.

The intensity of each of the four allowed transitions (two occurring at ν_A and at ν_X) will be the same because the Boltzmann distribution of spins (cf. Eq. 1-7) between states differing in F_z by unity will be the same. This might be more apparent if it is noted that Fig. 4-2A is not drawn to scale. The separation between states 3 and 4 may be 100 MHz, whereas the separation between states 2 and 3 will be more on the order of 100 Hz. The energy difference between A transitions and X transitions, being only about 100 Hz, is not sufficiently large to cause any measureable difference in intensities.

4.2.3. The Two-Spin AX System with Coupling

In this case the form of the Hamiltonian is

$$\mathcal{H} = \mathcal{H}_A + \mathcal{H}_{AX} + \mathcal{H}_X \qquad (4\text{-}15)$$

where $\mathcal{H}_{AX}$ is the additional factor caused by interaction of nucleus A with

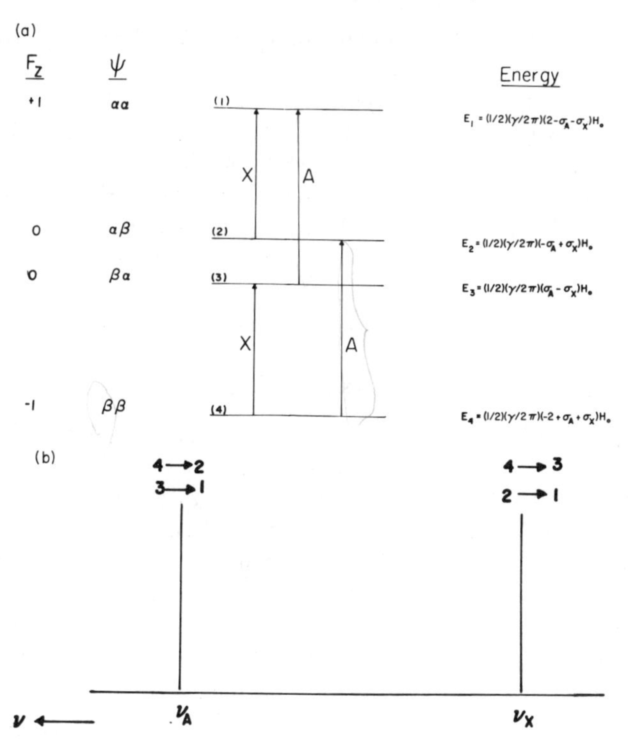

FIG. 4-2. (a) Energy level diagram and transitions for the AX case in the absence of any spin–spin interaction. (b) Spectrum for the AX case in the absence of spin–spin interaction.

nucleus X. $\mathcal{H}_{AX}$ is given by

$$\begin{aligned}\mathcal{H}_{AX} &= J\vec{I}_A \cdot \vec{I}_X \\ &= J[I_x(A)I_x(X) + I_y(A)I_y(X) + I_z(A)I_z(X)] \qquad (4\text{-}16)\end{aligned}$$

where $\vec{I}_A$ and $\vec{I}_X$ are the spin angular moments (denoted $\vec{J}'$ in Section 1.1) of nuclei A and X, respectively, and J is the scalar coupling constant between A and X. For the AX case, J is much less than the chemical shift between A and X.

If we solve Eq. 4-2 for the energy eigenvalue of the $\alpha\alpha$ state incorporating the spin–spin interaction Hamiltonian $\mathcal{H}_{AX}$ in addition to the Zeeman Hamiltonian terms, $\mathcal{H}_A$ and $\mathcal{H}_X$, we have

$$E_1 = \tfrac{1}{2}(\gamma/2\pi)(2 - \sigma_A - \sigma_X)H_0 + \tfrac{1}{4}J \qquad (4\text{-}17)$$

The $\alpha\beta$, $\beta\alpha$, and $\beta\beta$ energy eigenvalues are modified from the values for the noncoupling case given in Fig. 4-2A by $-\frac{1}{4}J$, $-\frac{1}{4}J$, and $\frac{1}{4}J$, respectively. The new energy levels under the influence of spin–spin coupling are shown in Fig. 4-3. It is apparent from the figure that the $4 \rightarrow 3$ (X) and $2 \rightarrow 4$ (A) transitions are decreased in energy by $J/2$ and the $2 \rightarrow 1$ (X) and $3 \rightarrow 1$ (A) transitions are increased in energy by $J/2$. In the absence of coupling, the two A transitions were degenerate and the two X transitions were degenerate. Coupling removes the degeneracy so each transition gives rise to a separate line in the spectrum, as shown in Fig. 4-4 for the AX system. The splitting between each of the A lines and each of the X lines is seen to be given by the magnitude of the coupling constant J. It is obvious from Fig. 4-3 that a change in the sign of J will change the energy levels, but the splitting observed in the spectrum will be the same.

4.2.4. The Two-Spin AB System with Coupling

The AB system is distinguished from the AX system by an additional effect of the spin–spin coupling when the chemical shift difference becomes sufficiently small that the shift difference and coupling constant become comparable. When states 2 and 3 in Fig. 4-3 come very close in energy, it is no longer satisfactory to describe them simply with spin functions $\alpha\beta$ and $\beta\alpha$, respectively. As the states approach each other in energy, it becomes necessary to describe each state by a mixture of $\alpha\beta$ and $\beta\alpha$, with greater mixing as the states come closer. We can describe the states by the wave functions

$$\psi_2 = a(\alpha\beta) + b(\beta\alpha) \tag{4-18}$$

and

$$\psi_3 = -b(\alpha\beta) + a(\beta\alpha) \tag{4-19}$$

The $\alpha\alpha$ and $\beta\beta$ states clearly do not enter into the mixture because their energies are much different from $\alpha\beta$ and $\beta\alpha$. This leads to a nonmixing rule; namely, states with different F_z values do not mix. The mixing coefficients a and b will depend on the value of $J/\Delta\delta$, such that b becomes larger and a becomes smaller as the two states come closer in energy. The coefficients will be related by

$$a^2 + b^2 = 1 \tag{4-20}$$

which is a consequence of the requirement that ψ_2 and ψ_3 be normalized wave functions (cf. eq. 4-9). Equations 4-18 and 4-19 are also formulated in such manner that the wave functions are orthonormal, i.e.,

$$\int \psi_n \psi_m \, dV = \delta_{nm} \tag{4-21}$$

where δ_{nm} is the Kronecker delta ($\delta_{nm} = 1$ when $m = n$ and is zero otherwise).

The four energy eigenvalues can be obtained with the use of Eq. 4-1,

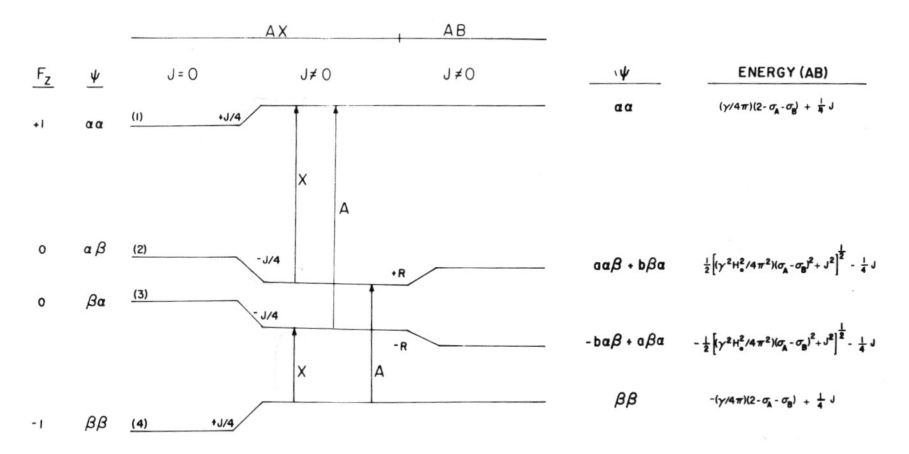

FIG. 4-3. Energy level diagram and transitions in the AX system with and without coupling and in the AB system with coupling.

setting up a secular determinant with the four resulting linear equations (11). (See Eq. 4-22, p. 104). The determinant can be factored into two 1×1 determinants and a 2×2 determinant, as shown by the dashed lines in Eq. 4-22. As illustrated in Fig. 4-3, the energy eigenvalues for the $\alpha\alpha$ and $\beta\beta$ energy states are unchanged from those which obtained in the case of the AX system with coupling. The 2×2 determinant is a quadratic equation with two roots for E, E_2 and E_3, which are given in Fig. 4-3. The change in the energy levels of E_2 and E_3 going from the AX system to the AB system is also indicated in that figure as $+R$ for the E_2 level and $-R$ for the E_3 level.

Transitions in the AB system will still be between the same levels as in the AX system. However, as indicated in Fig. 4-5, the chemical shift ν_A is not obtained simply as the half-way point between the resonance lines for the $3 \rightarrow 1$ and $4 \rightarrow 2$ transitions. The intensities of the transitions have also changed. The chemical shift for the A resonance therefore occurs at

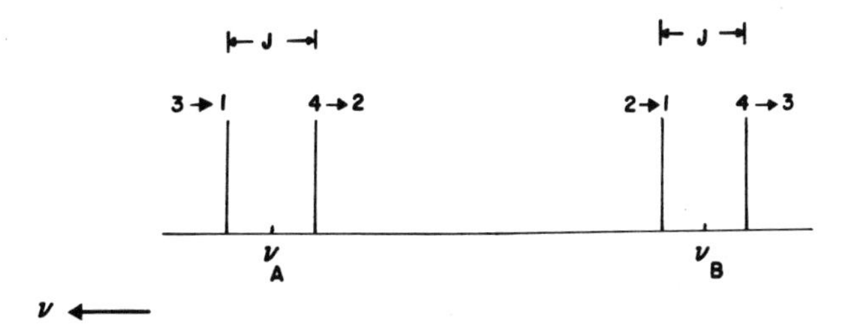

FIG. 4-4. Spectrum for the AX case in the presence of spin–spin coupling.

$$
\begin{vmatrix}
\frac{\gamma}{4\pi}(2-\sigma_A-\alpha_B)H_0+\frac{J}{4}-E & 0 & 0 & 0 \\
0 & \frac{\gamma}{4\pi}(\sigma_B-\sigma_A)H_0-\frac{J}{4}-E & J/2 & 0 \\
0 & J/2 & \frac{\gamma}{4\pi}(\sigma_A-\sigma_B)H_0-\frac{J}{4}-E & 0 \\
0 & 0 & 0 & \frac{\gamma}{4\pi}(-2+\sigma_A+\sigma_B)+\frac{J}{4}-E
\end{vmatrix} = 0
$$

Equation 4-22

the "center of gravity" of the two A lines and the chemical shift for the B resonance occurs at the "center of gravity" of the two B lines. The chemical shift difference (in Hz) between $3 \rightarrow 1$ and $2 \rightarrow 1$ transitions and between the $4 \rightarrow 3$ and $4 \rightarrow 2$ transitions may be expressed as

$$\Delta\nu = \left[\frac{\gamma^2 H_0^2}{4\pi^2}(\sigma_A - \sigma_B)^2 + J^2\right]^{1/2} \tag{4-23}$$

For the AB case, J may be obtained directly from the splitting. The resonance positions of the A and B protons, ν_A and ν_B, are $\pm 1/2\Delta\nu$ from the midpoint of the symmetrical four-line spectrum.

It has been found to be convenient to introduce an angular parameter θ, which can vary between 0 and π. The mixing coefficients can then be expressed as $a = \cos\theta$ and $b = \sin\theta$ and the relative intensities of the transitions expressed as in Table 4-1. The expression defining θ is

$$\sin 2\theta = J \Big/ \left[\frac{\gamma^2 H_0^2}{4\pi^2}(\sigma_A - \sigma_B)^2 + J^2\right]^{1/2} \tag{4-24}$$

The line positions and intensities for theoretical AB spectra are shown in Fig. 4-6, which illustrates the symmetry of the AB pattern and the increasing intensity of the inner peaks at the expense of the outer peaks as the $J/\Delta\delta$ ratio increases.

Although the AB spectra do exemplify some of the characteristics of non-first-order spectra, they do not show the dependence of the spectra on the sign of the coupling constant nor the second-order splitting that is evident in the A_2B_2 spectra of Fig. 4-1. Both Roberts (7) and Corio (8) discuss complex spin–spin interactions in systems with many spins in some detail.

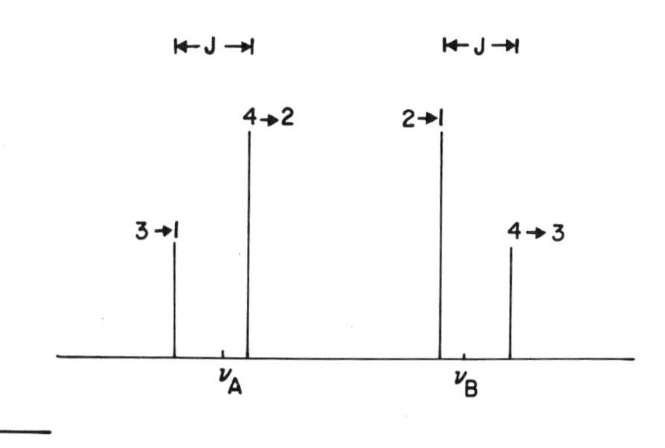

Fig. 4-5. Spectrum for the AB case.

TABLE 4-1

RESONANCE ENERGY AND RELATIVE PEAK INTENSITY FOR THE AB SYSTEM

TRANSITION	PEAK POSITION	RELATIVE INTENSITY
$3 \rightarrow 1$	$\frac{1}{2}J\,[1 + (\sin 2\theta)^{-1}]$	$1 - \sin 2\theta$
$4 \rightarrow 2$	$\frac{1}{2}J\,[-1 + (\sin 2\theta)^{-1}]$	$1 + \sin 2\theta$
$2 \rightarrow 1$	$\frac{1}{2}J\,[1 - (\sin 2\theta)^{-1}]$	$1 + \sin 2\theta$
$4 \rightarrow 3$	$\frac{1}{2}J\,[-1 - (\sin 2\theta)^{-1}]$	$1 - \sin 2\theta$

4.2.5. AIDS IN ANALYSIS

A. *Spin Decoupling*

In some cases it will be apparent from the J splitting which multiplets arise from two proximate chemical groups because the coupling constant will result in equal splitting for the two multiplets, as shown in Fig. 1-4. However, in a more complicated spectrum, involving several coupling constants or second-order coupling, it may be quite helpful to apply a second rf field at the resonance frequency of a multiplet. As discussed in Section 2.5.1, application of a strong rf field at one multiplet resonance will

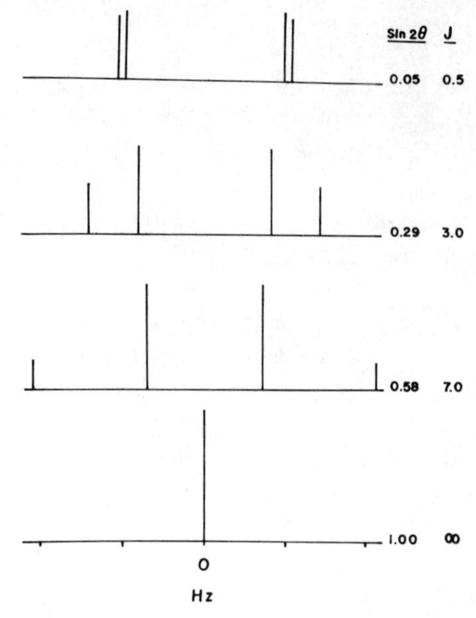

FIG. 4-6. Theoretical AB spectra for two nuclei with a chemical shift difference of 10 Hz.

cause the J splitting of the resonance of a second nucleus coupled to the irradiated resonance to collapse. In practice, the nuclei to be decoupled should be separated by at least 10 Hz. Spin decoupling by application of a strong field at one resonance frequency is illustrated in Fig. 2-17 for a simple system.

In complicated systems it is usually possible to identify coupled resonances by such double resonance techniques as homonuclear (e.g., both nuclei are protons) or heteronuclear (e.g., one nucleus is a proton and the other is a nitrogen) decoupling. The analytical advantages of decoupling are: (*a*) The splitting effects of certain nuclei can be removed, thus simplifying the spectrum; (*b*) the proximity of chemical groups may be determined by noting which nuclear resonances are decoupled; (*c*) lines in the spectrum can be sharpened by strongly irradiating the resonance of a nucleus possessing an electric quadrupole moment, such as ^{14}N, which is coupled to nuclei of interest in the spectrum; (*d*) the relative sign of the coupling constants in a weakly coupled system, e.g., AMX, can be determined. A moderately strong $\vec{H}_2$ field applied at a frequency between two of the four peaks (two peaks split by coupling to X) in the A resonance will cause a collapse of two of the peaks in the X resonance. The peaks that collapse in the X resonance will be those peaks which were split by coupling to A and which also correspond to the same energy state of M as the irradiated A peaks if both J_{AM} and J_{MX} have the same sign. If J_{AM} and J_{MX} have opposite signs, the collapsed peaks will be those from the other energy state of M. By irradiating at different positions, it is possible to determine the relative signs of the three coupling constants. These advantages (*a* through *d*) can be extended by using an additional rf field as in triple resonance. The uses and complications of spin decoupling have been thoroughly reviewed (12–14).

B. *Magnetic Field Increase*

The coupling constant results solely from spin–spin interactions and is therefore independent of the applied magnetic field, but the chemical shift is linearly dependent on the applied magnetic field. By going to higher fields, it is possible to reduce a complicated spectrum to one that is more easily recognized and analyzed. For example, the complicated A_2B_2 spectrum in Fig. 4-1 at $J/\Delta\delta = 0.4$ is not immediately recognizable. However, a fourfold increase in field strength will reduce the spectrum to that in Fig. 4-1 at $J/\Delta\delta = 0.1$, which can easily be analyzed for δ_A, δ_B, and J_{AB}. In fact, a good approximate analysis could be made treating the spectrum as A_2X_2. The fourfold increase is almost achieved by comparing proton NMR spectra obtained at 60 MHz and 220 MHz. This is illustrated in

Fig. 4-7 for an AB_2 spin system. Greater simplification, of course, would be achieved by running the spectrum on a 360 MHz instrument.

C. *Shift Reagents and Changes in Solvent Conditions*

As discussed in Section 3.3.3, the addition of lanthanide shift reagents to solutions can cause large chemical shifts in the nuclear resonances of compounds in the solution. The use of a shift reagent will have about the same effect as an increase in applied magnetic field strength. When a shift reagent is added to a solution containing a compound possessing a complex splitting pattern, the splitting pattern may be simplified. Because the different nuclei will not all be affected to the same extent by the shift reagent and the coupling constant will generally be unaffected by the shift reagent (15) (assuming no conformation changes), the $J/\Delta\delta$ ratio will decrease, leading to a simpler spectrum.

Shift reagents can cause dramatic changes in chemical shifts, resulting in a much simpler spectrum. Any other factor discussed in Chapter 3 that affects chemical shifts can also lead to simplification of a spectrum if the nuclear resonances are differentially affected. In Section 3.1.2, A, the effect of the solvent on chemical shifts was discussed. The solvent contributions to the shielding constant that can differentially affect the nuclear resonances of a solute molecule are σ_a, arising from anisotropy in the molecular susceptibility of the solvent molecules; σ_w, caused by van der Waals forces between solvent and solute molecules; σ_E, the polar effect resulting from the charge distribution in the solvent molecules setting up an electric dipole or quadrupole field affecting the solute molecules; and specific molecular interactions. The specific molecular interactions may entail self-association of solute molecules, e.g., polar solute molecules in a nonpolar solvent. Specific solute–solvent interactions may involve molecular complexation to a greater extent than is usually encountered with normal solvation. The greatest specific effects, however, will entail hydrogen bond formation or protonation, involving an electron donor site on either a solute or solvent molecule, as discussed in Section 3.1.2.

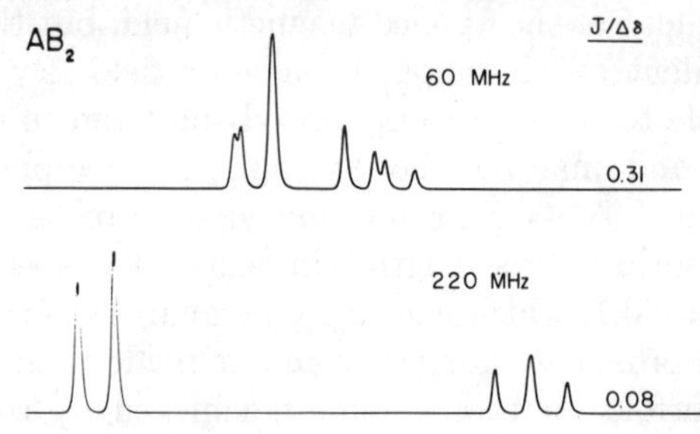

FIG. 4-7. AB_2 splitting pattern in 60 MHz spectrum simplified in 220 HMz spectrum.

D. *Isotopic Substitution*

A complicated spectrum can be simplified considerably by appropriate isotopic substitution. Deuterium is commonly used to replace hydrogen in simplifying a spectrum, e.g., changing an ABC spectrum to an ABX spectrum, where C is a proton and X is a deuteron. In addition the effects of removing the C resonance from the proton NMR spectrum, the coupling constants J_{AC} and J_{BC} will be much smaller by a factor of $\gamma_D/\gamma_H = 0.154$ such that $J_{AX} = 0.154\ J_{AC}$ and $J_{BX} = 0.154\ J_{BC}$. In certain ABC cases, the substitution of a deuteron for a proton is tantamount to reducing the spectrum to AB, from which the spectral parameters are easily obtained.

It is also useful at times to substitute a ^{13}C for a nonmagnetic ^{12}C nucleus. For example, substitution of ^{13}C in

$$\begin{matrix} H_a & & H_c \\ & C{=}^{13}C & \\ H_b & & H_d \end{matrix}$$

renders the protons nonequivalent.

A ^{13}C substitution was used to help assign the proton NMR spectrum of phosphoenol pyruvate,

$$\begin{matrix} H_B & & OPO_3^{2-} \\ & C{=}C & \\ H_A & & COO^- \end{matrix}$$

which is shown in Fig. 4-8 (16). The resonance centered at 4.67 ppm (τ) is split by J_{AB} and J_{AP}, and the resonance centered at 4.85 ppm (τ) is split by J_{AB} and J_{BP} such that each resonance is composed of two overlapping doublets with $J_{AB} = 1.55$ Hz, $J_{AP} = 1.45$ Hz, and $J_{BP} = 1.15$ Hz. The coupling constants are all small and quite similar in value. As to be discussed in Section 4.3, molecular geometry plays in important role in determining the magnitude of coupling constants. It was known that the trans ^{13}CH coupling constant of acrylic acid is larger than the cis coupling constant. This knowledge was used to assign the phosphoenol pyruvate resonances by examining the proton NMR spectrum of [1 − ^{13}C] phosphoenol pyruvate (60%) shown in Fig. 4-9. The upfield peak at 4.85 ppm clearly has a larger ^{13}CH coupling constant as the two ^{13}C satellite triplets are clearly resolved from the central triplet of the ^{12}C component. This evidence confirms other data implying that the upfield peak at 4.85 ppm is caused by H_B, the proton cis to the phosphate group.

The incorporation of deuterium from deuterium oxide in phosphoenol pyruvate in the reversible enolase-catalyzed reaction

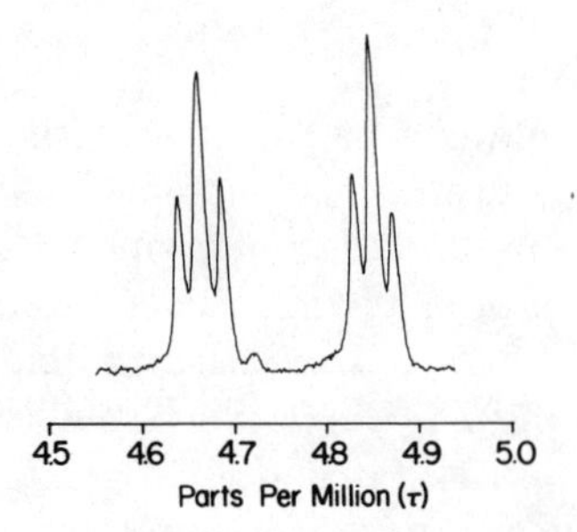

FIG. 4-8. 60 MHz proton NMR spectrum of 0.3 *M* phosphoenol pyruvate in D_2O at approximately pH 7.5 and 28°C (16).

$$\begin{array}{c} COO^- \\ | \\ HCOPO_3^{2-} \\ | \\ HCH \\ | \\ OH \end{array} \rightleftharpoons \begin{array}{c} {}^-OOC \quad OPO_3^{2-} \\ \diagdown C \diagup \\ \| \\ C \\ \diagup \quad \diagdown \\ H_A \qquad H_B \end{array} + H_2O$$

provides information about the stereochemistry of the enzyme-catalyzed reaction. The proton NMR spectrum of the phosphoenol pyruvate-3-*d* formed in the reaction is shown in Fig. 4-10. Comparison with the spectrum of protonated phosphoenol pyruvate in Fig. 4-8 reveals that the spectrum of the deuterated species contains only a doublet at 4.86 ppm caused by ^{31}P coupling, in contrast to the two groups of overlapping doublets at 4.67 and 4.85 ppm in the spectrum of the protonated species. The conclusion is that deuterium substitutes for H_A trans to the phosphate group and the enolase-catalyzed elimination of water from 2-phosphoglycerate is specifically anti.

E. *Computer Calculation of Spectra*

Probably the most valuable aid for analyzing a very complex spectrum is a digital computer. The general procedure is essentially the following. The user provides the computer with a first guess of all the coupling constants and chemical shifts derived from knowledge of similar compounds or, perhaps, from the observed spectrum. The program calculates the Hamiltonian matrix from the input NMR parameters and compares the energy eigenvalues and calculated intensities with the line positions and intensities determined from the observed spectrum. The comparison is used to obtain new values for the coupling constants and chemical shifts, and the procedure is continued in an iterative fashion until a satisfactory fit of experimental and calculated values of the energies and intensities is obtained with the correct values of the NMR parameters.

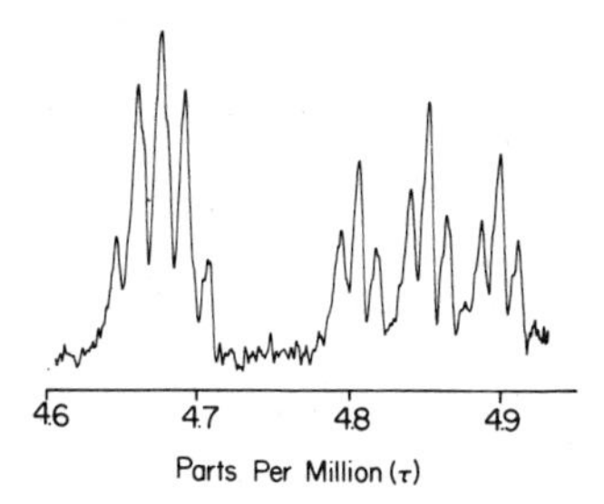

FIG. 4-9. 100 MHz proton NMR spectrum of 0.28 *M* phosphoenol pyruvate—60% 1-^{13}C, 40% 1-^{12}C in D_2O at approximately pH 7.6 and 31°C (16).

Various options and modifications of the above procedure are commonly used. In general practice, most spectra can be treated adequately with one or more of the programs available from the Quantum Chemistry Program Exchange* (17–19).

Calculation of a spectrum is also possible for a complicated system containing exchange-broadened resonance lines (20). A compilation of complex spectra generated from computer calculations and a rationale of the theoretical treatment have been given in the text by Wiberg and Nist (21), and the use of computer programs in the analysis of NMR spectra has been reviewed by Haigh (21a).

4.3. Molecular Structure

The sign and magnitude of the coupling constant J depend on the number of bonds and the type of bonds through which nuclei interact, in

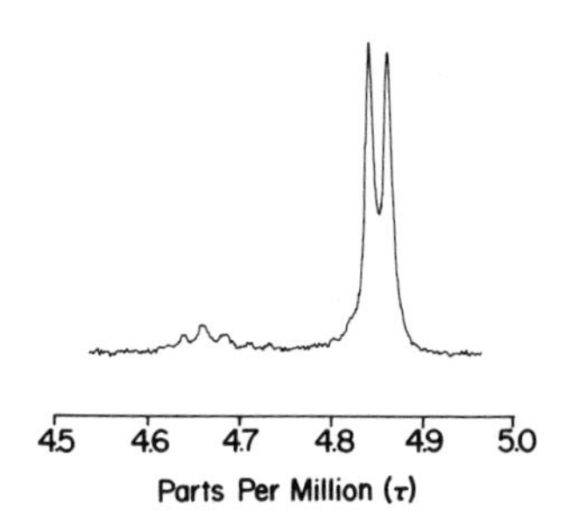

FIG. 4-10. 60 MHz proton NMR spectrum of 0.3 *M* phosphoenol pyruvate-3-*d* in D_2O at approximately pH 7.5 and 28°C. The weak triplet at 4.67 ppm is caused by a small amount of the protonated phosphoenol pyruvate as a contaminant (16).

* For information, contact: Quantum Chemistry Program Exchange, Chemistry Department, Indiana University, Bloomington, Indiana 47401.

addition to the geometrical orientation of those nuclei. The structural and geometrical aspects of proton–proton coupling constants have been reviewed by Bothner-By (22).

4.3.1. Geminal Proton–Proton Coupling Constants

It has been shown that an adjacent π electron system will cause a decrease in the geminal coupling constant of CH_2 (i.e., the coupling constant will become more negative) (23–25). The π electron system has a nodal plane; the probability of an electron being in the nodal plane is zero. If the nodal plane of the π electrons intersects the internuclear vector connecting the geminal protons, the coupling will be decreased. This is illustrated in Fig. 4-11, showing the larger negative coupling constant (-21.5 Hz) for structure A, in which the nodal plane of each carbonyl group bisects the geminal proton–proton axis in the rigid molecule. A similar structure B, in which there is internal rotation, has a less negative geminal coupling constant (± 12.7 Hz).

Geminal coupling constants are also affected by substituent groups. The electronegativity of neighboring atoms appears to be one important factor, although back donation of electrons from the neighboring atoms to bonding orbitals in the CH_2 group is also possible. The latter effect depends on geometry and both effects will tend to make the coupling constant more positive. A correlation of the geminal coupling constant with the electronegativity of the atom attached to the CH_2 group is possible with the rigid three-membered rings shown in Table 4-2.

4.3.2. Vicinal Coupling Constants

Many parameters enter into the case of vicinal coupling constants. The substituents will have some effect, as will the various bond lengths and

(A) O, H, O, H $J_{gem} = -21.5$ Hz

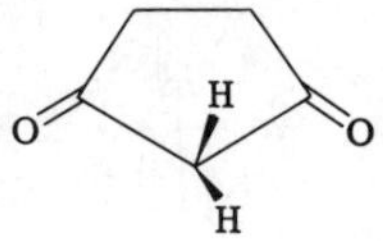

(B) HOOC, OH, $(CH_3)_2HC$, HO, H, O, H $J_{gem} = \pm 12.7$ Hz

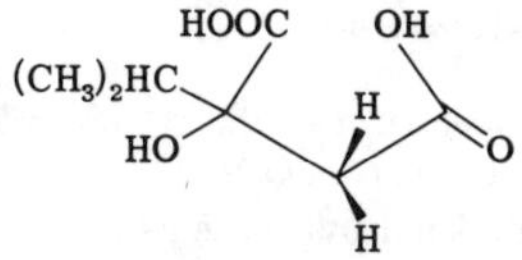

Fig. 4-11. Geminal coupling constants for CH_2 adjacent to unsaturated groups.

bond angles (22). The primary concern here will be with the effect of the dihedral angle θ (cf. Fig. 4-12) on vicinal coupling constants and the use of coupling constant values in determining molecular structures.

Karplus (26) used valence bond theory to calculate vicinal coupling constants. It was found that the calculated coupling constant could be approximately related to the dihedral angle by

$$J = A + B \cos \theta + C \cos 2\theta \qquad (4\text{-}25)$$

The coefficients are $A = 4.22$, $B = -0.5$, and $C = 4.5$ Hz in the case of vicinal proton–proton coupling. A plot of Eq. 4-25 for proton–proton coupling is given in Fig. 4-13. The implication is that adjacent axial–axial protons, having a dihedral angle of 180°, are strongly coupled, but axial–equatorial protons are only moderately coupled. Equation 4-25 has proved to be useful experimentally. However, Karplus (27) has pointed out that a relation such as Eq. 4-25 has limitations and should primarily prove useful comparing closely related chemical species rather than determining a dihedral angle to within a few degrees. The primary limiting factor is ignorance of the variation in other contributions to the coupling constant from substituents, other bond angles, and bond lengths, and the interrelationships among these factors.

A. *Sugar Conformation*

Vicinal proton–proton coupling constants with corroborating chemical shift data were used to establish the chair conformations and the conformaional equilibria in aqueous sugar solutions (28). The possible chair forms are shown in Fig. 4-14. The 100 MHz proton NMR spectra of D-xylose (A), D-lyxose (B), D-arabinose (C), and D-ribose (D) at equilibrium in deuterium oxide at 35°C are shown in Fig. 4-15. The hydroxyl groups are all deuterated and do not contribute to the NMR signal. The mutarotation could be followed after a sugar was dissolved by observing the anomeric proton peaks until equilibrium was achieved.

For example, dissolution of α-D-xylopyranose in deuterium oxide initially showed a sharp doublet at 5.26 ppm (from TMS) with a one-proton intensity. The intensity of the 5.26 ppm peak, ascribed to the α-anomeric proton, gradually decreased and a second doublet appeared at 4.65 ppm that

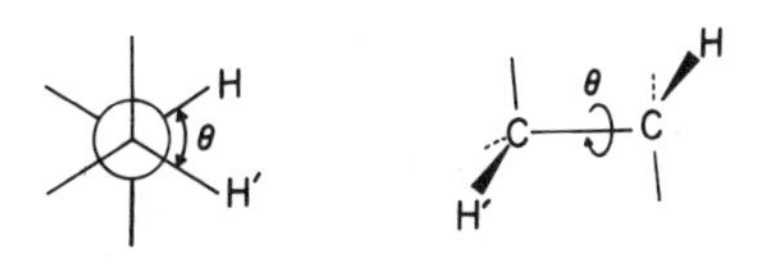

FIG. 12. The dihedral angle θ.

TABLE 4-2

EFFECT OF NEIGHBORING ATOMS ON GEMINAL PROTON–PROTON COUPLING CONSTANTS IN THREE-MEMBERED RINGS

COMPOUND	STRUCTURE	J_{gem}(Hz)
A	Br, H, H	−5.9
B	H_3C, H, H	$-0.4 < J_{gem} < 0.4$
C	H, N, H, H	±20
D	O, H, H	+5.5

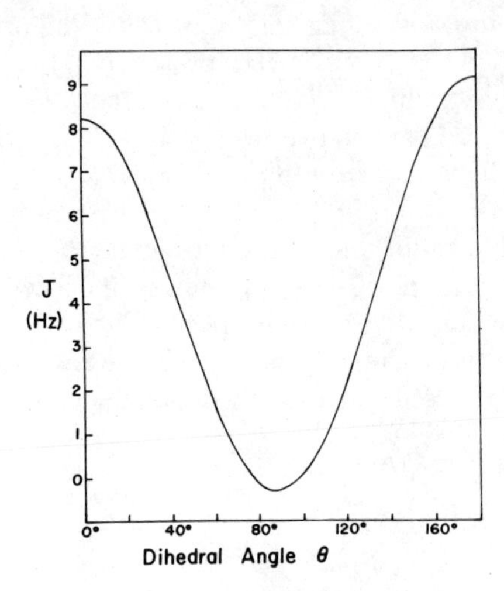

FIG. 4-13. Vicinal proton–proton coupling constant dependence on the dihedral angle according to Eq. 4-25.

gradually increased in intensity. The sum of the two peak intensities (one from the α anomer and one from the β anomer) corresponded to one proton. The intensity ratio indicated 33% α anomer and 66% β anomer at equilibrium, in good agreement with the ratio calculated from optical rotation and bromine oxidation. The splittings of the anomeric proton peaks at 5.46 and 4.65 ppm caused by coupling with the C2 proton are 3.1 and 7.4 Hz, respectively. According to the plot in Fig. 4-13, these values correspond to a small dihedral angle (i.e., axial–equatorial, $\sim 60°$) for the α anomer and a large dihedral angle (i.e., axial–axial, $\sim 180°$) for the β anomer. The α anomer must therefore have the 1C chair conformation of Fig. 4-14 and the β anomer must have the C1 chair conformation. Chemical shift data are in agreement with this conformational assignment (28).

The splitting of the anomeric proton resonance by coupling with C2 protons is listed in Table 4-3 for some other sugars (28). The coupling constants and, by implication, the dihedral angles indicate that the C1 conformation is appropriate for β-D-ribopyranose and β-D-lyxopyranose, and the 1C conformation is appropriate for α-D-arabinopyranose and α-D-ribopyranose, but the conformational forms of α-D-lyxopyranose and β-D-arabinopyranose cannot be determined from their vicinal coupling constants. Chemical shift data are also in agreement with these conformational assignments.

Spectrum D in Fig. 4-15 exhibits evidence for four tautomeric forms of D-ribose. The two downfield doublets are ascribed to the furanose anomers. The spectrum of 2-deoxy-D-ribose also indicates four tautomers are present at equilibrium.

Further refinements and extensions have been made for these and others sugars, including studies in organic solvents (29, 30). The conformation of the sugar moiety in other biochemical compounds has also been investigated. In particular, nucleotides have been studied. Phosphorus-31 NMR studies of the dinucleotides ApA and UpU have implied that J_{PH} depends on the dihedral angle (31). The ^{1}H—^{31}P and ^{1}H—^{1}H coupling constants were used in a study of the conformations of oxidized and reduced nicotinamide mononucleotide (32). The relationship between the

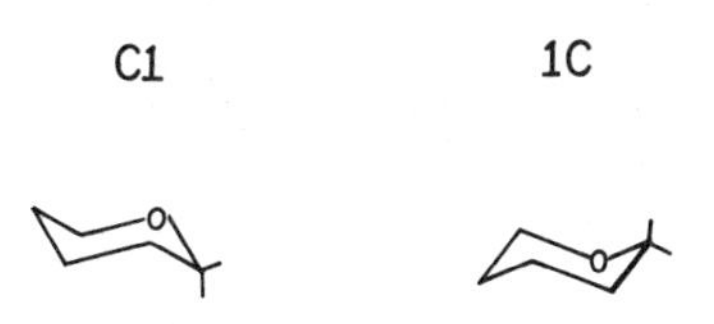

FIG. 4-14. Possible chair conformations of sugar pyranose rings. All bonds are shown for the anomeric carbon.

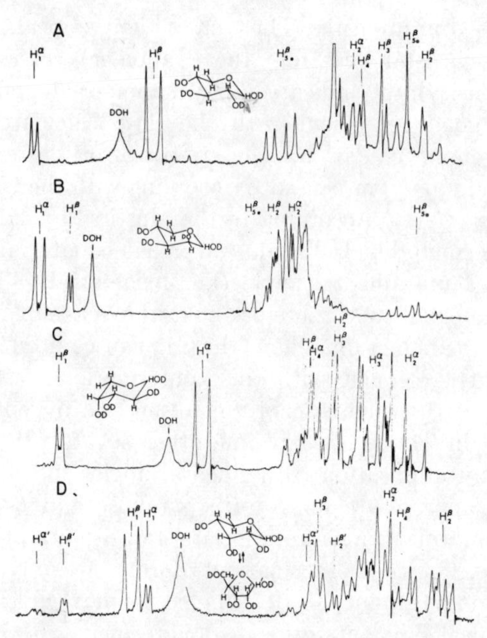

FIG. 4-15. 100 MHz proton NMR spectra of the aldopentoses D-xylose (A), D-lyxose (B), D-arabinose (C), and D-ribose (D) at tautomeric equilibrium at 35°C in deuterium oxide (28).

dihedral angle and the $^{1}H—^{31}P$ coupling constant was also examined in a proton NMR study of 3′,5′-cyclic nucleotides (33) and 2′,3′-cyclic nucleotides (34). The utility of $^{31}POC^{13}C$ coupling constants for estimating the importance of the trans and gauche conformations from the ^{13}C NMR of 3′,5′-cyclic nucleotides (35) and 2′,3′-cyclic nucleotides (34) has also been investigated.

The $^{13}C—^{31}P$ coupling constants obtained from ^{13}C NMR spectra (cf. Fig. 4-16) were used in conjunction with the ^{13}C chemical shifts of D-fructose 6-phosphate and D-fructose 1,6-diphosphate to assign ^{13}C resonances to the α and β anomers (36). The tautomeric composition was estimated to be 19% α anomer and 81% β anomer for F6P, and 23% α anomer and 77% β anomer for FDP. A comprehensive review of NMR studies on carbohydrate structure has been published (37).

B. *Peptide Conformation*

As with sugars, the dihedral angle dependence of vicinal coupling constants has proved useful in studies of peptide conformation (38, 39). The dihedral angles ϕ and ψ, which define the peptide backbone conformation, are shown in Fig. 4-17. ϕ is the dihedral angle between the C^1–N^1–C^α plane and the N^1–C^α–C^2 plane, and θ is the dihedral angle between the H^1–N^1–C^α and N^1–C^α–H^α planes. The vicinal proton–proton coupling constant $J_{H^1H\alpha}$ is related to the angle θ, which in turn is related to the standard peptide dihedral angle ϕ by $\theta = |\phi - 60°|$ for an L-peptide unit. ϕ is defined as zero for the cis conformation of the peptide unit.

The dependence of vicinal proton–proton coupling on the angle θ may be given by a variation of Eq. 4-25, viz.,

$$J = a \cos^2 \theta + b \cos \theta + c \sin^2 \theta \tag{4-26}$$

The peptides have a certain amount of freedom in internal rotation so that more than one conformation is energetically plausible, and all these conformations must be accounted for in a theoretical treatment. The vicinal coupling constants for H^1—H^α were measured for several amides (J_{obs}) in dimethylsulfoxide and corrected for electronegativity or bond angle distortion (J_{corr}) (39). These values are compared in Table 4-4 with values calculated (J_{cal}) from the mean value of the trigonometric functions of θ for a distribution of θ. The mean value of the dihedral angle $\langle\theta\rangle$ used in obtaining J_{cal} is also given in Table 4-4. A least-squares analysis of J_{corr} for the compounds in Table 4-4 yielded values for the constants a, b, and c of Eq. 4-26 such that

$$J = 7.9 \cos^2 \theta - 1.55 \cos \theta + 1.35 \sin^2 \theta \tag{4-27}$$

TABLE 4-3

VICINAL COUPLING CONSTANTS FOR THE ANOMERIC PROTON RESONANCES OF THE PYRANOSE FORMS OF THE D-SUGARS AT TAUTOMERIC EQUILIBRIUM IN D_2O AT 35°C[a]

	OBSERVED SPLITTING (Hz)		TAUTOMERIC COMPOSITION (%)	
SUGAR	α	β	α	β
Xylose	3.1	7.4	33	67
Lyxose	4.2	1.5	71	29
Arabinose	2.7	7.2	63	37
Ribose	2.1 (~5)	6.4 (~1)	20 (6)	56 (18)
Glucose	3.5	7.5	36	64
Mannose	1.7	1.0	67	33
Allose	—	8.2	—	>90
Galactose	2.8	7.1	27	73

[a] The values in parentheses refer to the furanose forms where present. Composition was determined from anomeric proton resonance intensities (28).

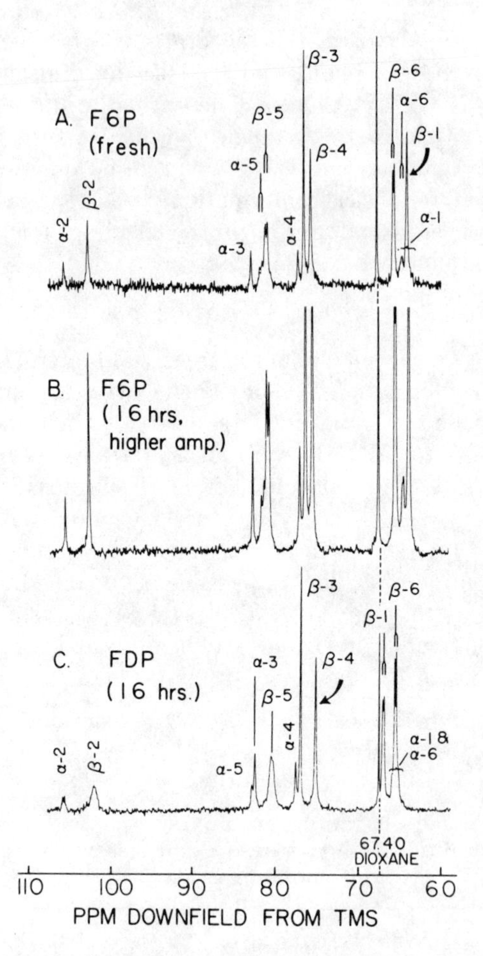

FIG. 4-16. Natural abundance proton-decoupled ^{13}C NMR spectra obtained in D_2O at 35°C. (A) 0.70 *M* D-Fructose 6-phosphate, disodium salt (F6P), pD = 8.6, immediately after dissolution (fresh) and (B) at equilibrium (16 hr) with higher amplification; and (C) 0.70 *M* D-fructose 1,6-diphosphate, tetrasodium salt (FDP), pD = 7.9, at equilibrium (16 hr). The three spectra from top to bottom are the result of 1.8×10^3, 19.4×10^3, and 13.6×10^3 accumulations. A slight time dependence is observed in the spectrum of F6P, with 5% more β anomer present at equilibrium than immediately after dissolution (36).

with an average deviation of 0.05 Hz for the correlation (39). The coupling constant values calculated from the dihedral angle distribution agree fairly well with the observed values for the compounds listed in Table 4-4 as well as for some other aliphatic dipeptides. In other cases, notably aromatic dipeptides, the calculated and observed values do not agree as well, varying by as much as 0.75 Hz. Other factors evidently contribute in these cases.

Equation 4-27 will still be useful with certain classes of peptides. In particular, peptides with a rigid structure, such as cyclic oligopeptides, should fit the correlation summarized in Eq. 4-27. Ramachandran *et al.* (39) found Eq. 4-27 could correlate quite well the published J values with the dihedral angles determined from x-ray studies for ferrichromes, valinomycin, and gramicidin S-A. In particular, NMR data (40) on alumichrome, an aluminum analog of the iron-containing cyclic hexapeptide ferrichrome A,

Orn 3
Orn 2 — Ser 1
Orn 1 — Ser 2
Gly

have been compared with information derived from the x-ray crystallographic study of ferrichrome A (41). In addition to a change in metal ion, alumichrome has two residues of glycine in place of the serines in ferrichrome A. Nevertheless, the NMR and x-ray data on the dihedral angle θ of the three ornithine residues agree quite well, as shown in Table 4-5.

The relationship between the dihedral angle θ and the proton–proton coupling constant has been theoretically calculated for the model compound N-methylacetamide and the resulting mathematical model applied to peptide conformations (42). Reference (42) also contains several liter-

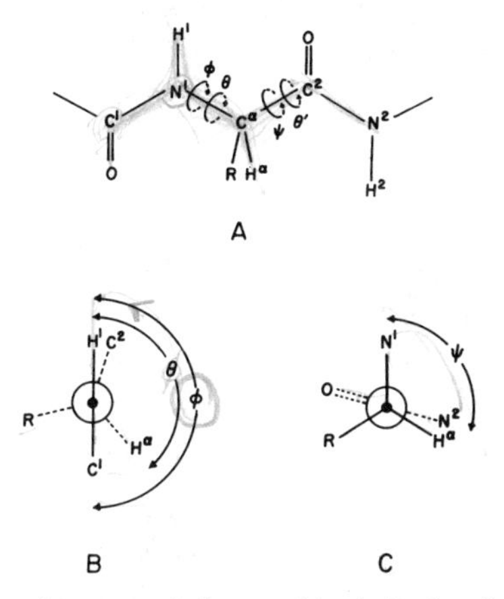

FIG. 4-17. (A) Segment of an extended L-peptide chain showing the relationship of the dihedral angles θ, ϕ, and ψ. ϕ = C^1–N^1–C^α–C^2. ϕ = H^1–N^1–C^α–H^α. ψ = N^1–C^α–C^2–N^2. (B) Sighting along the N^1–C^α bond. (C) Sighting along the C^α–C^2 bond.

TABLE 4-4

OBSERVED AND CALCULATED H^1–N^1–C^α–H^α PROTON–PROTON COUPLING CONSTANTS FOR PEPTIDE MODEL SYSTEMS IN DIMETHYL SULFOXIDE (~35°C)[a]

COMPOUND	$<\theta>$[b]	J_{obs} (Hz)	J_{corr} (Hz)[c]	J_{cal} (Hz)	θ_{cal}[d]
(1) *N*-Methylacetamide	100°	4.8	4.5	4.63	127°
(2) *N*-Ethylacetamide	102°	5.7	5.5	5.54	133°
(3) *N*-Isopropylacetamide	148°	7.7	7.7	7.63	150°
(4) Glycylglycine	100°	6.0	5.8	5.76	135°
(5) Alanylalanine	154°	7.8	7.8	7.80	152°
(6) Dihydrothymine (equatorial)	45°	3.7	3.6	3.53	45°
(7) Dihydrothymine (axial)	75°	1.5	1.5	1.39	75°
(8) Diketopiperazine	60°	2.2	2.2	2.21	60°

[a] From Ramachandran *et al.* (39).
[b] The mean value of the dihedral angle distribution used in computing J_{cal}.
[c] Corrected for electronegativity variations and bond angle distortions.
[d] Determined from J_{cal} and Eq. 4-27 in the text.

ature references to other applications of vicinal coupling constants to the study of peptide conformation.

Recent work has indicated that the nitrogen–hydrogen coupling constant J_{NH} between H^α and N^2 may provide information about the other peptide dihedral angle ψ (cf. Fig. 4-17) (43), where ψ is the angle between the N^1–C^α–C^2 plane and the C^α–C^2–N^2 plane. In this case, the dihedral

TABLE 4-5

COMPARISON OF THE OBSERVED VICINAL H–N–C^α–H PROTON–PROTON COUPLING CONSTANTS OF THE ORNITHINE RESIDUES OF ALUMICHROME (40) WITH THE DIHEDRAL ANGLES OF FERRICHROME A (41)[a]

RESIDUE	COUPLING CONSTANT J (Hz)			DIHEDRAL ANGLE, ϕ (°)	
	IN DMSO	IN WATER	AVERAGE	FROM J	FROM X-RAY
Orn 1	7.4	7.7	7.6	−150	−145
Orn 2	5.2	5.5	5.4	− 73	− 77
Orn 3	8.6	8.5	8.6	−100	−104

[a] From Ramachandran *et al.* (39).

angle θ' pertinent to the coupling constant is related to ψ by $\theta' = |\psi - 120°|$. Using a valence bond treatment, the vicinal J_{NH} coupling constant was related to the $J_{HH'}$ coupling constant of the analogous hydrocarbon by the approximate relationships

$$J_{^{15}NH}(\theta') \approx -1.514\, a^2 J_{HH'}(\theta') \tag{4-28a}$$

$$J_{^{14}NH}(\theta') \approx -1.079\, a^2 J_{HH'}(\theta') \tag{4-28b}$$

where the constant a is the coefficient of the nitrogen atom Hartree-Fock 2s function used to compose the normalized hybrid orbital on the nitrogen atom (43). Equations 4-28 illustrate the dihedral angle dependence of the coupling constant and also show that the s character of the hybrid orbital on nitrogen can be an important factor. The observed values of J_{NH} were compared with values of J_{NH} calculated from Eqs. 4-28 using $J_{HH'}$ values from the same or similar compounds; the comparison is shown in Table 4-6 (43–47) for compounds that are not amides. The hybridization was assumed to be sp^3, making $a^2 = \frac{1}{4}$, for the quaternary nitrogen and sp, making $a^2 = \frac{1}{2}$, for the isocyanide nitrogen.

TABLE 4-6

COMPARISON OF OBSERVED AND CALCULATED NITROGEN–HYDROGEN COUPLING CONSTANTS[a]

COMPOUND (REFERENCE)	θ'(°)	$J'_{HH'}$ (Hz)	$J_{^{14}NH}$ (Hz) EXP.	$J_{^{14}NH}$ (Hz) CALC.	$J_{^{15}NH}$ (Hz) EXP.	$J_{^{15}NH}$ (Hz) CALC.
Amino acids (44)	60	2.6	—	0.7	(−) 1.8 ± 0.8	−1.0
	180	13.6	—	3.7	(−) 5.1 ± 1.2	−5.2
Trimethylvinylammonium bromide (45, 46)	0 (cis)	8.3, 8.5	2.6	2.3	—	−3.2
	180 (trans)	14.8 15.0	5.5	4.0	—	−5.6
3,3-Dimethylbutylisocyanide (47)	60	2.3	0.7	1.2	—	—
	180	16.0	6.9	8.6	—	—
Vinyl isocyanide (47)	0 (cis)	8.5	3.2	4.6	—	—
	180 (trans)	15.6	6.3	8.3	—	—

[a] From Karplus and Karplus (43).

TABLE 4-7

COUPLING CONSTANTS IN MONOSUBSTITUTED ETHYLENES[a]

COMPOUND	J_{cis} (Hz)	J_{trans} (Hz)	J_{gem} (Hz)
$H_2C=CH_2$	+11.6	+19.1	+2.5
$H_2C=CHCO_2C_2H_5$	+11.7	+19.0	+0.24
$H_2C=CHCO_2H$	+10.2	+17.2	+1.7
$H_2C=CHCN$	+11.75	+17.92	+0.91
$H_2C=CHCH_3$	+10.02	+16.81	+2.08
$H_2C=CHOt$-Bu	+ 6.2	+13.2	−0.1
$H_2C=CHCl$	+ 7.3	+14.6	−1.4
$H_2C=CHF$	+ 4.65	+12.75	−3.2

[a] From Bothner-By (22).

More germane to an interest in peptides, however, is the use of [^{15}N]-acetylglycine in D_2O as a peptide model (43). The methyl group splitting yielded J_{15NH} (obs) = (−)1.5 Hz. The theoretical J_{15NH} coupling constant was calculated using $J_{HH'}$ from acetaladehyde (2.85 Hz) and an sp^2 hybrid orbital on nitrogen ($a^2 = \frac{1}{3}$) giving a calculated value of −1.44 Hz in excellent agreement with the observed coupling constant. Further correlations with peptide models using Eqs. 4-28 may show vicinal nitrogen–hydrogen coupling constants to be useful in determining the dihedral angle ψ.

4.3.3. Coupling in Unsaturated Compounds

The geminal proton–proton coupling constants of olefins are surprisingly small, especially when compared with J_{cis} and J_{trans}, which involve spin–spin interactions through an additional bond. It is also observed that J_{cis} is smaller than J_{trans}. Correlation of the coupling constants with the electronegativity of the substituent group in monosubstituted ethylenes may be expressed by (6)

$$J_{\text{cis}} = 11.7(1 - 0.34\,\Delta X) \tag{4.29a}$$

$$J_{\text{trans}} = 19.0(1 - 0.17\,\Delta X) \tag{4.29b}$$

where ΔX is the difference in electronegativity between hydrogen and the substituent. We see from Eqs. 4-29 that $J_{\text{cis}} \approx \frac{2}{3} J_{\text{trans}}$. Some examples of coupling constants in monosubstituted ethylenes are given in Table 4-7. A more comprehensive list was tabulated by Bothner-By (22).

The coupling constants of the aromatic protons in substituted benzenes exhibit little dependence on the substituents, and the coupling constants can be generalized as: J_{ortho}, 6.5–9.5 Hz; J_{meta}, 1–3 Hz; and J_{para}, 0.3–1 Hz.

References

1. P. L. Corio, *Chem. Rev.* **60,** 363 (1960).
2. N. F. Ramsey and E. M. Purcell, *Phys. Rev.* **85,** 143 (1952).
3. N. F. Ramsey, *Phys. Rev.* **91,** 303 (1953).
4. M. Barfield and D. M. Grant, *Advan. Magn. Resonance* **1,** 149 (1965).
5. J. N. Murrell, *Progr. NMR Spectrosc.* **6,** 1 (1971).
6. C. N. Banwell and N. Sheppard, *Discuss. Faraday Soc.* **34,** 115 (1962).
7. J. D. Roberts, "An Introduction to the Analysis of Spin-spin Splitting in High-Resolution Nuclear Magnetic Resonance Spectra." Benjamin, New York, 1962.
8. P. L. Corio, "Structure of High-Resolution NMR Spectra." Academic Press, New York, 1966.
9. R. G. Jones, *Nucl. Magn. Resonance* **1,** 191 (1972).
10. A. Messiah, "Quantum Mechanics," Vol. I, p. 72. Wiley, New York, 1959.
11. J. A. Pople, W. G. Scheider, and H. J. Bernstein, "High-Resolution Nuclear Magnetic Resonance," pp. 105 and 120. McGraw-Hill, New York, 1959.
12. W. McFarlane, *Annu. Rev. NMR (Nucl. Magn. Resonance) Spectrosc.* **1,** 135 (1968).
13. J. D. Baldeschwieler and E. W. Randall, *Chem. Rev.* **63,** 81 (1963).
14. B. D. Nageswara Rao, *Advan. Magn. Resonance* **4,** 271 (1970).
15. J. K. M. Sanders and D. H. Williams, *J. Amer. Chem. Soc.* **93,** 641 (1971).
16. M. Cohn, J. E. Pearson, E. L. O'Connell, and I. A. Rose, *J. Amer. Chem. Soc.* **92,** 4095 (1970).

17. A. A. Bothner-By and S. Castellano, Program 111, Quantum Chemistry Program Exchange, Indiana University, Bloomington.
18. J. W. Cooper, Program 126, Quantum Chemistry Program Exchange, Indiana University, Bloomington.
19. J. W. Cooper, Program 127, Quantum Chemistry Program Exchange, Indiana University, Bloomington.
20. D. A. Kleier and G. Binsch, Program 165, Quantum Chemistry Program Exchange, Indiana University, Bloomington.
21. K. B. Wiberg and B. J. Nist, "The Interpretation of NMR Spectra." Benjamin, New York, 1962.
21a. C. W. Haigh, *Annu. Rep. NMR* (*Nucl. Magn. Resonance*) *Spectrosc.* **4,** 311 (1971).
22. A. A. Bothner-By, *Advan. Magn. Resonance* **1,** 195 (1965).
23. M. Barfield and D. M. Grant, *J. Amer. Chem. Soc.* **85,** 1901 (1963).
24. M. Barfield and D. M. Grant, *J. Chem. Phys.* **36,** 2054 (1962).
25. S. Castellano and C. Sun, *J. Amer. Chem. Soc.* **88,** 4741 (1966).
26. M. Karplus, *J. Chem. Phys.* **30,** 11 (1959).
27. M. Karplus, *J. Amer. Chem. Soc.* **85,** 2870 (1963).
28. R. U. Lemieux and J. D. Stevens, *Can. J. Chem.* **44,** 249 (1966).
29. P. L. Durette and D. Horton, *Chem. Commun.* p. 1608 (1970).
30. P. L. Durette and D. Horton, *Carbohyd. Res.* **18,** 403 (1971).
31. M. Tsuboi, S. Takahashi, Y. Kyogoku, H. Hayatsu, T. Ukita, and M. Kainosho, *Science* **166,** 1504 (1969).
32. R. H. Sarma and R. J. Mynott, *J. Amer. Chem. Soc.* **95,** 1641 (1973).
33. B. J. Blackburn, R. D. Lapper, and I. C. P. Smith, *J. Amer. Chem. Soc.* **95,** 2873 (1973).
34. R. D. Lapper and I. C. P. Smith, *J. Amer. Chem. Soc.* **95,** 2880 (1973).
35. R. D. Lapper, H. H. Mantsch, and I. C. P. Smith, *J. Amer. Chem. Soc.* **95,** 2878 (1973).
36. T. A. W. Koerner, Jr., L. W. Cary, N. S. Bhacca, and E. S. Younathan, *Biochem. Biophys. Res. Commun.* **51,** 543 (1973).
37. T. D. Inch, *Annu. Rev. NMR* (*Nucl. Magn. Resonance*) *Spectrosc.* **2,** 35 (1969).
38. G. N. Ramachandran and R. Chandrasekaran, *Biopolymers* **10,** 935 (1971).
39. G. N. Ramachandran, R. Chandrasekaran, and K. D. Kopple, *Biopolymers* **10,** 2113 (1971).
40. M. Llinas, M. P. Klein, and J. B. Neilands, *J. Mol. Biol.* **52,** 399 (1970).
41. A. Zalkin, J. D. Forrester, and D. H. Templeton, *J. Amer. Chem. Soc.* **88,** 1810 (1966).
42. M. Barfield and H. L. Gearhart, *J. Amer. Chem. Soc.* **95,** 641 (1973).
43. S. Karplus and M. Karplus, *Proc. Nat. Acad. Sci. U.S.* **69,** 3204 (1972).
44. R. L. Lichter and J. D. Roberts, *J. Org. Chem.* **35,** 2806 (1970).
45. M. Ohtsuru and K. Tori, *Chem. Commun.* p. 750 (1966).
46. J. M. Lehn and R. Seher, *Chem. Commun.* p. 847 (1966).
47. A. A. Bothner-By and R. H. Cox, *J. Phys. Chem.* **73,** 1830 (1969).

CHAPTER 5

EXPERIMENTAL METHODS

To this point we have been discussing the principles and some uses of nuclear magnetic resonance while ignoring a vital question. How is the NMR spectrum actually obtained or an NMR relaxation time of a sample measured? In this chapter we will attempt to answer these questions with an emphasis on the problems encountered with samples of biochemical origin.

5.1. Detection of Nuclear Magnetic Resonance

The simple schematic diagram of an NMR spectrometer in Fig. 5-1 illustrates the experimental set up necessary for observing the NMR phenomenon. The spectrometer shown in Fig. 5-1a is the double-coil instrument first used by Bloch *et al.* (1, 2). A sample is placed in a cylindrical glass tube between the pole pieces of the magnet M. The sample tube is placed inside a receiver coil R, which is oriented so that the receiver coil axis, the transmitter coil axis, and the magnetic field are at right angles to one another. The rf transmitter applies an oscillating rf field to the sample via the transmitter coil at a frequency appropriate to achieve resonance at

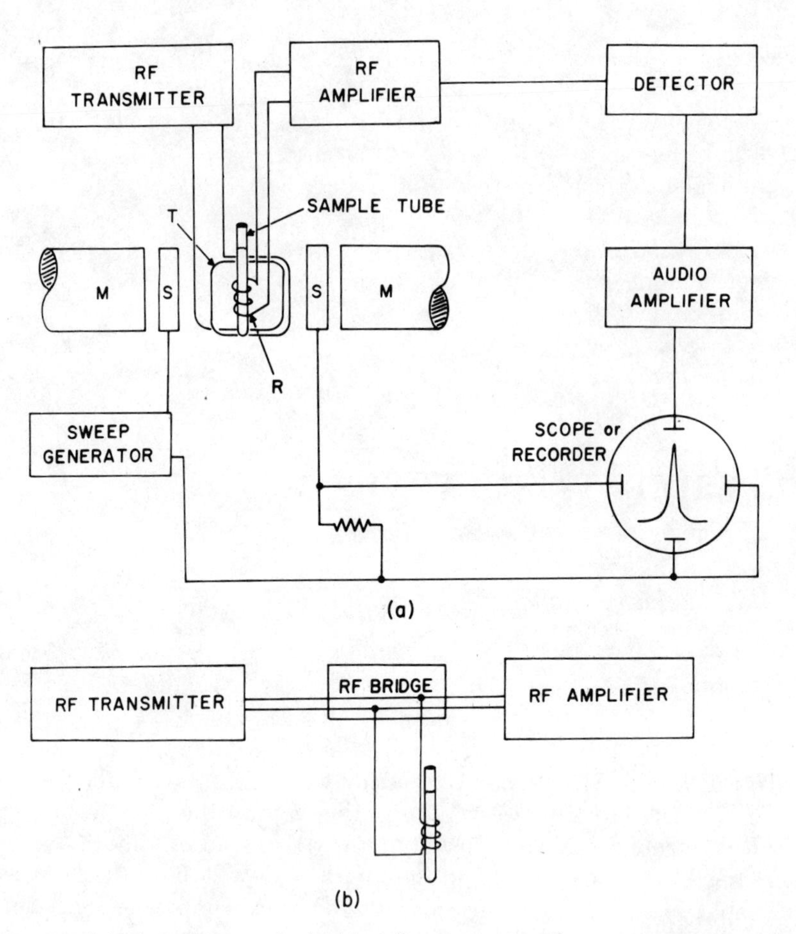

FIG. 5-1. (a) Block diagram of a double-coil nuclear magnetic resonance spectrometer. M = magnet, S = sweep coils, T = transmitter coil, R = receiver coil. (b) Modifications for a single-coil spectrometer.

the field strength of the magnet. The magnetic field is swept through the resonance condition, $\omega = \gamma H_0$, by varying the current in the magnet's sweep coils S with the sweep generator. The magnet's field strength and the frequency are very close to the resonance condition before the sweep current is applied; the field is swept over a very small range. It is also possible to sweep the frequency over a small range while holding the magnetic field strength constant.

When resonance is achieved, the nuclei in the sample flip (as described in Sections 1.2 and 2.1), inducing a voltage in the receiver coil. This double-coil process is termed "nuclear induction." The induced voltage is amplified, electronically manipulated, and displayed on an oscilloscope or recorder as the NMR signal.

A single-coil instrument (3, 4) was first used to detect an NMR signal. As shown in Fig. 5-1b, the single coil is wound around the sample tube at right angles to the magnetic field, the single coil acting as an inductor in a tuned circuit. When the resonance condition is achieved the nuclei absorb energy with a resultant voltage drop across the coil. This voltage drop is detected, amplified, and displayed as the NMR signal. Both single-coil and double-coil instruments are widely used.

5.2. High-Resolution Continuous Wave Spectrometer

In this section some aspects of the components of a high-resolution continuous wave NMR spectrometer are discussed. Many of the points are also applicable to pulsed and Fourier transform spectrometers.

5.2.1. Magnetic Field

Whether a permanent magnet, an electromagnet, or a superconducting magnet is used, the magnetic field necessary for high-resolution NMR experiments must be stable and must be the same over the sample volume (i.e., it must be homogeneous). A permanent magnet has advantages in economy, stability, fewer maintenance problems, and lack of noise pollution from a heat exchanger. An electromagnet has advantages in that the magnetic field may be varied (necessary for studying some nuclei) and, for a given gap size, a slightly higher magnetic field, ~25 kGauss, may be obtained. The maximum value of the magnetic field of an electromagnet is limited by saturation of the iron in the pole pieces of the magnet. To enjoy the advantages of very high magnetic fields, it is necessary to use a superconducting magnet that operates in liquid helium. Superconducting magnets having fields of ~85 kGauss are being built for NMR purposes and in the future even more powerful magnets will be used.

A. *Stability Requirements*

The magnetic field strength needs to be maintained to better than 1 part in 10^8 for most high-resolution purposes. A permanent magnet and a superconducting magnet have fewer problems in this regard because adequate thermostatting and magnetic shielding of the magnet are all that is required. Superconducting magnets are actually fairly sensitive to movement of nearby magnetic materials. An electromagnet requires more elaborate instrumentation to maintain stability. The magnet power supply has a voltage regulator to maintain the line voltage. In addition, a magnet flux stabilizer is used to guard against short-term changes (<10 sec) in the magnetic field. A pair of coils, the pickup and buckout coils, are used with the flux stabilizer to maintain a stable field. Any rapid change in magnetic field induces a current in the pickup coils that is proportional to the rate of change of magnetic field. The current actuates a balanced photocell circuit that produces a current in the buckout coils. The current in the buckout coils produces a small opposing magnetic field to "buckout" the magnetic field change. A compensation for long-term drift may also be incorporated by having a small amount of current continuously flowing in the buckout coils. However, this will not protect against changes in the direction of the field drift. The buckout coils may also be used for sweeping the field.

Even with these devices, a field sufficiently stable for high-resolution NMR is usually not attained. External or internal locking, discussed in Section 5.2.2, provides the relative stability necessary for a high-resolution instrument.

B. *Homogeneity*

To resolve lines separated by 1 Hz on a 100 MHz instrument requires that the magnetic field vary less than one part in 10^8 over the sample volume ($\sim$0.1 ml within the receiver coil for proton samples in 5 mm NMR tubes). To achieve this, the magnet pole pieces must be metallurgically uniform, parallel, and polished to optical flatness. Electromagnets (and superconducting magnets, although for another reason) are "cycled." Cycling entails increasing the magnet current (and thus the field) above the operating value for a short time and then lowering it to the operating value. This helps achieve a constant field over a region around the center of the magnet gap. For this reason, the sample is positioned carefully in the magnet gap. Improvement in homogeneity may also be attained by use of smaller volumes.

Homogeneity is also improved by using small coils, called "homogeneity" or "shim coils," placed on the pole faces of the magnet. Each of these coils is designed to affect different types of field gradients. There are usually at least three shim coils (the x, y, and z coils). Currents are passed through these coils to counteract field gradients in the x, y, and z directions (z is the direction of the $\vec{H}_0$ field). In addition, there is a curvature coil that provides for second-order corrections along the field direction. Additional higher order correction coils (e.g., x^2, xy, xz, etc.) are used with some instruments.

Finally, a considerable improvement in homogeneity is achieved by spinning the sample as illustrated in Fig. 5-2. Suppose the field varies over a

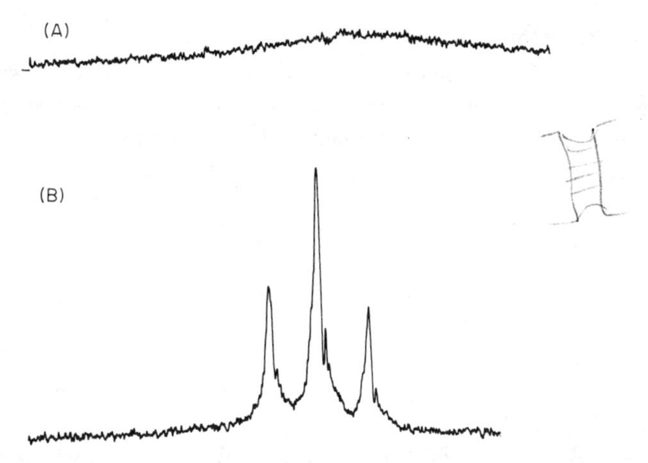

FIG. 5-2. Effect of sample spinning on the proton NMR spectrum of the methyl peak of 1% ethylbenzene. (A) Nonspinning. (B) Spinning.

sample by an amount ΔH. If the sample motion is sufficiently rapid, each nucleus in the sample will behave as though it experiences the average field of ΔH instead of each nucleus experiencing a different field over the range ΔH. For the motion to be sufficiently rapid, the spinning rate

$$\nu_s \geq 5\sqrt{2}\,\gamma\Delta H/\pi^2 \tag{5-1}$$

Spinning the cylindrical sample tube with an air turbine averages the field in two directions; however, inhomogeneity is still left in the direction along the tube axis. Adjustment of the homogeneity coils in this direction is therefore particularly important, and second-, third-, and fourth-order shim coils exist for the z direction. Values for the spinning rate are typically 20–30 rps, although in some superconducting magnets spinning rates an order of magnitude larger are needed.

An undesirable side effect of spinning, however, is the creation of spinning side bands, which appear equally spaced on each side of the resonance line. They are produced by modulation resulting from spinning the sample nuclei through different values of the magnetic field caused by inhomogeneity. The more inhomogeneous the field is, the more intense the side bands will be. Although they can complicate a spectrum, spinning side bands are easily distinguished by their dependence on the spinning rate; the spinning side bands will be located at $\pm\nu_s$ from the resonance line. A finely polished completely symmetrical sample tube is also a necessity if good resolution is desired.

5.2.2. Modulation and Locking

A more detailed block diagram of a high-resolution NMR spectrometer is shown in Fig. 5-3. For improved base line stability and capabilities for field locking, the field is modulated at an audiofrequency ν_m (typically 5–40 kHz). [Haworth and Richards (5) have reviewed the use of modulation in magnetic resonance.] The field modulation results in side bands being produced at $\nu_r + \nu_m$ and $\nu_r - \nu_m$, each flanking the center band at frequency ν_r, which is the frequency established by the quartz crystal and multipliers in the transmitter (e.g., 220 MHz). An NMR signal can be observed if any one of the three frequencies meets the resonance condition, $\omega = \gamma H_0$. In practice the first upper side band is often used, e.g., the NMR frequency for observation with a 220 MHz instrument modulated at 20 kHz is 220.02 MHz.

Figure 5-3 shows an additional NMR tube in the magnetic field besides the sample tube. The additional tube is physically placed close to the analytical sample tube and is used for external locking. It contains a control sample (usually paramagnetically "doped" water for proton NMR). By "locking" on the resonance frequency of the control sample, the stability of the spectrometer may be increased by more than an order of magnitude. "Locking" on the control signal entails closing a circuit containing the coil that is wound around the control sample. Any change in the magnetic field will be manifest as a change in the control sample's resonance frequency,

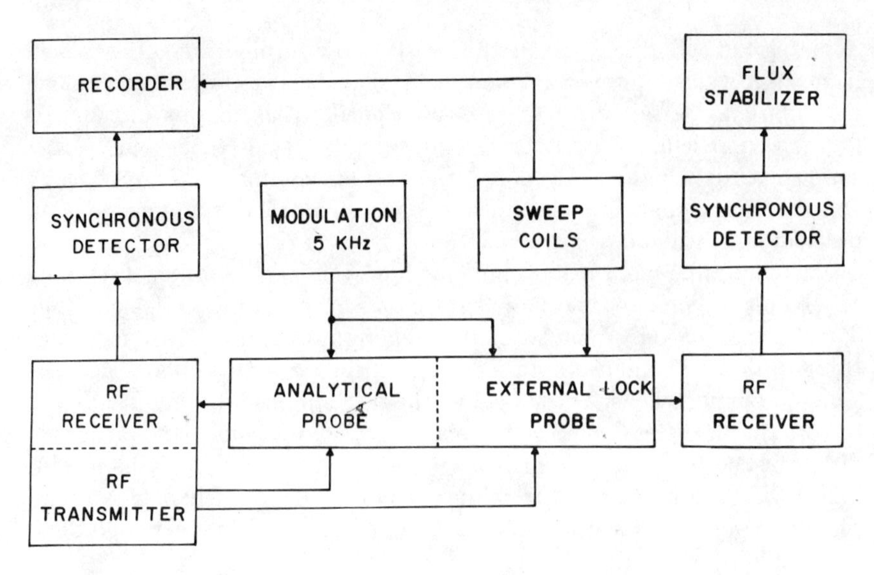

Fig. 5-3. Schematic diagram of an NMR spectrometer with external lock.

which ultimately alters the current in the lock circuit. That current is used to counteract the magnetic field change. The process for locking is as follows. The magnetic field is adjusted to the resonance condition at the upper sideband of the frequency-modulated signal for the external lock ("doped" water). Frequency modulation has essentially the same effect as field modulation. The audiofrequency phase detector is adjusted so that the dispersion (in-phase) mode of the lock signal is obtained (cf. Fig. 2-4). The magnetic field is adjusted until the external lock sample is exactly on resonance. At that point, a control loop, including an inductor coil around the lock sample, is "locked on." With the dispersion mode, there is no voltage in the coil at the exact center of resonance. Any small variation in the magnetic field will result in the lock signal moving slightly off resonance. This results in an induced positive or negative voltage in the external control coil, which is part of a feedback loop serving to correct any drift or field fluctuation. The feedback may operate through the magnetic power supply, the flux stabilizer, or alteration of the audiofrequency. The purpose it serves, however it is accomplished, is to provide the necessary stability for the high-resolution NMR experiment. The external lock system succeeds in stabilizing the field so that the observed resonance line will appear to shift less than 0.005 ppm/hr.

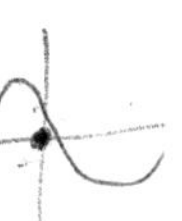

Because the external lock signal does not experience exactly the same field variations as the analytical sample, some instability persists. An advancement over external locking is provided by internal locking, which gives an order of magnitude greater stability. The reason for this improvement is that the voltage used in the feedback loop is derived from changes in the NMR signal of a nucleus in the analytical sample tube. The lock nucleus may be on a compound in the sample specifically added for that purpose (e.g., tetramethylsilane) or it may be on the solvent. The lock may be homonuclear or heteronuclear with respect to the observed NMR spectrum. For running ^{1}H, ^{13}C, and ^{31}P spectra, it is most convenient to use a heteronuclear deuterium lock derived from deuterated solvents.

The most accurate spectral calibration involves generating side bands at a precisely known frequency. The chemical shift of a sample resonance can then be measured from the chart paper thus calibrated. With external and internal locking, there is sufficient stability for most purposes that precalibrated chart paper can be used with only an occasional check on the calibration.

5.2.3. Probes

The probe refers to the unit that holds the sample tube in the magnet gap and includes the plastic sample spinner (air turbine), the receiver coil (and transmitter coil for a double-coil instrument), the sweep coils, and a preamplifier. The probe may also entail a Dewar system with temperature sensor, a heater for temperature control, a decoupling coil, and an external lock coil. All parts are manufactured from nonmagnetic material, chiefly aluminum.

A double-coil probe for a commercial instrument is shown in Fig. 5-4. Note that the receiver coil wound around the sample tube is perpendicular to both the transmitter coil and the stationary magnetic field H_0 in order to minimize pickup from these sources. Nevertheless, a small amount of "leakage" between the transmitter and receiver coils still persists. To minimize the "leakage," metallic devices called "paddles" (both coarse and fine) are moved in or out of the induced fields to optimize the coupling between the coils.

In a single-coil instrument, the sample probe contains a single coil around the sample tube that acts as an inductor in an LC circuit tuned to the resonance condition $[\omega = (LC)^{-1/2}]$, where L is the inductance of the coil and C is the capacitance in parallel with the coil. The single coil, operating both for the transmitter and the receiver, has some advantage over the double coil in that the probe configuration is simplified and the rf power from the transmitter is not wasted by putting the transmitter coil outside the receiver coil. The latter point is not too vital for CW NMR but is an important consideration in pulsed and Fourier transform NMR,

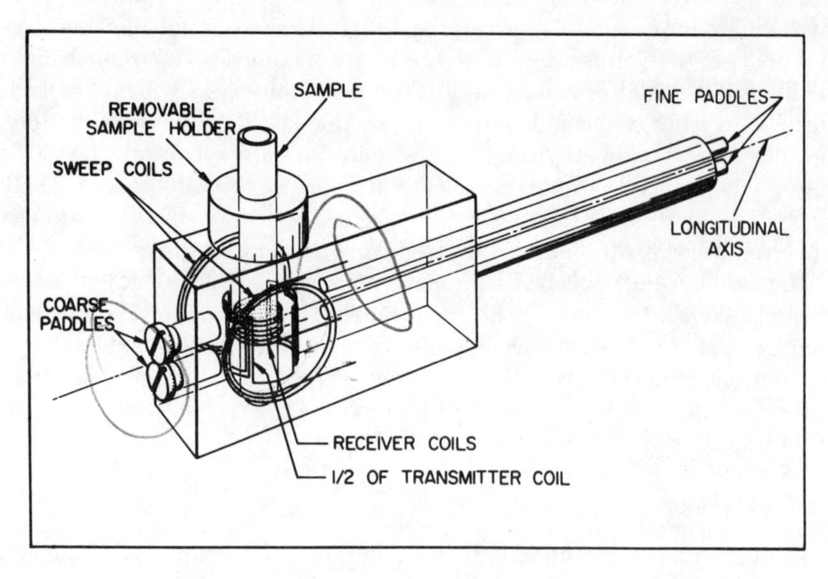

FIG. 5-4. Double-coil probe. (Courtesy of Varian Associates.)

where very high power levels are desired. The double-coil configuration, in contrast, has an advantage in that the transmitter and receiver coils can be tuned independently.

Abragam (6) indicates that for thermal noise generated in the receiver coil, the signal-to-noise ratio for the coil may be represented by

$$S/N \propto f V_c^{1/2} Q^{1/2} \tag{5-2}$$

where f is the filling factor (V_s/V_c), V_c is the volume of the coil, V_s is the volume of the sample, and Q is the quality factor for the coil. The implications of Eq. 5-2 are: (*1*) The filling factor should be as large as possible. This is accomplished by using thin-walled, precision-polished sample tubes that just fit inside the coil. (*2*) The signal-to-noise can be improved via the electronic circuitry to a certain point. (*3*) At first glance, increasing the coil volume would seem to give an improvement. However, increasing V_c leads to a lower Q factor for the coil.

The first point above, concerning the filling factor, is related to an item of some concern in biological applications of NMR, that of a limited amount of sample. Although a spherical sample shape is a better model of an ideal Lorentzian cavity, most NMR tubes are cylindrical. Cylindrical tubes are more convenient and can be used with virtually no distortion if the sample in the cylinder is "infinitely" long (actually a length of sample five times as long as the coil diameter will suffice). Most of the sample in the cylinder is not in the field of the coil. For a 5 mm tube, the volume within the coil may be less than 0.1 ml, but the required sample volume is about 0.5 ml. For many biochemical samples, only a small quantity of sample may be available. In such cases, it is obvious that a greater amount of the sample will be detected by using a spherical cavity, which just fills the coil, rather than a long cylindrical cavity. For such purposes, microcells are used. These may contain as little as 25 μl of sample. The spherical microcells may be fabricated entirely out of one cylindrical piece of glass or may involve simply a glass bulb that can be centered in a regular sample tube. Both types of microcell are commercially available and make possible the study of small quantities of compounds (as little as 100 μg). Microcells are inconvenient for most purposes because the technique of filling and cleaning cells is time consuming, and the cell must be very carefully positioned in the coil region to achieve good signal-to-noise and homogeneity.

To avoid the inconvenience of the spherical microcell with only a small decrease in signal-to-noise ratio, it is possible to use a commercially available probe with a 1 mm receiver coil accommodating a 1 mm cylindrical tube. With this tube it is possible to observe useful spectra with 5 μl of sample routinely and 2.5 μl with more effort. The proton Fourier transform NMR spectrum of a 10 μg sample can therefore be obtained in less than 15 min (including time for filling the 1 mm sample tube).

For many purposes, it is desirable to work at high or low temperatures. Commercial spectrometers are available that have temperature probes capable of working in the range $-150°C$ to $+200°C$. The temperature can usually be maintained within $\pm 1°C$. The temperature is controlled by passing nitrogen gas through a coil in a cooling bath (if low temperatures are desired), over a heater, and around the sample tube in the Dewar-jacketed probe insert. A thermistor or thermocouple senses the temperature and activates a proportional controller to maintain the temperature at a selected value. To minimize response time and provide better control, the heater and sensor should be as close to the sample as possible; this is not done on all commercial instruments.

5.2.4. Spectral Integration

As pointed out in Section 2.1.2,C, the area under a resonance signal is directly related to the number of nuclei contributing to the signal. For quantitative studies we are therefore interested in the integrated intensities of resonance peaks. Nuclear Overhauser effect measurements also require knowledge of resonance intensities. Most commercial instruments are equipped for electronic integration of signals. Problems with electronic integration arise from samples with poor signal-to-noise, baseline drift, and very close resonance peaks. The latter is especially troublesome for multiple-irradiation spectra where the strongly irradiated peak is near the peak to be integrated. Saturating the peaks by applying too powerful an H_1 field should also be avoided because (as described in Section 2.1.2,A) saturation will lead to an underestimation of the number of nuclei contributing to the integrated peak intensity.

In certain situations it may be better to determine the area under the peak by cutting and weighing the spectrum chart paper or by use of a planimeter. If the spectrum is stored on a digital signal averager or computer, the peaks may conveniently be integrated digitally.

5.2.5. Sensitivity Improvement by Signal Averaging

Probably the greatest boon to sensitivity enhancement is the use of time-averaging techniques. These have been reviewed (7, 8). The instrumentation necessary for accumulating and averaging several spectra entails either a large-scale computer, a general laboratory computer, or a special purpose time-averaging computer. The latter lacks versatility but has been most widely used. A small special purpose signal averager of this type is commonly referred to as a "computer of average transients" or CAT. The use of a CAT to accumulate several spectra over a long period of time (e.g., overnight) is made possible by the stability improvement engendered by external or internal locking.

The basis for time averaging is as follows. As a number N of spectral scans are added, the accumulated signal for any peak will be n times as large as the signal for one scan. However the noise, which is inclined to be random in nature, tends to average out and the accumulated noise from N scans is $N^{1/2}$ times as large as the noise from one scan. The signal-to-noise ratio for the accumulated spectral scans will then be $N^{1/2}$ $(= N/N^{1/2})$ times the signal-to-noise ratio for a single scan. This means 100 scans will theoretically give a sensitivity enhancement of 10. It also means that there is very little improvement in obtaining 2500 scans rather than 2000 scans; efficiency in use of time is also important. The signal-to-noise improvement by use of time averaging is illustrated in Fig. 5-5.

The method of time averaging is as follows. The spectrum is divided into a certain number of evenly spaced channels (which number is set by the computer's capability and the operator's desired scanning time) across the spectrum width. The number of channels might typically be 1024. As the spectrum is swept starting with the first channel, the digital response to the analog voltage signal from the NMR sample is stored in the computer's memory core at the site allocated for channel 1. When the spectrum reaches

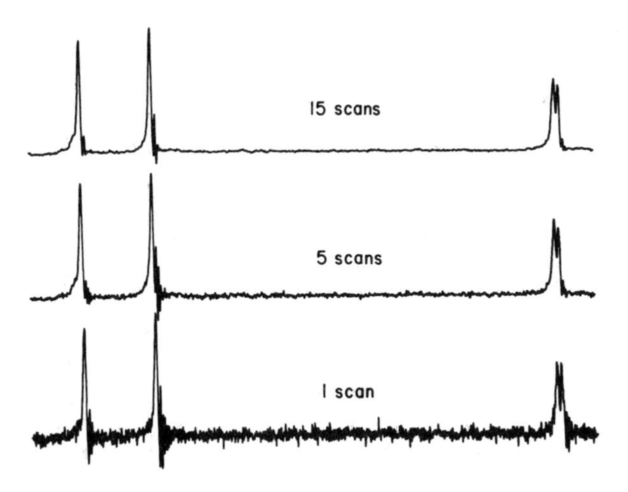

FIG. 5-5. Improvement in signal-to-noise ratio by accumulation of several frequency-swept spectra with a laboratory computer. The H-8′, H-2′, and H-1′ peaks of the proton NMR spectrum of 0.1 M ADP in D_2O are shown.

the frequency for channel 2, the signal for that frequency will be stored in core at the site for channel 2. This process will continue for the remaining channels until the spectrum has been scanned. At that point, the spectral scan is begun again at the frequency for channel 1. This time the signal from channel 1 is added to the channel 1 signal of the first scan, which is stored in the computer's memory core. The signals for the remaining channels and for the remaining scans are likewise added and stored in the appropriate place in memory. When desired, the signals can be recalled from memory, scaled to fit the chart paper, and plotted with a smooth line connecting the points from the channels.

Aside from the stability required, signal averaging will have limits in the resolution attainable. For a 1000 Hz sweepwidth using a computer with 1024 channels, a resolution of little better than 1 Hz may be attained. If 0.5 Hz resolution is desired, a computer with 2048 channels will be necessary.

In practice, at least for macromolecule spectra, it is found that the best signal-to-noise improvement for a given amount of time is achieved by using a fairly strong rf field (just less than saturating), sweeping at a fairly quick rate across the spectrum (e.g., 25–50 Hz/sec), and accumulating several spectra.

5.3. Pulsed NMR Spectrometer

A pulsed NMR spectrometer, of course, has many features in common with a continuous wave NMR spectrometer. This is particularly true for pulsed operation in which the transient signal is Fourier transformed to yield a high-resolution frequency spectrum. Discussion of Fourier transform spectrometers is deferred to the next section. This section will be concerned with a "pulsed" or "spin echo" NMR spectrometer, which is used predominantly to measure relaxation times of a single component, as discussed in Section 2.1.3, and which is much less expensive than a Fourier transform system. For these purposes, the magnetic field need not be very homogeneous. Because inhomogeneity is not a problem, much larger sample tubes can be used to increase signal-to-noise within limits of the instrument's rf power output; the power decreases as r^{-2} as the transmitter coil radius r increases. Such a spectrometer has several uses that will be described in later chapters, such as relaxation time and self-diffusion coefficient measurements of water protons in enzyme solutions and in cellular systems, and relaxation time measurements of quadrupolar ions, such as $^{23}Na^{+}$, in solution.

The construction of pulsed NMR spectrometers has been described by Buchta *et al.* (9). The pulsed spectrometer, diagrammed in Fig. 5-6, operates in the following manner. The magnetic field and rf frequency are adjusted so that the nucleus of interest achieves the resonance condition. The 15 MHz rf field, generated by a 7.5 MHz crystal oscillator with a frequency doubler, runs continuously. The pulse programmer, which establishes the pulse sequence and phase to be used, triggers the gate on the gate amplifier, permitting an rf pulse of several microseconds duration to be amplified by the power amplifier. The output from this transmitter is coupled to the coil in the probe. The intensity (voltage) of the resulting NMR signal is enhanced with a broad-band amplifier. The signal is then phase detected, using a 15 MHz reference from the crystal oscillator and doubler. The phase-sensitive detector, which has intrinsically linear response, yields an increase in signal-to-noise ratio because the pseudorandom noise will not produce a coherent component at the frequency and phase of the detector. The phase-detected output is displayed as desired (e.g., oscilloscope).

Besides the timing and gating, the pulsed spectrometer differs from a CW spectrometer in the requirement for high-power rf output (100–1000 watts) from the transmitter so short pulses (1–50 μsec) may be used. The pulse shapes must also be square, having rise and recovery times <1 μsec. The probe and receiver also must be able to handle the high-voltage pulses and recover quickly (<10 μsec).

Because pulsed NMR methods using strong rf fields can cause transverse nuclear magnetizations comparable to the equilibrium magnetization (cf. Section 2.1.3), it might be expected that better signal-to-noise ratios would be obtained than with CW NMR, which uses low-power rf to avoid saturation. Unfortunately, this is not the case because the duration of the rf pulses is so short that a broad-band amplifier must be used in the receiver to avoid a nonlinear response. The broad-band amplifier admits more noise and decreases the signal-to-noise sufficiently to more than compensate for the large transverse magnetization. For optimum sensitivity, the bandwidth of the receiver should be no larger than necessary.

For many uses, the spectrometer diagrammed in Fig. 5-6 is satisfactory. Some studies, however, necessitate sensitivity enhancement. Fast-response computers will perform this task admirably. A less expensive way of achieving this sensitivity enhancement (within a factor of two to three of that possible with a computer) is provided by use of a gated amplifier or, as it is often called, a "boxcar integrator." A boxcar integrator is a device for averaging repetitive signals; it is most useful for signals of short duration that repeat infrequently compared with their durations, i.e., signals with a

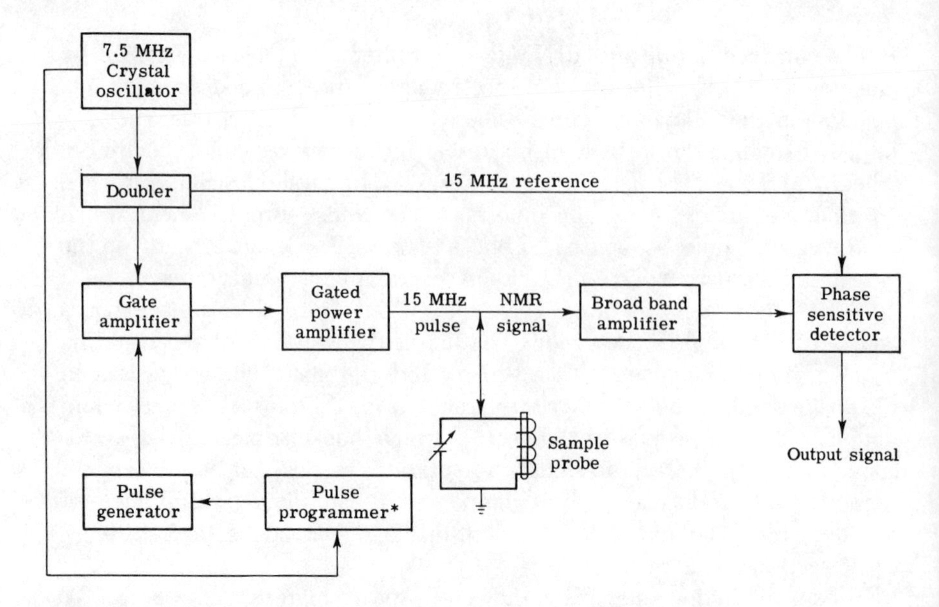

FIG. 5-6. Functional diagram of a pulsed NMR spectrometer.

low duty factor. The signal is averaged on an RC low-pass circuit that is gated so that it is connected to the signal source only during the actual signal. Between signals the RC circuit is disconnected and stores the previous signals to be averaged with future signals.

A diagram is given in Fig. 5-7 for a pulsed NMR spectrometer modified for use with a boxcar integrator. Possible modes of utilizing the boxcar integrator for relaxation time measurements and the theoretical basis for signal-to-noise enchancement have been presented (10). In the 180°–τ–90° method for determining T_1 (cf. Section 2.1.3,B), the signal-to-noise improvement in averaging several free induction decay signals is

$$\approx 0.25(t_{av}/T_1)^{1/2} \tag{5-3}$$

if a time equal to $7T_1$ is used to allow the nuclear spin system to return to equilibrium before the following two-pulse sequence commences. The total length of time used for averaging is t_{av}. In practice this method will give improvements of nearly two orders of magnitude in sensitivity for a nucleus with $T_1 = 50$ msec. Longer T_1 values will give less improvement with a given time of averaging.

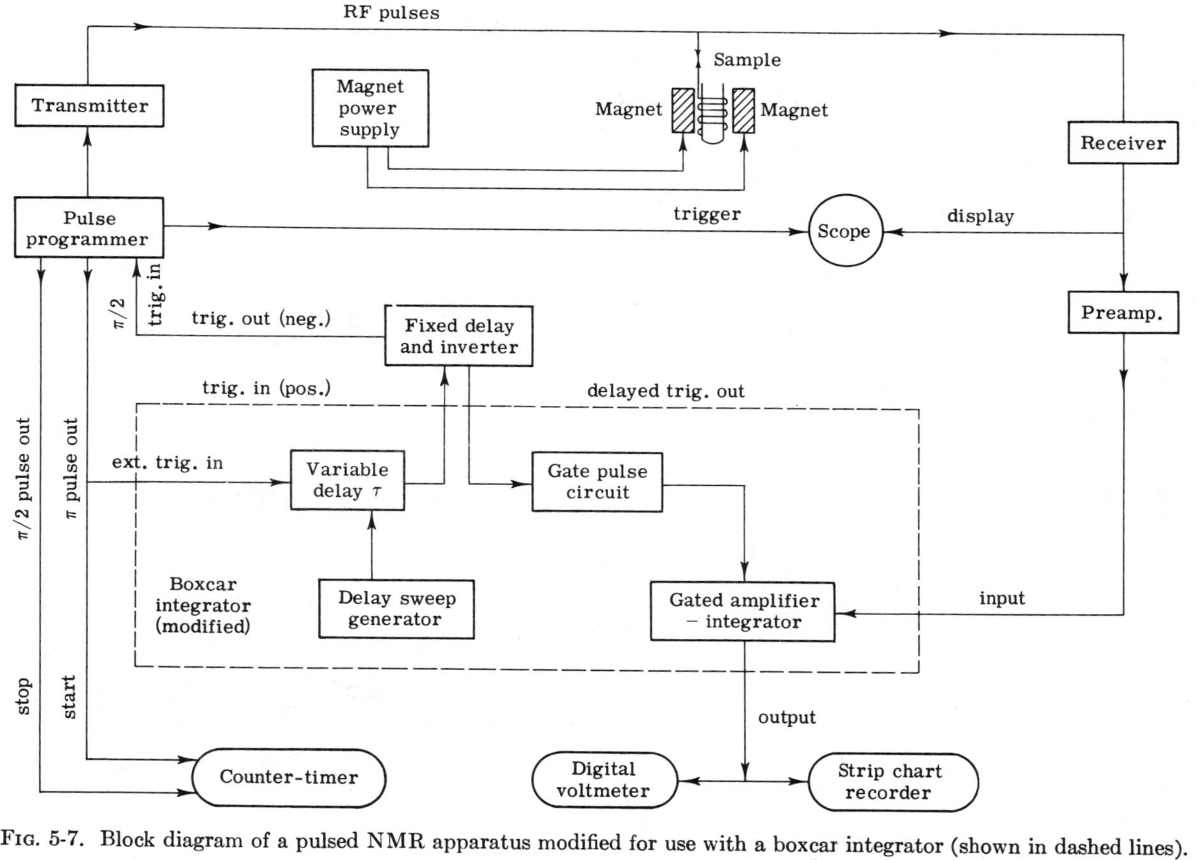

FIG. 5-7. Block diagram of a pulsed NMR apparatus modified for use with a boxcar integrator (shown in dashed lines).

5.4. Fourier Transform NMR Spectrometer

High-resolution Fourier transform NMR operation combines the requirements for pulsed NMR and continuous wave NMR operation. The transmitter power, probe, and receiver requirements are at least as demanding as those of ordinary pulsed NMR, and the homogeneity and stability requirements of CW NMR are ordinarily also required. In addition, a computer is necessary. The principles of operation, advantages, and limitations of the Fourier transform method are given in Section 2.1.4. A typical multinuclei Fourier transform NMR spectrometer with associated computer hardware and software is shown in Fig. 5-8.

5.4.1. Spectrometer

The essential components of a Fourier transform NMR spectrometer system with multinuclei capabilities (^{1}H, ^{13}C, ^{31}P, etc.) are (*a*) magnet, (*b*) lock system, (*c*) pulse programmer and power amplifier, (*d*) probe, (*e*) pseudorandom noise decoupler, (*f*) broad-band receiver, (*g*) frequency synthesizer to tie all components to a single master crystal oscillator, and (*h*) data accumulation and reduction system. These essentials are embodied in the schematic diagram of the Fourier transform NMR spectrometer system shown in Fig. 5-9.

For field stability, a heteronuclear internal lock system is very desirable. A convenient choice is a deuterium lock for use with deuterated solvents. It is necessary to have a strong lock signal. If a homonuclear lock is used, the limited dynamic range of the computer makes it difficult to detect weak

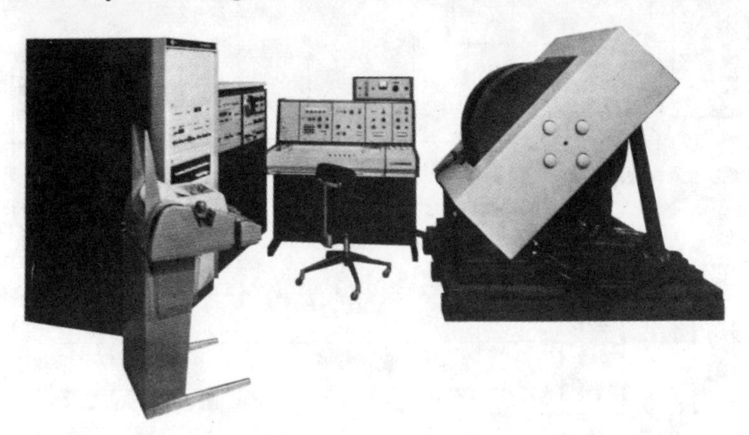

Fig. 5-8. Multinuclei Fourier transform NMR spectrometer system. (Courtesy of Varian Associates.)

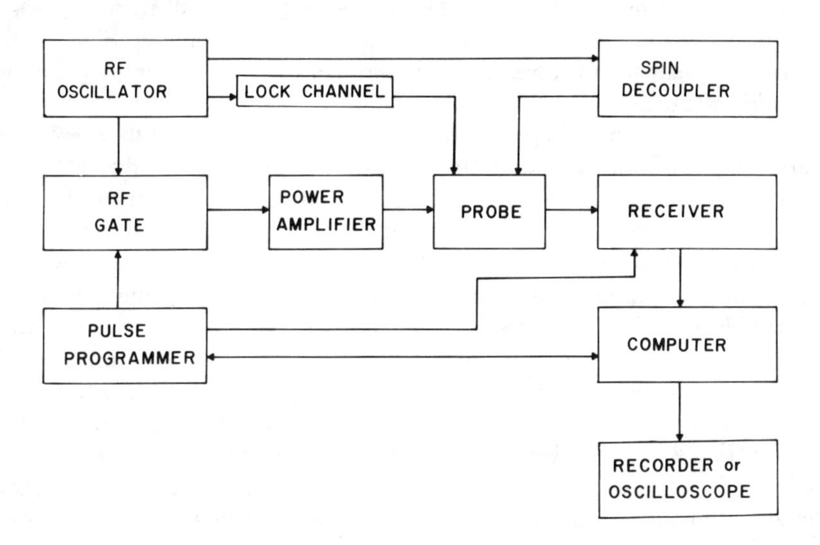

FIG. 5-9. Schematic diagram of a Fourier transform NMR spectrometer.

signals in the presence of the strong lock signal. An additional problem with a homonuclear lock is that the strong rf pulses will also perturb the lock signal. Although it is possible to gate the lock system around the rf pulse, it is still best to avoid the problems by using a heteronuclear lock.

The noise decoupler is very useful for decoupling over a broad frequency range (e.g., all protons). This can be extremely valuable for ^{13}C NMR. Decoupling protons removes the ^{1}H–^{13}C splitting, thus simplifying the ^{13}C spectrum, and can lead to increased signal-to-noise caused by the nuclear Overhauser effect (cf. Section 2.5.2).

A very important consideration is the amount of rf power (H_1) available from the power amplifier. For the entire spectrum to be excited uniformly, it is necessary that the field strength of the rf pulse (in frequency units) be greater than the width of the spectrum SW (11):

$$\gamma H_1/2\pi > SW \tag{5-4}$$

If condition 5-4 is not achieved, the following problems are encountered: (*1*) appreciable phase shifts (*post hoc* correctable) of off-resonance lines; (*2*) large frequency-dependent line intensity variations that cannot be corrected; and (*3*) systematic errors in relaxation time measurements.

The practical aspects of modifying an existing CW spectrometer for Fourier transform operation have been presented (12, 13). The extent and expense of the modification depends on the particular CW instrument and the desired capabilities of the Fourier transform spectrometer. Some of the newer CW instruments already fulfill many of the requirements for Fourier transform operation.

It might be assumed *a priori* that 90° pulses (i.e., $\theta = \gamma H_1 t_w = \pi/2$ from Eq. 2-20) are used and that the pulses are repeated at sufficiently long times ($\sim 7T_1$) that the nuclear spin system returns to equilibrium between pulses. Such an experiment would produce the largest free induction decay (FID) following any individual pulse. However, it is found that a greater signal-to-noise enhancement for a given period of time averaging is achieved by rapid repetition of pulses having pulse widths such that θ is less than 90° (7, 14). When a series of pulses is applied to a spin system, a steady state is created, with the pulses tilting the macroscopic magnetization being balanced by the relaxation of the magnetization back toward the z direction (cf. Section 2.1.3,A). Maximum sensitivity enhancement requires that the balance be optimized. It is apparent that a nucleus with a long T_1 will have less magnetization relaxed back along the z axis when the following pulse is applied, resulting in a decreased signal. For smaller values of θ, the equilibrium magnetization will be restored along the z axis more quickly, which more than compensates for the smaller signal produced when a small value of θ is used. The appropriate balance is usually obtained empirically. The rf field H_1 is generally held constant and the pulse width t_w is varied. The optimum value for t_w may be obtained by noting the maximum signal obtained among those recorded for several different settings of t_w. Of course, a compromise must be reached for samples containing nuclei with widely varying values of T_1. The different values of T_1 provide the explanation of why signal intensities cannot be meaningfully compared in ordinary Fourier transform NMR. For an equal number of nuclei, a resonance having a shorter T_1 will be more intense than a resonance having a longer T_1 because more of the magnetization will be restored along the z axis before the subsequent pulse is applied. For accurate peak intensity comparisons, the time between pulses should be about five times the longest T_1.

Certain multiple-pulse techniques provide an increase in sensitivity that is especially helpful for samples with long T_1 values, more specifically $T_1 \gg T_2^*$. The "driven equilibrium Fourier transform" (DEFT) technique, using a repetitive 90°–τ–180°–τ–90° three-pulse sequence, provides such a sensitivity increase (15). Data acquisition occurs during the free induction decay following the first 90° pulse. The 180° pulse refocuses the individual spin moments (just as in the Hahn sequence described in Section 2.1.3,C). The final 90° pulse "drives" the magnetization back along the z axis to augment the signal from the next pulse. For a given period of time averaging, DEFT has a signal-to-noise advantage of 1.28 over ordinary Fourier transform NMR when $T_1 \gtrsim 2.5T_2^*$ (16).

As mentioned in the previous section, for pulsed NMR spectrometers, the receiver for a Fourier transform spectrometer must have a fast recovery time from the effects of the pulse ($t_{rec} \ll \tau$). Data acquisition is delayed until the receiver has recovered. The receiver must also have a large linear range. A sample having a very strong signal can "saturate" the receiver if the receiver gain is too high. If there are additional strong signals, lines at combination frequencies may appear.

The free induction decay signals following the pulse cover a frequency range F, which is the frequency difference between the resonance lines occurring at the highest and the lowest fields, as shown in Fig. 5-10. The heterodyne detection process in the receiver compares the frequency range F with the carrier frequency ν (frequency of the applied H_1 field) converting the signal to a band of audiofrequencies determined by ν and F. If the carrier frequency is at ν' as shown in Fig. 5-10, the spectral lines 3 and 4 will be "folded" about ν' and will appear in the positions designated 3′ and 4′ because only the difference, and not the sign of the difference, between the carrier frequency and the spectral line frequency is important. To avoid complications from folding, the carrier frequency is usually set at one end or the other of the spectrum as shown for carrier frequency ν in Fig. 5-10.

Just as signal frequencies are folded about the carrier frequency, high-frequency noise is folded back and appears in the spectrum. To minimize the noise, a sharp low-pass filter, usually set to equal the spectral width, is utilized. A filter set less than the spectral width will strongly affect the intensities and phases of lines near the edges of the spectrum. In addition to filtering out high-frequency noise components, the filter may be used to

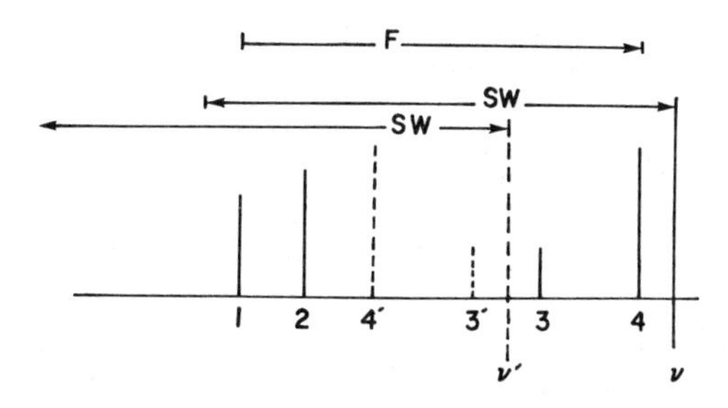

FIG. 5-10. Four-line spectrum from a Fourier transform NMR experiment. The spectral positions if the applied H_1 field is at frequency ν are indicated by unprimed numbers. The spectral positions observed if the applied H_1 field is at frequency ν' are indicated by 1, 2, 3′, and 4′.

cut out (or decrease) certain peaks (e.g., a solvent peak) if only a portion of a spectrum is desired. A filter is not infinitely sharp at its cutoff frequency, so the line to be reduced must be distant from the lines of interest and care must be taken to recognize any folded back line (albeit reduced in intensity) caused by the filtered resonance if the spectral width is also set to exclude the line.

5.4.2. Data Acquisition

For data processing, the free induction decay is divided into discrete segments, each characterized by the average voltage in the segment. The analog-to-digital converter, or digitizer, converts the analog signal (voltage) in each segment to a digital (numerical) signal, and the digital signal for each segment is stored in the computer's memory at a specific location allocated for that segment. For averaging purposes, the signals of subsequent FID's are also divided into segments, with the digital signals for any particular segment being added to the signal already stored at the location for that segment in the computer memory.

In order to uniquely define the digital signal from any waveform, it is necessary to sample twice per segment. The data sampling rate is the number N of data samples taken per unit time. The sampling theorem then requires that the sampling rate be twice the spectral width SW, or the number of data points

$$N = 2(SW)t_{ac} \tag{5-5}$$

where the acquisition time t_{ac} is the total time the data from a single FID is sampled. For a given spectral width and acquisition time, the number of required words or channels in computer memory for data storage is N.

The ultimate linewidth of a resonance is determined by the T_2 relaxation processes or, for nuclei with long T_2, by magnetic field inhomogeneity (T_2^*). It is also possible to artificially broaden lines in the Fourier transform spectrum if the FID is not sampled long enough. If the acquisition time is t_{ac}, the best resolution possible is $1/t_{ac}$. For high resolution, it is necessary to sample the FID for as long as possible. To obtain very high-resolution spectra from samples requiring a large spectral width, the computer must have a large memory. For example, ^{13}C spectra at 25 MHz typically require a 5000 Hz spectral width (covering shifts of 200 ppm). To achieve a resolution of 0.5 Hz, the acquisition time must be 2 sec and the computer memory allocated for data storage must be (from Eq. 5-5) $2(5000)(2) = 20{,}000$ words. In general, a resolution of 1 Hz and a 10 K (i.e., 10,000 words) memory will suffice. If a combination of very high resolution and large sweep widths is necessary, it is recommended that an interactive disk system, having typically 1200 K storage, be used to augment computer memory. Cassettes and disks are also invaluable aids in software manipulations of acquired data.

As with so many cases in spectroscopy, a tradeoff between resolution and sensitivity must be made. To improve resolution, a long acquisition time is used. However, at the tail end of the FID, a proportionally larger amount of noise relative to signal is being sampled. The resulting spectrum will have good resolution with relatively low signal-to-noise. By truncating the FID early (e.g., after a period of one T_2*) and pulsing again, better signal-to-noise is obtained at the expense of resolution. Even when high resolution is desired, there is nothing to be gained by extending the acquisition period past $5T_2$* because magnetic field inhomogeneity (T_2* processes) will ultimately limit the resolution anyway.

The analog-to-digital converter (digitizer) used must be fast and must have a word length sufficient to handle the strongest FID to be used. The digitizing accuracy of a digitizer with 12 bit words is usually sufficient to handle all but a few proton NMR spectra having such strong FID signals that the signal-to-noise ratio of a single FID is greater than 2^{12} (= 4096).

5.4.3. Data Reduction

In the preceding section, one function of the digital computer, acquisition and averaging of the several FID's, was considered. In this section, another computer function, that of reduction of the accumulated data into useful forms, is considered. This requires typically 4 K (actually 4096) to 8 K (8192) words of computer memory in addition to the memory necessary for data storage discussed in the preceding section. Whether 4 K or 8 K is necessary depends on the programs for data reduction and the amount of peripheral software used.

A strong impetus to Fourier transform NMR was provided by the fast Fourier transform (FFT) algorithm of Cooley and Tukey (17). This speeded up the Fourier transform calculations by two or three orders of magnitude. The NMR frequency spectrum is obtained from the averaged FID, generally using the FFT modification developed by Berglund (18). With many modern laboratory computers the transformation can be completed in less than 5 sec for an 8 K transform. Restrictions on computer memory usually necessitate the destruction of the FID data when the transform is made unless the FID is saved on a disk.

In addition to the vital task of computing the frequency spectrum, the computer also performs several tasks to improve the appearance of the spectrum, including smoothing, filtering, and sensitivity or resolution enhancement.

Some additional factors concerning truncation of the FID after a time t_{ac} should be mentioned. The resulting frequency spectrum following Fourier transformation of the truncated FID is multiplied by a function $\sin(\omega t_{ac})/(\omega t_{ac})$, showing damped oscillations on each side of the peaks in the spectrum. These oscillations may be decreased by multiplying the FID by a triangular apodization function, which decreases to zero at the Nth point if N points are used in the Fourier transformation.

By truncating the FID, even after a fairly long time, all of the computer memory may not necessarily be utilized. In certain cases, improved spectral appearance from a greater number of points in the frequency spectrum (although with no greater resolution) can be obtained by "zero filling," a process by which the remaining unfilled channels in the computer memory following truncation are filled with zeroes. For example, if a computer has 8 K of memory available for data storage but the acquisition time is set so that the FID is truncated after only 4 K of memory is used, the remaining 4 K of memory is filled with zeroes and the resultant spectrum has the same resolution as a 4 K transform but contains 8192 points.

Various convolution processes (14) can be used for improved spectral quality. These processes involve multiplication of the FID signal by a weighting function. One possibility is to use an exponential weighting function. Multiplying the FID by a positive exponential results in a spectrum with enhanced sensitivity but broader lines. A negative exponential gives enhanced resolution but more noise. This is illustrated in Fig. 5-11, where spectrum A was obtained without weighting and spectra B and C were

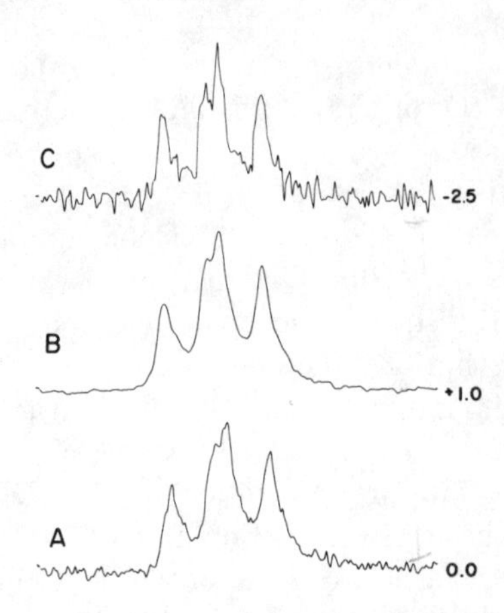

FIG. 5-11. Effect of exponential weighting of the methylene resonance of tryptophan. The argument of the exponential weighting factor is given beside the spectrum. (A) No weighting. (B) Sensitivity enhancement. (C) Resolution enhancement.

obtained with the exponential arguments +1.0 and −2.5, respectively. The optimum weighting function (not used in Fig. 5-11) utilizes the time constant characteristic of the FID.

A number of other manipulations available via the computer include baseline correction, drift correction, other multipoint smoothing and filtering techniques, integration, automatic homogeneity control, and automatic phase correction.

The Fourier transform of a free induction decay is composed of a real and an imaginary part, i.e., the cosine and sine transforms (cf. Section 2.1.4). The cosine and sine transforms theoretically correspond to the absorption (v) and dispersion (u) spectra, respectively. However, certain factors result in a mixing of the absorption and dispersion spectra. The three primary causes are: (*a*) setting of the spectrometer phase detector, (*b*) delay between the end of the pulse and the start of data acquisition, and (*c*) electronic filtering. Item (*a*) is adjusted using a strong line in the spectrum prior to pulsing, just as in the case of CW spectra. Compensation for items (*b*) and (*c*) are made after the Fourier transform is obtained, with the spectrometer operator visually adjusting the "mix" between cosine and sine transforms until a satisfactory absorption or dispersion spectrum is observed on an oscilloscope.

It is also possible to display the power spectrum $(u^2 + v^2)^{1/2}$. In a few instances the power spectrum may be useful because no phase corrections are necessary. In general, however, the power spectrum is of little use because the peaks have very broad "tails" on each side, permitting weak signals to be hidden in the tails of strong signals. Peak overlap in the power spectrum is not additive and perturbations in intensity and lineshape occur.

The use of a computer to control display form is convenient. Other computer-controlled features commonly employed in display of a spectrum include: (*1*) expansion of a particular region, (*2*) adjustment of signal amplitude, (*3*) constant velocity plotting, (*4*) peak frequency printout, (*5*) intensity printout, (*6*) intensification of a region on the oscilloscope display, (*7*) subtracting or adding spectra, and (*8*) automatic accumulation and display of a number of spectra for T_1 measurements.

5.4.4. Elimination of the "Water Problem"

Most biological studies entail use of aqueous solutions. However, it is difficult to detect the proton NMR signals of compounds of interest in the presence of H_2O. The dynamic range of the computer limits the detection of weak signals in the presence of strong signals, as mentioned in Section 2.1.3,A. The dynamic range of the computer is expressed by its word length, 12–20 bit words being typical for most dedicated laboratory computers. In averaging with a computer having 12 bit words, as soon as the accumulated signal-to-noise of the FID reaches 2^{12} (= 4096) the signals are normalized and little further improvement is made by averaging more FID signals. If the FID was composed primarily of signal from the solvent, any weak signals (10^{-4}–10^{-3} M) may still not have enough signal-to-noise for detection. A computer with 20 bit words permits an FID signal-to-noise of 2^{20} (= 1,048,576) before being limited, providing a significant advantage over 12 bit words. Even a computer with 20 bit words will limit long-term averaging in aqueous solutions, however. To partially alleviate the problem, 99.9+% D_2O is used. Nevertheless, the residual HDO peak may still be two or three orders of magnitude larger than the peaks of interest. In some cases, the effect of using D_2O rather than H_2O will eradicate or perturb the proposed experiment, e.g., when certain sites of interest on a molecule become deuterated or when deuterium affects kinetics of interest. For several reasons then, it is desirable to sidestep the dynamic range problem so studies can be conducted in H_2O solution. We discuss here a few of the methods proposed for overcoming the problem.

Redfield and Gupta (19) have used a long, weak rf pulse containing the frequencies necessary to promote resonance of the spins of interest. The rf pulse, however, is off resonance for H_2O. The spectral density of the pulse looks like the curve in Fig. 2-14; a longer pulse width t_w results in a narrower range of frequencies excited $(\pi t_w)^{-1}$. The rf field strength H_1 is kept low to maintain the proper flip angle θ $(= \gamma H_1 t_w)$. If the H_2O resonance frequency is outside the range of excited frequencies, the signal from H_2O will be minimized and studies of exchangeable proton resonances in H_2O solutions can be made. A major problem with this approach is that the resonances of interest must be quite far removed from the H_2O resonance, even with use of electrical filtering, which will aid in diminishing the effects of H_2O.

Patt and Sykes (20) have examined proton NMR spectra of proteins in solutions containing HDO. By taking advantage of the longer spin–lattice relaxation time (T_1) of the HDO resonance compared with the protein resonances, they were able to eliminate the HDO signal. Using the inversion recovery pulse sequence $(\pi\text{–}\tau\text{–}\pi/2\text{–}T_d)_n$ (cf. Section 2.1.4,B) and setting the pulse spacing $\tau = T_1$ (HDO) ln 2, there will be no z component of the HDO magnetization at the time of the 90° pulse. Consequently, there is no contribution of HDO to the sampled FID following the 90° pulse. The advantage of the method is clearly the elimination of an HDO (or H_2O) resonance so that lines near (or even under) the HDO peak may be examined. One disadvantage is that studies are limited to resonances having a T_1 different from that of HDO (or H_2O); that is not a major problem for most macromolecules. This method also is not adaptable to relaxation time measurements. Another disadvantage is the increased time necessary for time averaging. To permit the HDO nuclear spins to return to equilibrium, the delay time T_d is set to be $\sim$5–7 T_1 (HDO); a sizeable share of the time is spent waiting for the system to relax. Some improvement on this long time is possible if shorter values of τ and T_d are used.

Benz *et al.* (21) also examined some of the ramifications of the inversion recovery method and compared it to the usual steady-state rapid pulse technique which, in fact, has a smaller HDO signal than might be expected on the basis of concentration because the HDO protons relax slowly and are still far from equilibrium when the next pulse hits.

It would appear that another feasible method of eliminating (or diminishing) the water problem lies in saturating the water resonance while examining the nuclei of interest. The possibility of eliminating a single peak in a Fourier transform spectrum by saturation has been recently demonstrated by the elimination of the 2% TMS resonance from CCl_4 solution by homonuclear decoupling (22). This technique would retain the flexibility necessary for making relaxation time measurements.

The method of block averaging deals directly with the computer. At some point before the accumulated FID's exceed the dynamic range, they are transformed and stored in double precision (two words are used for one number) or in the floating point mode in a particular region of memory. The process is repeated with the Fourier transform spectrum from each block of FID's being added to the preceding ones just as in the case of CW spectrum time averaging. It is also possible to compare each transformed block with a reference peak position to cancel long-term drifts. The block averaging technique has some disadvantages in that data manipulation is slower and the amount of memory available is decreased. The latter disadvantage can be circumvented to a large extent by storing the transformed blocks on a disk system. Unlike earlier block averaging programs, those presently available from instrument manufacturers permit phase adjustment of the final averaged spectrum.

Another promising technique for suppressing the water peak (as well as homonuclear decoupling) via the use of a synthesized rf excitation source has been described (23). In essence, the rf pulse used in the Fourier transform NMR spectrometer is created such that the frequency spectrum of the pulse has a "hole" at the frequency corresponding to the water proton resonance frequency. The idea is then that the water proton resonance will not be induced when the rf pulse is applied.

The block averaging technique, still using 99.9+% D_2O as solvent, is the method of choice for most long-term averaging. Relaxation time measurements can also be made using block averaging. If it is necessary to use H_2O rather than D_2O as solvent, a computer with a long word length (at least 16 bits, preferably 20 bits) and perhaps a 15 bit digitizer (rather than the usual 12 bit) should be used in conjunction with a block averaging program. However, if it is possible to circumvent the problem by saturating the water resonance, the computer and digitizer requirements may be eased. For examining peaks very near (or under) the HDO resonance the inversion recovery method is necessary because the block averaging method will still permit a sizeable HDO resonance. Measurement of relaxation times of weak proton resonances very near the HDO resonance is presently not possible.

5.5. Multiple Irradiation

The use of an additional strong rf field H_2 extends the utility of NMR, as described in Sections 2.5 and 4.2.5,A. Here we deal with some of the experimental aspects of multiple irradiation.

Any commercial high-resolution NMR spectrometer can be adapted for double resonance. In fact, most manufacturers offer optional (or non-optional) double irradiation units for their instruments. The second field may be derived from audiofrequency modulations (24, 25) or simply by applying a second rf field (26–28). The former is generally used in the case of homonuclear double resonance and the latter generally in the case of heteronuclear double resonance.

5.5.1. Homonuclear Double Resonance

It is generally desirable to stabilize operation by using a field/frequency lock. Using the audiomodulation technique for homonuclear double irradiation necessitates three modulation frequencies:

1. For observation: ν_1 with amplitude H_1
2. For locking: ν_L with amplitude H_L
3. For saturation: ν_2 with amplitude H_2

These audiomodulation frequencies, typically in the range 1–50 kHz, result in the sample being subjected to fields not only at the centerband frequency ν, but also at the sidebands $\nu \pm \nu_1$, $\nu \pm \nu_L$, and $\nu \pm \nu_2$ as shown in Fig. 5-12. Typically, the field is locked on the upper sideband $\nu + \nu_L$ and the phase sensitive detector set to observe only the upper sidebands.

There are two possible ways of decoupling while the spectrum is being scanned. With the field sweep method, the H_0 field is swept, effectively providing a compensatory sweep of ν_L. This method maintains $\nu_1 - \nu_2$

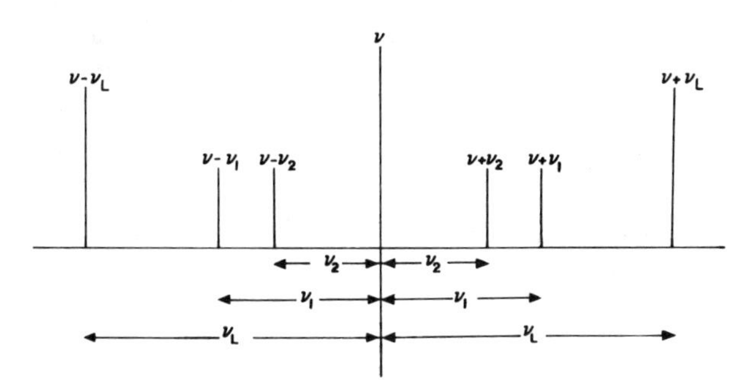

FIG. 5-12. Audiomodulation in the operation of an internally locked double-resonance spectrometer: audiofrequencies ν_1 for observation, ν_2 for strong second field, and ν_L for locking.

constant as the field sweeps through the resonance at ν_1. The experimental technique is simple but has the disadvantage that a new sweep must be made, resetting $\nu_1 - \nu_2$, for each resonance suspected of coupling with the observed resonance at ν_1. This inconvenience can be obviated by using the frequency sweep method, maintaining a constant field H_0 and constant ν_2 while ν_1 is varied.

For most experiments, the H_2 field must be strong enough to saturate a resonance. However, there are curbs on the H_2 field strength. If H_2 is too strong, receiver overload may result. Also, distortion of the observed resonance may occur if $| \nu_1 - \nu_2 |$ is not much greater than $(\gamma H_2/2\pi)(T_1/T_2)^{1/2}$. Fortunately, most resonances can be saturated without serious perturbation of neighboring resonances. However, although distortion effects depend on the relaxation times, whenever $\nu_1 - \nu_2 \lesssim 10$ Hz, distortion may result. This figure goes up, of course, as it becomes necessary to saturate all components of a multiplet having large splittings.

It is also possible to perform homonuclear double resonance experiments in the Fourier transform mode (29). The second rf field (H_2) is generally shut off just before the rf pulse (H_1) inducing the FID is applied. The long duration of H_2 permits a narrow range of frequencies that are directly affected by this strong field.

5.5.2. Heteronuclear Double Resonance

Heteronuclear double irradiation generally entails using a second rf field. A most convenient and versatile means of providing such a field is to use a frequency synthesizer with an rf power amplifier. For fixed frequency use, a cheaper alternative is to replace the synthesizer by a crystal clock with necessary frequency multipliers and dividers (27).

With double-coil probes, the transmitter coil is double tuned for both the observing and irradiating frequencies (27). With single-coil probes, a second coil orthogonal to the first is introduced to carry the H_2 field.

It has been noted in Section 2.5 that decoupling all protons provides a significant improvement in signal-to-noise ratio in ^{13}C NMR spectra. Besides removing the splitting, proton decoupling can increase the signal-to-noise by up to 1.988 from the nuclear Overhauser effect. An effective means of decoupling all protons is noise decoupling, i.e., modulating the saturating field with white noise (30).

Straightforward application of noise decoupling results in loss of the information inherent in spin–spin coupling. There are various techniques by which it is possible to enjoy the sensitivity enhancement provided by the NOE with retention of splitting. One of these techniques is off-resonance decoupling (31, 32), which can also be used in the Fourier transform mode. If the noise decoupling is limited to a small or moderate bandwidth ($\sim$100–500 Hz) and ν_2 is positioned $\sim$500–1000 Hz away from the center of resonance, the ^{13}C resonances with large J_{CH} will not be completely collapsed; those resonances will be broadened, however. The ^{13}C resonances with small J_{CH} (in practice, nonprotonated carbons) will be sharp and will get a small intensity boost from the NOE. By suitable adjustment of the frequency offset and bandwidth, it is possible to obtain some NOE gain while splitting is retained. However, off-resonance decoupling distorts lineshapes and gives unreliable coupling constant values. Reliable coupling constants with sizeable NOE gain may be obtained by alternately pulsing the ^{13}C and proton resonances, obtaining the Fourier transform spectrum from accumulated ^{13}C FID's (33). With the alternate pulse technique, the proton decoupling pulse is on for a relatively long time ($\sim$1 sec), followed by a short ^{13}C pulse ($\sim$30 μsec) and FID.

5.6. Correlation Spectroscopy

The use of correlation analysis (34, 35) may be billed as the "ultimate filter" for improvement of signal-to-noise in spectroscopy. Although it is possible to employ autocorrelation analysis, the brief discussion here will be restricted to two applications of cross-correlation analysis. Sensitivity enhancement via cross-correlation is based on the postulate that signals are periodic or coherent, and therefore predictable, but noise is random. By correlating between a signal and a reference, it is possible to distinguish the signal from any random noise present.

Ernst (36) developed a form of cross-correlation spectroscopy that he termed "stochastic magnetic resonance." The process is schematically diagrammed in Fig. 5-13. Gaussian pseudorandom noise $s(t)$, stimulated by a binary noise generator and filtered with a low-pass filter, is continuously applied to the sample. The response $v(t + \tau)$ contains both the signal and the noise information. The cross-correlation function

$$R_{sv}(\tau) = \overline{s(t)v(t + \tau)} \tag{5-6}$$

contains the signal of interest with the noise considerably diminished because the noise is incoherent and the resulting noise correlation has a time average of zero. $R_{sv}(\tau)$ may be obtained from a special purpose correlator or a digital computer. The usual NMR frequency spectrum is obtained by Fourier transformation of $R_{sv}(\tau)$. This correlation function and some of its properties may be implied from the discussion in Section 2.3.

It was shown that the sensitivity enhancement achievable with stochastic resonance and pulsed Fourier transform NMR is identical (36). In Section 5.4.2 the "tradeoff" of sensitivity and resolution in Fourier transform NMR was discussed. With the stochastic resonance technique, sensitivity and resolution may be optimized independently. This technique does not require the high power levels of Fourier transform NMR and should prove especially useful for such nuclei as ^{13}C, which have large chemical shifts.

A second method of cross-correlation spectroscopy has recently been presented (37). The process involves cross-correlation of the ringing of a rapidly scanned CW spectrum with the ringing from a strong reference line scanned under the same conditions. The rapidly scanned acetaldehyde quartet is shown as spectrum I of Fig. 5-14. The chloroform reference spectrum is shown as spectrum II obtained under the same conditions, and the result of cross-correlation is shown as spectrum III.

The cross-correlated spectrum contains information about the frequencies that are common to the signal response and the reference response. The noise on the signal will be incoherent with the noise on the reference,

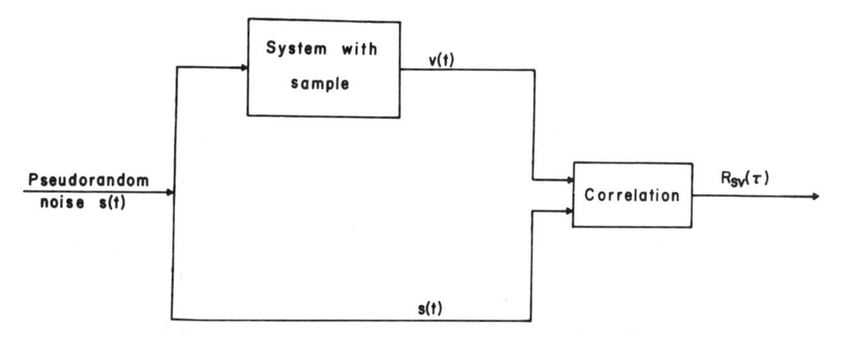

FIG. 5-13. Schematic diagram of the stochastic NMR process.

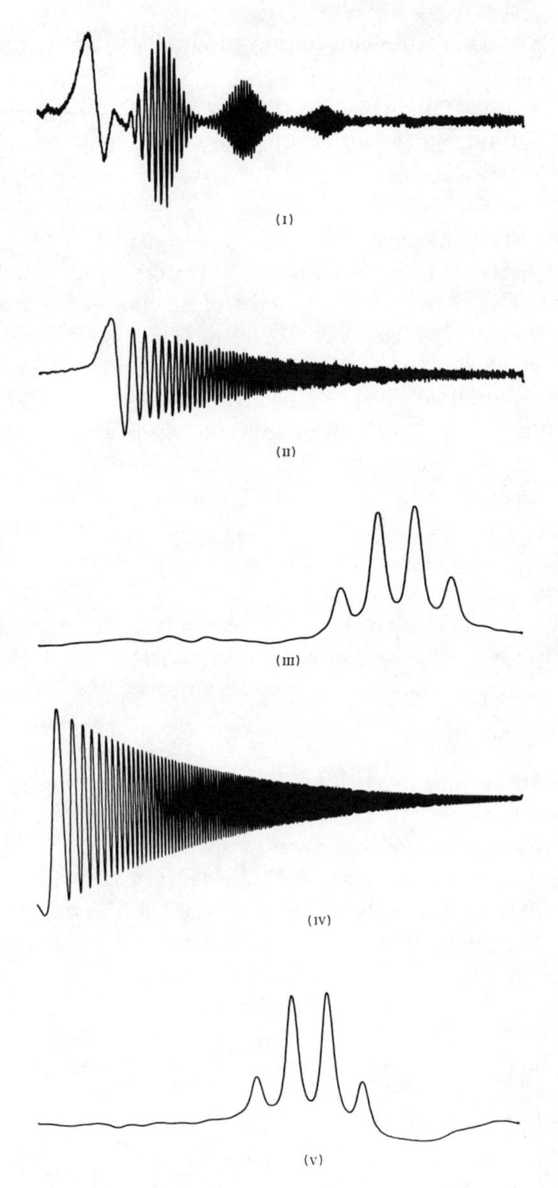

FIG. 5-14. Cross-correlation spectroscopy of the acetaldehyde quartet. Spectrum I: the acetaldehyde quartet scanned over 500 Hz at 125 Hz/sec with 4 K data points accumulated. Spectrum II: the chloroform reference line scanned under the same conditions. Spectrum III: cross-correlation of spectra I and II. Spectrum IV: calculated reference spectrum. Spectrum V: cross-correlation of spectra I and IV. (Courtesy of J. Kisslinger and J. Cooper, Nicolet Instrument Corp.)

so the time average of the correlation product will have much diminished noise.

It should be apparent that cross-correlation between a noisy signal and a calculated reference spectrum (having no noise) should also result in a relatively clean spectrum. This is illustrated by spectrum V of Fig. 5-14, which results from cross-correlation of the sample spectrum I with the calculated reference spectrum IV.

This correlation technique provides improved sensitivity, with advantages over Fourier transform NMR in that a CW spectrometer with low rf power may be used; a selected portion of the spectrum is obtained (i.e., no water problem or foldback); and the resolution, signal-to-noise and phase are adjusted after the rapidly scanned spectrum is obtained.

5.7. Relaxation Time Measurements

There are various methods for measuring spin–lattice (T_1) and spin–spin (T_2) relaxation times. These methods can be divided into three general domains: (*1*) continuous wave methods, (*2*) pulsed methods, and (*3*) Fourier transform methods. Pulsed and Fourier transform methods have greater accuracy, e.g., T_1 can be measured to within ±2–5%. CW methods are capable of measuring T_1 to only ±10–40% but currently have the advantage of more widely available instruments.

5.7.1. Continuous Wave Methods

A. *T_1 Determination; Saturation Recovery*

With this method (38–40), the H_0 field is maintained exactly at the center of the resonance and the H_1 field is increased beyond saturation until the signal disappears. The H_1 field is quickly decreased to a value much below saturating and the signal is monitored on an oscilloscope or recorder as it exponentially recovers to its equilibrium value with a time constant T_1. In the absence of distortion effects (39, 40), the shape of the recovery curve is $1 - \exp[-t/T_1]$. To maintain the field exactly at the center of resonance, it is necessary to use field locking (preferably internal). Alternatively, for narrow lines with long T_1, the resonance may be scanned repetitively as the signal recovers. With some effort it is possible to measure T_1 values as low as 0.2 sec. T_1 values on the order of 1–5 sec can be measured to ±15–20% with care.

Progressive Saturation. This method (41) permits the measurement of shorter T_1 values (0.01–1 sec) than the saturation recovery technique but suffers from inaccuracy on the order of ± 20–40%. If progressive saturation is used, the calculated T_1 values should probably be used only as an estimate of the order of magnitude of T_1. As discussed in Section 2.1.2,A, saturation of a resonance depends on T_1 as embodied in the saturation factor $(1 + \gamma^2 H_1^2 T_1 T_2)$. Progressive saturation involves increasing the H_1 field while measuring the signal amplitude (Eq. 2-12b). The amplitude passes through a maximum when $(\gamma H_1)^2 = (T_1 T_2)^{-1}$. T_1 may be calculated if T_2 and H_1 are known. T_2 may be estimated from the linewidth in the absence of saturation if T_2 is sufficiently short that magnetic field inhomogeneities do not control the linewidth. H_1 may be estimated independently (42).

Adiabatic Rapid Passage. Adiabatic rapid passage (43) may be used to measure spin–lattice relaxation times longer than 0.01 sec with fair accuracy ($\sim 10\%$). Measurement of the ratio of the signal amplitudes for two rapid passages separated by a time τ enables T_1 to be determined. If a sufficiently strong H_1 field is swept through resonance, at the end of the passage, M_z will be oriented in the negative z direction but will have the magnitude of the equilibrium magnetization, $M_z = M_0$. Relaxation processes result in a decreased magnitude of M_z measured during a second passage through resonance, $M_z = M_0\,(1 - 2\exp[-\tau/T_1])$.

B. *T_2 Determination; T_2 from the Linewidth*

Resonance linewidths can be used to measure T_2 values $\lesssim 1.0$ sec. The method and precautions have been discussed in Sections 1.3 and 2.1.2,A.

Decay of Ringing. The ringing (or wiggles) following passage of the swept field through the resonance frequency decays with a time constant T_2 if magnetic field inhomogeneity does not dominate. This phenomenon was discussed in Section 2.1.2,B.

Adiabatic Half-Passage. This technique (44, 45) overcomes the limitations of magnetic field inhomogeneity, permitting measurement of long T_2 values. Using a strong H_1 and starting with the H_0 field off resonance, the H_0 field is swept precisely to the center of resonance and stopped. The transverse magnetization has been polarized and commences to decay to zero with a time constant T_2; $u = M_0 \exp[-t/T_2]$.

5.7.2. Pulsed Methods

The pulsed methods described in this section provide the most accurate means of T_1 and T_2 determination, but they are generally limited to only one constituent in a sample. (In certain cases it is possible to resolve exponential decays into two components and thus to obtain information about two constituents.) It may also be possible in particular cases to selectively measure relaxation times of constituents in separate experiments if their relaxation times are long. The principal advantages of these techniques in comparison with those to be discussed in the Fourier transform section are the much less expensive instrumentation involved and the comparative ease with which T_1 and especially T_2 are measured.

A. T_1 *Determination; Two-Pulse Sequences*

The $180°$–τ–$90°$ and $90°$–τ–$90°$ pulse sequences used for T_1 determinations were described in Section 2.1.3,B. The $180°$–τ–$90°$ sequence is most commonly used. By measuring the magnitude of the FID following the $90°$ pulse over values of τ extending to $\sim 3T_1$, T_1 may be determined to $\pm 2\%$ from the slope $[1/(2.3T_1)]$ of a plot of $\log(M_0 - M_z)$ vs τ (cf. Eq. 2-23). With a sacrifice in accuracy, the "null" variation of the $180°$–τ–$90°$ technique may be used. It is apparent in Fig. 2-8A that the FID passes through zero as τ is increased. By adjusting τ such that $M_z = 0$, we obtain

$$T_1 = \tau_{\text{null}}/\ln 2 = \tau_{\text{null}}/0.69 \tag{5-7}$$

from Eq. 2-23.

The magnitude of the rf power puts a lower limit on the T_1 value which can be measured. The larger the H_1 field is, the smaller the pulse width t_w need be to give $180°$ and $90°$ pulses. In discussing the pulsed NMR experiment, it was assumed that the time involved for nutation (i.e., rotation) of the magnetization vector through $90°$ or $180°$ was infinitely small and the equations describing the decay of the magnetization were derived from the Block equation in the absence of a magnetic field. In fact, the field is on for a finite time t_w during the $90°$ pulse. In general, the time t_w can be ignored if $\tau \gtrsim 10t_w$. If t_w is 5 μsec, values of τ down to 50 μsec can be used and T_1 values down to 0.1 msec can be measured. However it is not always possible to obtain sufficient power for t_w to be ≤ 5 μsec. If one is pressed, it is still possible to obtain a reasonable value of T_1 using τ values as low as $3t_w$ if the length of time t_w is included in τ. In such cases it is best to place more emphasis on the measurements at longer τ values when evaluating T_1.

Several sources of experimental error should be considered. Incorrect adjustment of pulse width; instabilities in the H_0 field, pulse spacing, gating, and pulse widths; and an inhomogeneous H_1 field may all lead to errors. In particular, an inhomogeneous H_1 field will lead to sizeable errors if the "null" method is used. If H_1 is inhomogeneous, the magnetization vector will not be nutated through exactly $180°$ and Eq. 5-6 will give a value of T_1 that is too short. Using the full range of values, however, usually enables an accurate T_1 to be obtained from the slope of the plot. In general practice, it is wise to make T_1 measurements on samples of known T_1 to gauge the stability and accuracy of an unfamiliar instrument. A convenient standard for proton NMR is an aqueous solution of 0.10 mM $MnCl_2$, which has a water proton T_1 of 0.78 ± 0.03 sec at room temperature.

Triplet Pulse Sequences. Some improvement in the accuracy of T_1 measurements is possible if an additional 180° pulse follows the 180°–τ–90° sequence and the magnetization of the resultant spin echoes is monitored. The improvement results from the magnetization being monitored at some time remote from the pulse. More importantly, a series of triplet pulse sequences (46–48) prove advantageous in measurements of long T_1 values. The reason for this is that the longitudinal component of the magnetization is sampled several times during the course of relaxation following a 180° pulse and T_1 may be accurately calculated from a single scan. The best triplet sequence (48) is $180_y°–\tau_2–[(90_y°–\tau_1–180_x°–\tau_1–90_{-y}°) - 2\tau_2 -]$. The initial $180_y°$ pulse inverts M_z. In each of the triplet sequences, the $90_y°$ pulse nutates the magnetization into the x direction and the FID is measured. The spreading of the individual spin vectors in the xy plane of the rotating coordinate system ($x'y'z$) is reversed with the $180_x°$ pulse. The $90_{-y}°$ pulse, at the moment phase coherence is regained, nutates the magnetization again in the z direction, where relaxation continues. The magnetization is therefore sampled N times if N triplet sequences are used in the course of relaxation following the initial $180_y°$ pulse and the magnetization has the value $M_z = \{1 - 2 \exp[-2(N - 1)\tau_2/T_1]\}$ if $\tau_1 \ll T_1$.

Selective Determination of T_1. It is possible to use the 180°–τ–90° pulse sequence to selectively measure the T_1 of one resonance if that resonance is separated from others and has a long T_1 (49, 50). Using a weak irradiation level ($\gamma H_1/2\pi = 1$ Hz) and long pulse durations (0.25–1 sec), T_1 values >2 sec could be measured for a selected resonance without serious degradation from nearby lines (50). The low power level required makes it possible to perform these measurements on a CW spectrometer with only moderate modification. This selective technique is also applicable to spin echo measurements of T_2 and to relaxation measurements in the rotating frame (50).

B. T_2 *Determination*

The Hahn 90°–τ–180° sequence, the Carr-Purcell sequence, and the Carr-Purcell sequence with the Meiboom-Gill modification were discussed in Section 2.1.3,C, and methods are illustrated in Figs. 2-8B and 2-8C. To an even greater extent than for T_1 measurements, T_2 measurements require a very stable magnetic field ($\sim$1 part in 10^8). The most important consideration in T_2 measurements is the H_0 magnetic field inhomogeneity. It is desirable to have a field sufficiently inhomogeneous that $T_2^* < T_2$, but if the field is too inhomogeneous molecular diffusion terms enter (cf. Eq. 2-24) and can lead to serious errors if not recognized. With the Hahn two-pulse technique, the effects of diffusion will be apparent in the nonlinearity in the plot of log $M_\perp$ vs τ. The Carr-Purcell sequence is not as susceptible to the effects of diffusion. Nevertheless, a very inhomogeneous field will lead to diffusion errors that are noticeable if the calculated T_2 depends on the pulse separation 2τ. Slow chemical exchange can also lead to T_2 measurements dependent on pulse separation (51). Use of a pulse separation small compared to T_2 will usually be sufficient to obtain a reliable value of T_2.

Other sources of error for T_2 measurements are the same as for T_1 measurements: inexact adjustment of pulse widths, instrumental instabilities in timing sequences, and slow rise time in the transmitter. The Meiboom-Gill modification minimizes these limitations and can generally result in T_2 values accurate to ± 5–10%.

For the measurement of short T_2 values, it is sufficient to use a homogeneous magnetic field and to simply measure the time constant of the exponential FID following a single 90° pulse.

C. $T_{1\rho}$ *Determination*

The usual (or laboratory frame) spin–lattice relaxation time T_1 provides information about molecular motions with $\tau_c \sim 1/\omega_0$, i.e., $\sim 10^{-8}$ sec. It is possible to obtain information about slower motions from measurements of $T_{1\rho}$, the spin–lattice relaxation time in the rotating frame (52–55). The dependence of $T_{1\rho}$ on molecular motion (τ_c) is illustrated in Fig. 5-15 and

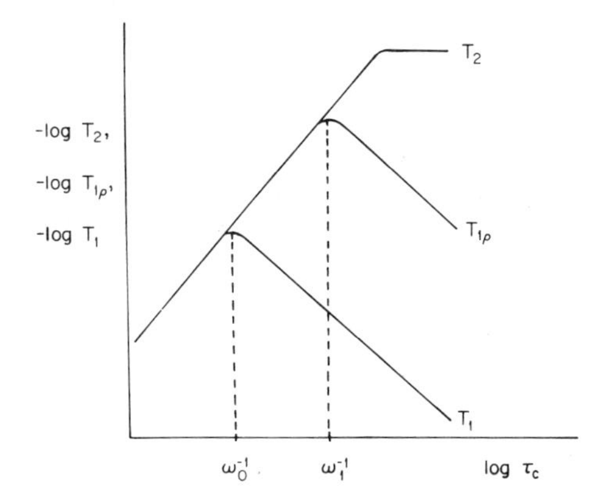

FIG. 5-15. Dependence of T_2, $T_{1\rho}$, and T_1 on correlation time. The maximum in the T_1 curve occurs when $\tau_c = 1/\omega_0 = 1/(\gamma H_0)$ and in the $T_{1\rho}$ curve when $\tau_c = 1/\omega_1 = 1/(\gamma H_1)$.

compared with T_1 and T_2. $T_{1\rho}$ will be sensitive to molecular motions having correlation times $\tau_c \sim 1/\omega_1 = 1/(\gamma H_1)$ (i.e., $\sim 10^{-5}$–10^{-4} sec) as long as γH_1 is greater than the local dipolar fields. Such slow motions may be of interest in studies of membranes and very viscous biological fluids.

We can describe the $T_{1\rho}$ experiment in the rotating coordinate system in the following manner. The magnetization is initially at equilibrium oriented in the z direction defined by the stationary magnetic field $\vec{H}_0$. A resonant 90° rf pulse $\vec{H}_1$ is applied, nutating the magnetization $\vec{M}$ into the y' direction. This is immediately followed by a second pulse, of duration τ, which is phase shifted by 90° with respect to the 90° pulse such that $\vec{M}$ is aligned with $\vec{H}_1$ in the x' direction. Because $\vec{M}$ and $\vec{H}_1$ are parallel, no torque is exerted on $\vec{M}$ by $\vec{H}_1$, so $\vec{M}$ remains parallel to $\vec{H}_1$ and the nuclear spin system is said to be spin locked. This procedure is often referred to as "forced transitory precession." While the $\vec{H}_1$ field is on during the time τ, the spin system will come to internal thermal equilibrium in $\vec{H}_1$, with a characteristic time constant $T_{1\rho}$. We see from Fig. 5-15 that this relaxation in the rotating frame will be faster than T_1 relaxation if $\tau_c > 1/\omega_0$. The amplitude of $\vec{M}$ in the rotating frame may be monitored after a time τ by quickly turning the $\vec{H}_1$ field off and measuring the height of the resulting FID. The magnitude of the magnetization is simply

$$M = M(0)e^{-\tau/T_{1\rho}} \tag{5-8}$$

Other methods for setting up the forced transitory precession, such as use of a field pulse (55) and adiabatic half-passage (53), have also been employed.

In certain cases it is convenient to measure $T_{1\rho}$ rather than T_2. From Fig. 5-15, it is apparent that $T_{1\rho} = T_2$ over a large range of τ_c, at least if the motion is isotropic. The equality is probably valid for viscous fluids until the liquid crystal region is reached. Another use of $T_{1\rho}$ measurements has been in studying fast chemical exchange reactions (45, 56).

Another spin locking method enables the indirect detection of "rare spins," as little as 10^{-5} "rare spins" per "abundant spin." Hartmann and Hahn (57) proposed that an "abundant" nuclear spin system be spin locked. If a strong secondary rf field is applied at the resonance frequency of weak spins and there is strong dipolar coupling between the two spin systems, the weak spins may be detected by their effect on the amplitude of the FID of the abundant spins when the locking field is turned off. This spin locked double resonance technique has only been applied to inorganic crystals (58) so far, but the technique might be expected to have some application in detection of ^{33}S, ^{15}N, ^{2}H, and ^{13}C in biological systems.

5.7.3. Fourier Transform Methods

Fourier transform methods combine the best parts of the CW methods and the pulsed methods, making it possible to accurately obtain relaxation time measurements on individual resonances.

A. T_1 *Determination*

The essential ideas for obtaining T_1 values of individual resonances via Fourier transform NMR were presented in Section 2.1.4,B for the 180°–τ–90° pulse sequence (59–61). A variation of the procedure described in that section entails subtracting the signal following a 180°–τ–90° sequence from the FID signal of a single 90° pulse. T_1 is obtained by measuring the difference signal intensity

$$M(\tau) = 2M_\infty e^{-\tau/T_1} \tag{5-9}$$

as a function of τ. The 180°–τ–90° pulse sequences unfortunately require a long time if sensitivity enhancement is desired and signals from several pulse sequences are summed, because a delay time on the order of $5T_1$ must be used between pulse sequences to permit the magnetization to return to equilibrium.

There is a quick method of estimating T_1, analogous to the progressive saturation method of CW NMR. With this technique (62), a series of 90° pulses separated by a time τ are applied. A dynamic balance between the effect of the pulses and the relaxation processes is established. The deviation of the signal magnitude from its equilibrium value $[M_\infty - M(\tau)]$ is an exponential function of τ with time constant T_1. The accuracy of this technique is about ±10–20%.

Two similar methods have been used that permit accurate T_1 determinations in much less time than the 180°–τ–90° method (63, 64). These methods require no *a priori* guess of the magnitude of T_1, eliminate the long delay time following the pulse sequence (63, 64), and circumvent the difficulties in producing accurate 180° and 90° pulses (64). These methods are reminiscent of the saturation recovery method of CW NMR. Indeed, one method (63) uses a burst of nonselective pulses to saturate the system, followed after a time τ by a 90° pulse that samples the magnetization for Fourier transformation.

The sequence for the second method (64) is shown in Fig. 5-16. A 90° pulse nutates the magnetization into the y' direction. A pulse of current (homo spoil pulse) in the z homogeneity coils produces a field gradient that rapidly dephases the spins, reducing the magnetization to zero. The spin system commences to relax back toward equilibrium, producing a new

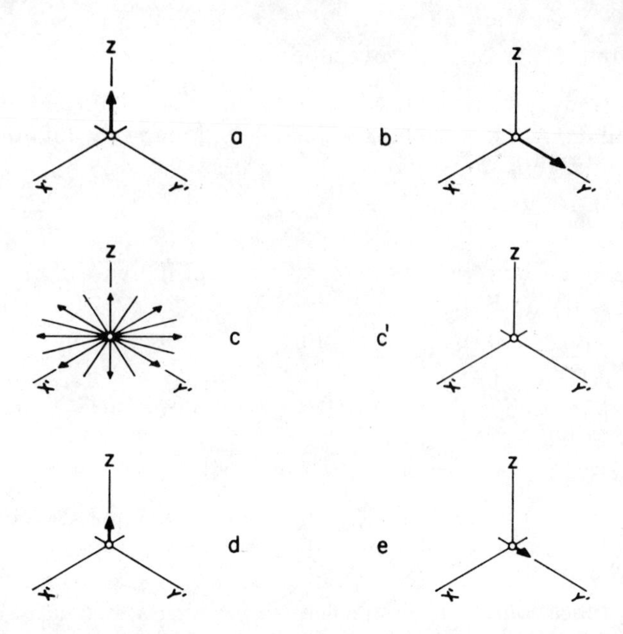

FIG. 5-16. Homogeneity spoil recovery (HSR) sequence for T_1 determination. (a) Net magnetization at equilibrium is along the z axis; (b) a 90° pulse rotates the magnetization vector along the y' axis; (c) application of a field gradient spoils the field homogeneity and rapidly dephases the spins, resulting (c') in zero magnetization; (d) after a decay time τ, the spin system has partially relaxed, resulting in a new magnetization along the z axis; (e) application of another 90° pulse rotates the magnetization vector along the y' axis, where it is detected (64).

magnetization along the z axis. After a time τ, a 90° pulse is applied, nutating the magnetization into the y' direction, where it is detected. Application of a homo spoil pulse eliminates any remaining magnetization after the FID signal has been sampled, and the pulse sequence may be immediately repeated. In both methods, the magnetization recovers exponentially with the decay time τ,

$$M_z(\tau) = M_\infty(1 - e^{-\tau/T_1}) \tag{5-10}$$

where M_∞ is the magnitude of the z magnetization when τ approaches infinity.

The saturation recovery technique proves to be faster than the 180°–τ–90° sequences in achieving the same signal-to-noise for systems with long T_1 values. The homo spoil recovery (HSR) technique, however, has the best time advantage. The advantage depends on T_1, a longer T_1 giving a greater advantage to the HSR technique compared with the 180°–τ–90° technique for a given number of transients N producing the same signal-to-noise (64).

That ratio is

$$\frac{T_{\mathrm{HSR}}}{T_{180°-90°}} = \frac{AE + \sum_{i=1}^{E} \tau_i}{2E(A + D) + \sum_{i=1}^{E} \tau_i} \tag{5-11}$$

where A is the data acquisition time; D, the delay time before repeating the sequence; E, the number of τ values; and τ_i, the variable decay time. To measure a relaxation time of 3.0 sec with the same signal-to-noise, it was found that the time necessary for the HSR method was less than one-fourth that required for the 180°–90° method and about 0.3 times that required for the (90°)–(180°–τ–90°) method (64).

B. T_2 *Determination*

If the linewidth is not dominated by magnetic field inhomogeneity or by premature truncation of the FID, as discussed in Section 5.4.2, T_2 may be determined from the linewidth, just as in CW NMR. Fourier transform operation has an advantage over CW NMR in that no line broadening from saturation occurs, in contrast to the saturation broadening with CW slow-passage experiments (13). We might also note at this point that the problems of line broadening and ringing in fast CW sweeps are also circumvented.

Under certain conditions, it is also possible to use a 90°–τ–180° sequence or a Carr-Purcell train to obtain T_2 values for individual resonances by Fourier transformation of spin echoes (65, 66). Vold (65) reported that individual T_2 values could be obtained on noncoupled or weakly coupled systems by subjecting the system to a Carr-Purcell sequence (with rf power covering all resonance lines), Fourier transforming the trailing edge of the last spin echo, and repeating the experiment for several different durations of the Carr-Purcell train. In general, the phase and amplitude of the peaks in the Fourier transform spectrum vary periodically as the duration of the echo train is changed, but these distortions can be minimized by using short pulse spacings (65).

Weakly coupled or noncoupled spin systems, analyzed using the 90°–τ–180° pulse sequence, only show evidence of a phase error related to the coupling constant, $\sum 4\pi J_{AX} m_I(X)\tau$ radians (66). J_{AX} is the scalar coupling constant between the observed spin A and the coupled spin X, $m_I(X)$ is the magnetic quantum number of spin X, and τ is the time between the 90° and 180° pulses. The phase error is illustrated in the partially

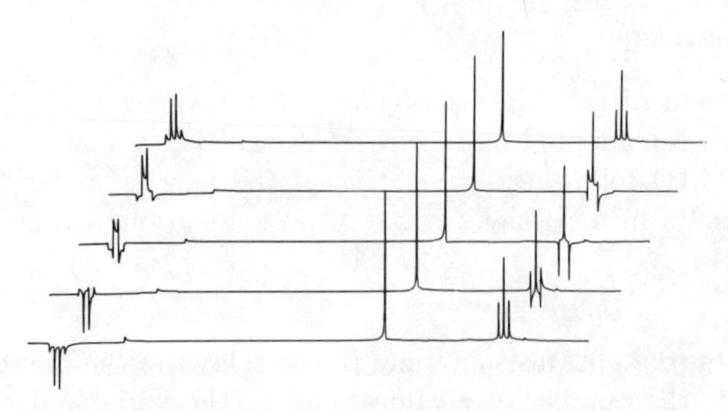

FIG. 5-17. Proton NMR spectra of ethyl acetate obtained from the Fourier transform of the spin echo using a 90°–τ–180° pulse sequence at times (from bottom to top) $2\tau = 1/J, 5/4J, 3/4J, 7/4J, 2/J$. The spectral width is 800 Hz. The downfield quartet is amplified by a factor of 2.4 (66).

relaxed spectra of ethyl acetate shown in Fig. 5-17. Note that the uncoupled acetate peak and the center peak of the methylene triplet, corresponding to $m_I(X) = 0$, have no phase errors. It is apparent from Fig. 5-17 that phase variations may be avoided by setting the pulse intervals to integral multiples of $1/J$. Problems with phase variations may also be obviated by using the absolute value of the Fourier transform as shown in Fig. 5-18 for ethyl acetate. The usual precautions concerning use of absolute value spectra apply (cf. Section 5.4.3). This method also suffers from the usual molecular diffusion complications of the 90°–τ–180° pulse method for long T_2 measurements. Use of the Fourier transform-modified Carr-Purcell sequence (65) with small pulse intervals will circumvent the molecular diffusion problem.

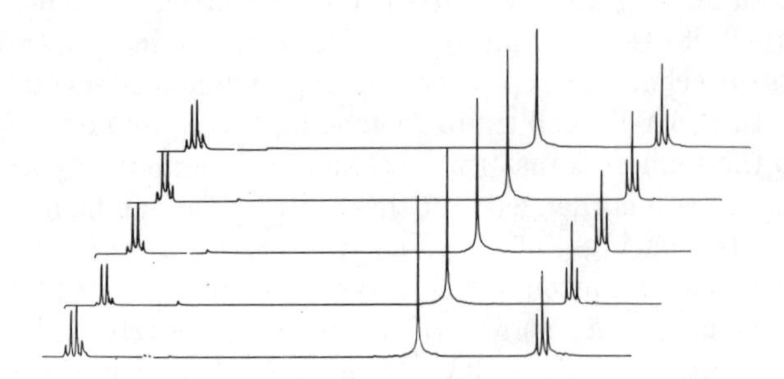

FIG. 5-18. Absolute value spectra for ethyl acetate obtained under the conditions listed in Fig. 5-17 (66).

C. $T_{1\rho}$ *Determination*

Measurement of the relaxation time in the rotating coordinate system, $T_{1\rho}$, has also been adapted to Fourier transform NMR. Fourier transformation of the FID following termination of the long H_1 pulse (cf. Section 5.6.2,C) results in a partially relaxed spectrum from which $T_{1\rho}$ may be estimated (67).

5.8. Self-Diffusion Coefficient Measurements

5.8.1. PULSED METHODS

The 90°–τ–180° pulse method may be used in the determination of the self-diffusion coefficient D with the aid of Eq. 2-24. The procedure involves measuring the magnitude of the spin echo $M(G)$ in the presence of a linear magnetic field gradient G along the z axis, and then measuring the echo magnitude $M(0)$ in the absence of the field gradient, for a given value of τ (68). The ratio of the echo magnitudes is

$$M(G)/M(0) = e^{(-2/3\gamma^2 G^2 D\tau^3)} \tag{5-12}$$

D may be determined from the slope of a plot of log $[M(G)/M(0)]$ vs τ^3 such as the one shown in Fig. 5-19. Discussion of the deviation from linearity observed in Fig. 5-19 for long τ^3 values is deferred for a moment. It is also possible to determine D by measuring the echo amplitude ratio as a function of G, keeping τ constant (69).

The linear field gradient G is produced by passing current through a pair

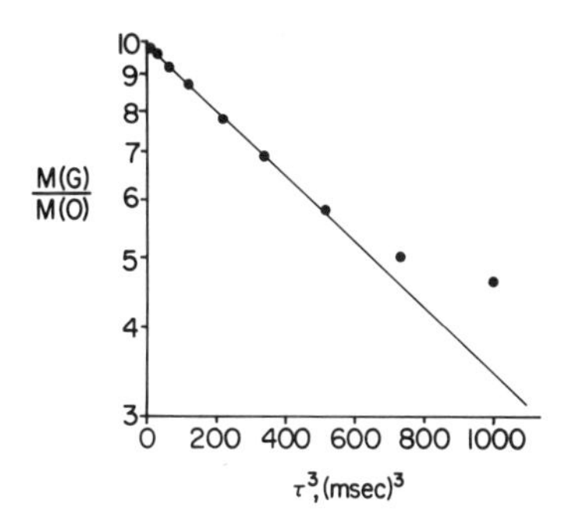

FIG. 5-19. Self-diffusion coefficient plot for water protons in frog eggs (*Rana pipiens*). Deviation from linearity is discussed in the text.

of coils the axes of which are aligned with the H_0 axis and which are connected in series. The gradient may be determined (*a*) by use of a substance with a known self-diffusion coefficient, (*b*) by direct calculation from the coil geometry and current if corrections for image currents in the pole faces of the magnet are included (70), or (*c*) from modulation of an FID following the 90° pulse or an echo following the 180° pulse (70). For a cylindrical sample tube, the modulated FID has the form of a first-order Bessel function from which the value of G may be obtained.

The lower limit of D that can be measured depends on the value of T_2 (cf. Eq. 2-24) and on the value of G, which is limited by the amount of current that can be passed through the gradient coils before they burn out. It is possible to increase the value of G employed by passing current through the coils for only a few seconds during the experiment. The lower limit of D for protons is nevertheless about 1×10^{-7} cm²/sec. This lower limit can be extended to about 10^{-9}, and with greater precision, by use of a pulsed gradient (71–73). The pulsed gradient technique permits the use of large gradients and a narrow-band amplifier in the receiver to obtain these improvements in range and precision. Even smaller values of D may be measured using a stimulated echo technique (74) or by the frequency dependence of $T_{1\rho}$ (75).

The deviation from linearity observed for larger values of τ^3 in Fig. 5-19 imply that measurements in this nonlinear range were in the region of restricted diffusion. Restricted or bounded diffusion arises when the sample cavity is very small and the size of the cavity restricts diffusion. For measurement of the self-diffusion of cellular water, the cell size may quite readily restrict diffusion. Restricted diffusion becomes a factor when $\gamma G a^3/D \gtrless 1$, where a is the dimension of the cavity. Restricted diffusion leads to a measured self-diffusion coefficient smaller than the true value of D (76, 77). The relationship of the observed values to the true self-diffusion coefficient has been explored for a few special cases: diffusion near an attractive center (71), diffusion within a laminar system (71, 77), and diffusion in a spherical cavity (78).

5.8.2. Fourier Transform Method

Just as Fourier transform NMR permits individual T_1 or T_2 values to be measured, it may be adapted for measurement of the self-diffusion coefficient of each component in a multicomponent system by obtaining the Fourier transform of the spin echo using the pulsed gradient technique (79). In addition to the capability of measuring individual D values, this technique has the usual advantages of Fourier transform NMR for signal-

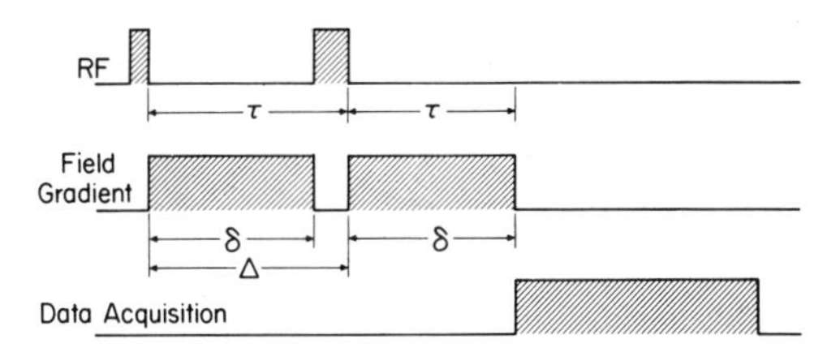

FIG. 5-20. Timing of 90°–τ–180° rf pulse sequence, pulsed field gradient, and data sampling. τ is the time between the 90° and 180° rf pulses, δ is the duration of the pulsed field gradient, and Δ is the time interval between the leading edges of the gradient pulses. The computer samples the data from the spin echo occurring at time 2τ (79).

to-noise enhancement and the advantages of a pulsed magnetic field gradient for precise study of diffusion.

Figure 5-20 shows the timing of the rf and field gradient pulses. A 90° pulse begins the process, followed after a time τ by a 180° pulse. A linear magnetic field gradient pulse of duration δ and magnitude G in the direction of the H_0 magnetic field is applied between the two rf pulses. A second field gradient pulse, identical to the first, is applied after the 180° pulse, with the time between the two gradient pulses being Δ. The frequency spectrum is obtained from the Fourier transform of the spin echo occurring at time 2τ. It should be noted that the field gradient is turned off when the rf pulses are applied and when the spin echo is monitored.

For the pulsed gradient experiment with no restrictions on diffusion, the signal magnitude in the presence of the field gradient relative to the signal magnitude in the absence of a gradient is given by

$$M(G)/M(0) = \exp\left[-\gamma^2 G^2 \delta^2 (\Delta - \delta/3) D\right] \tag{5-13}$$

For convenience, the experiment was set up as shown in Fig. 5-20 such that $\Delta = \tau$ and $\delta = \Delta -$ (rf pulse width) $\approx \tau$ because the 180° pulse width is $\lesssim$100 μsec. In that case, Eq. 5-13 reduces to that for a continuous field gradient, i.e., Eq. 5-12. Diffusion occurs as though a continuous field gradient were applied, but the advantages of the pulsed gradient method are retained.

The multicomponent technique is illustrated for the simple case of a 50% dimethyl sulfoxide–50% water solution in Fig. 5-21. The water peak and DMSO peak amplitudes can clearly be measured as a function of τ^3 from the Fourier transform of spin echoes in the absence and presence of a linear field gradient. The values of D for components in the solution at 25°C are $0.83 \pm 0.04 \times 10^{-5}$ cm^2/sec for water and $0.50 \pm 0.05 \times 10^{-5}$ cm^2/sec for DMSO.

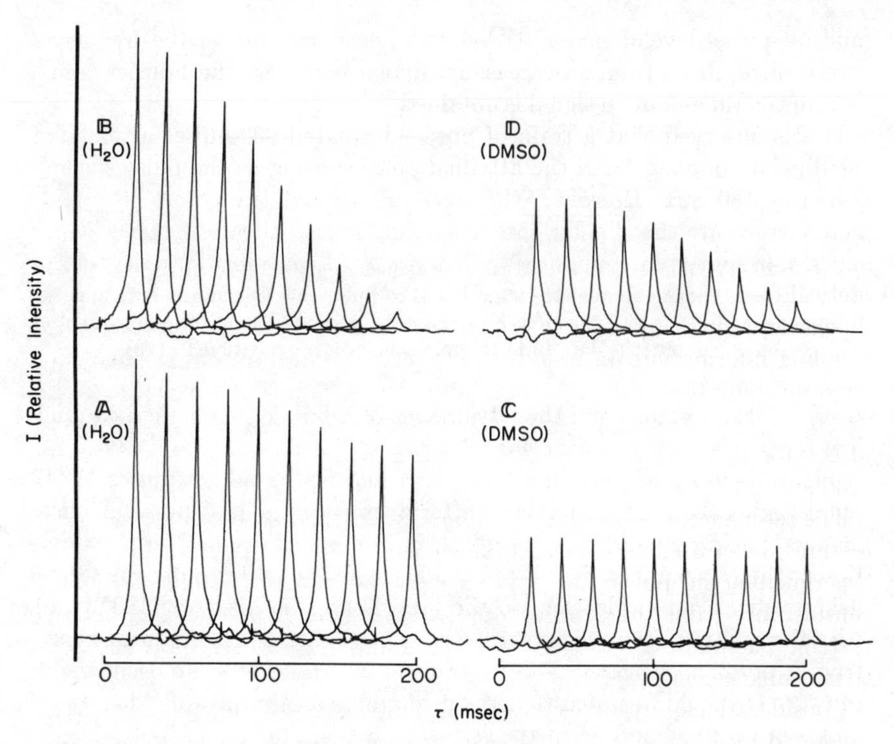

FIG. 5-21. Effects of diffusion on the Fourier transformed signal of each component in 50% dimethyl sulfoxide–50% water. (A) Water line in the absence of a pulsed field gradient; (B) water line in the presence of a pulsed field gradient; (C) dimethyl sulfoxide line in the absence of a field gradient; (D) dimethyl sulfoxide line in the presence of a field gradient. (A), (B), (C), and (D) are each composed of ten individual lines that were obtained with ten different values of τ, the time between the 90° and 180° rf pulses. Each spectrum shown is the power spectrum derived from the Fourier transform of four summed spin echoes (79).

5.9. Multiple-Pulse NMR Experiments

In experiments with solids, several pulse sequences have been employed in addition to those presented in preceding sections. Although these sequences have not yet been applied to biological or biochemical systems, it is safe to predict that some of them will find utility in NMR studies of membrane and model membrane systems as well as of certain biopolymer systems. The techniques will not be discussed in detail here but will be mentioned with reference to the original literature for the interested reader.

Much interest has centered on the elimination of dipolar or quadrupolar coupling effects that lead to broad lines and splitting in solids, although relaxation phenomena in the rotating frame have been the subject of some multiple-pulse developments. To see the line narrowing in the frequency spectrum resulting from a decrease in dipolar coupling, the Fourier transform of the time-domain signal is obtained.

It was observed that a train of phase-alternated 90° pulses attenuated the dipolar coupling, with the attenuation increasing as the pulse spacing decreases (80–82). However, the phase-alternated train does not completely eradicate the dipolar interation, and it also affects chemical shifts and scalar spin–spin coupling. Waugh *et al.* (83, 84) introduced the theoretical basis for more complicated pulse sequences capable of eliminating dipolar and quadrupolar coupling while retaining chemical shift and scalar coupling information. It would appear that the best-known sequence for accomplishing this is the four-pulse four-phase WAHUHA cycle: $\{\tau - 90_x^\circ - 2\tau - 90_{-x} - \tau - 90_y^\circ - 2\tau - 90_{-y}^\circ\}$ (85). Eight-pulse sequences have also found utility for reducing rf inhomogeneity effects and improving resolution in solid spectra (86).

The resonance offset technique is another method that has, among other features, been used to effect line narrowing (87, 88). In resonance offset experiments, the pulses are applied with the carrier rf frequency offset considerably from the Larmor precession frequency of the nuclear spin system. The phase-alternated 90° pulse train has also been used for sensitivity enhancement in solids (89).

Considerations in the design of multiple-pulse spectrometers have been presented by Ellet *et al.* (90). It will be noted that the effects of the multiple-pulse sequences require a rigid or nearly rigid sample. Therefore, it would appear that the presently known techniques will have little utility for biopolymer solutions but may be useful for solid or liquid crystalline materials.

5.10. Sample Considerations

Most sample considerations can be inferred from the discussion in preceding sections and chapters. Nevertheless, a few topics will be briefly mentioned here.

Many aspects of sensitivity have already been covered (cf. Section 5.2.3). Although the sensitivity depends on so many factors (magnetic field strength, instrument circuitry, probe design, sample tube size and shape, nuclear gyromagnetic ratio, relaxation time, time-averaging capabilities, etc.), it can be said that a single proton can be detected at concentrations less than 10^{-4} *M* in favorable cases. On an absolute scale, it is possible to obtain a useable NMR spectrum from 10 μg of sample using a microcell (cf. Section 5.2.3).

Chemical shift references were discussed in Section 1.4. The most accurate chemical shifts are determined with an internal standard, i.e., the reference is dissolved in the same solution as the compound of interest. However, sometimes it is necessary to use an external reference, e.g., in cases where the reference reacts (or interacts) with something in the solution. In such cases, the external reference is usually placed in a special coaxial capillary inside the NMR tube. There are two disadvantages to using such an external standard. First, it decreases signal-to-noise (by $\sim$25–50% in a 5 mm tube) because the capillary can take up a sizable amount of the volume inside the NMR tube. Second, for chemical shift determinations, it is necessary to apply a correction because of differences in the bulk diamagnetic susceptibilities in the contents of the reference capillary and the sample tube (91). The corrected chemical shift is

$$\delta_{\text{corr}} = \delta_{\text{obs}} + \left(\frac{2\pi}{3}\right)\Delta\chi \tag{5-14}$$

where $\Delta\chi$ is the difference between the bulk volume susceptibilities of the reference solution and the sample solution. However, many cases have been observed in which Eq. 5-14 was inadequate. If both the sample and the external reference compound are present as dilute solutions using the same solvent, it is satisfactory for many purposes to ignore the susceptibility correction.

A minor problem occasionally encountered with samples of biological origin is the presence of a small amount of insoluble material in the sample tube. These insoluble particles lead to poor field homogeneity in the sample, with resulting decreased spectral resolution. The insoluble material is best removed by centrifugation or filtration.

Another problem in those instances where long relaxation times ($\gtrsim$2 sec) are being measured is caused by dissolved oxygen. Molecular oxygen is paramagnetic and therefore provides an efficient relaxation mechanism. Molecular oxygen has special affinity for aromatic rings. When necessary, most of the oxygen can be removed by bubbling nitrogen through the solution for about 10 min and quickly capping the NMR tube. For accurate measurements of very long relaxation times it may be necessary to submit the sample to a few freeze–thaw cycles under reduced pressure, followed by sealing the NMR tube.

References

1. F. Bloch, W. W. Hansen, and M. E. Packard, *Phys. Rev.* **69,** 127 (1946).
2. F. Bloch, W. W. Hansen, and M. E. Packard, *Phys. Rev.* **70,** 474 (1946).
3. N. Bloembergen, E. M. Purcell, and R. V. Pound, *Phys. Rev.* **73,** 679 (1948).
4. E. M. Purcell, H. C. Torrey, and R. V. Pound, *Phys. Rev.* **69,** 37 (1946).
5. O. Haworth and R. E. Richards, *Progr. NMR Spectrosc.* **1,** 1 (1966).

6. A. Abragam, "Principles of Nuclear Magnetism," p. 82ff. Oxford Univ. Press, London and New York, 1961.
7. R. R. Ernst, *Advan. Magn. Resonance* **2,** 1 (1966).
8. G. E. Hall, *Annu. Rev. NMR* (*Nucl. Magn. Resonance*) *Spectrosc.* **1,** 227 (1968).
9. J. C. Buchta, H. S. Gutowsky, and D. E. Woessner, *Rev. Sci. Instrum.* **29,** 55 (1958).
10. T. L. James, Ph.D. Dissertation, University of Wisconsin, Madison (1969).
11. H. D. W. Hill and R. Freeman, "Introduction to Fourier Transform NMR." Varian Associates, Palo Alto, California, 1970.
12. H. Sternlicht and D. M. Zuckerman, *Rev. Sci. Instrum.* **43,** 525 (1972).
13. G. Kneissl and S. T. Dunn, *Amer. Lab.* **4,** 69 (1972).
14. R. R. Ernst and W. A. Anderson, *Rev. Sci. Instrum.* **37,** 93 (1966).
15. E. D. Becker, J. A. Ferretti, and T. C. Farrar, *J. Amer. Chem. Soc.* **91,** 7784 (1969).
16. P. Waldstein, *Rev. Sci. Instrum.* **42,** 437 (1971).
17. J. W. Cooley and J. W. Tukey, *Math. Comput.* **19,** 297 (1965).
18. G. D. Berglund, *Comm. Ass. Comput. Mach.* **11,** 703 (1968).
19. A. G. Redfield and R. K. Gupta, *J. Chem. Phys.* **54,** 1418 (1971).
20. S. L. Patt and B. D. Sykes, *J. Chem. Phys.* **56,** 3182 (1972).
21. F. W. Benz, J. Feeney, and G. C. K. Roberts, *J. Magn. Resonance* **8,** 114 (1972).
22. J. P. Jesson, P. Meakin, and G. Kneissel, *J. Amer. Chem. Soc.* **95,** 618 (1973).
23. B. L. Tomlinson and H. D. W. Hill, *J. Chem. Phys.* **59,** 1775 (1973).
24. R. Kaiser, *Rev. Sci. Instrum.* **31,** 963 (1960).
25. J. Itoh and S. Sato, *J. Phys. Soc. Jap.* **14,** 851 (1959).
26. V. Royden, *Phys. Rev.* **96,** 543 (1954).
27. R. C. Hopkins, *Rev. Sci. Instrum.* **35,** 1495 (1964).
28. A. Charles and W. McFarlane, *Mol. Phys.* **14,** 299 (1968).
29. R. Freeman, *J. Chem. Phys.* **53,** 457 (1970).
30. R. R. Ernst, *J. Chem. Phys.* **45,** 3845 (1966).
31. E. Wenkert, A. O. Clouse, D. W. Cochran, and D. Doddrell, *J. Amer. Chem. Soc.* **91,** 6879 (1969).
32. L. D. Hall and L. F. Johnson, *Chem. Commun.* p. 509 (1969).
33. O. A. Gansow and W. Shittenhelm, *J. Amer. Chem. Soc.* **93,** 4294 (1971).
34. I. H. F. H. Lang, "Correlation Techniques." Van Nostrand-Reinhold, Princeton, New Jersey, 1967.
35. R. M. Dagnall, B. L. Sharp, and T. S. West, *Nature* (*London*), *Phys. Sci.* **235,** 65 (1972).
36. R. R. Ernst, *J. Magn. Resonance* **3,** 10 (1970).
37. J. Dadok, R. F. Sprecher, and A. A. Bothner-By, *Abstr., 13th Exp. NMR Conf., 1972* Abstract 15.4; J. Dadok and R. F. Sprecher, *J. Magn. Resonance* **13,** 243 (1974).
38. H. C. Torrey, *Phys. Rev.* **76,** 1059 (1949).
39. J. E. Anderson and R. Ullman, *J. Phys. Chem.* **71,** 4133 (1967).
40. A. L. Van Geet, *Anal. Chem.* **40,** 304 (1968).
41. A. L. Van Geet and D. H. Hume, *Anal. Chem.* **37,** 979 (1965).
42. J. S. Leigh, Jr., *Rev. Sci. Instrum.* **39,** 1594 (1968).
43. A. Abragam, "Principles of Nuclear Magnetism," p. 66ff. Oxford Univ. Press, London and New York, 1961.
44. I. Solomon, *C. R. Acad. Sci.* **248,** 92 (1959).
45. S. Meiboom, *J. Chem. Phys.* **34,** 375 (1961).
46. A. Csaki and G. Bene, *C. R. Acad. Sci.* **251,** 228 (1960).
47. R. L. Streever and H. Y. Carr, *Phys. Rev.* **121,** 20 (1960).
48. Dinesh and M. T. Rogers, and G. D. Vickers, *Rev. Sci. Instrum.* **43,** 555 (1972).
49. S. Alexander, *Rev. Sci. Instrum.* **32,** 1066 (1961).

50. R. Freeman and S. Wittekoek, *J. Magn. Resonance* **1,** 238 (1969).
51. C. S. Johnson, Jr., *Advan. Magn. Resonance* **1,** 33 (1965).
52. D. C. Look and I. J. Lowe, "Fourth Omnibus Conference on the Experimental Aspects of Nuclear Magnetic Resonance Spectroscopy." Mellon Institute, Pittsburgh, Pennsylvania, 1963 (unpublished).
53. D. Ailion and C. P. Slichter, *Phys. Rev.* **135,** A1099 (1964); **137,** A235 (1965).
54. G. P. Jones, *Phys. Rev.* **148,** 332 (1966).
55. S. B. W. Roeder and D. C. Douglass, *J Chem. Phys.* **52,** 5525 (1970).
56. C. Deverell, R. E. Morgan, and J. H. Strange, *Mol. Phys.* **18,** 553 (1970).
57. S. R. Hartmann and E. L. Hahn, *Phys. Rev.* **128,** 2042 (1962).
58. A. Hartland, *Proc. Roy. Soc., Ser. A* **304,** 361 (1968).
59. R. L. Vold, J. S. Waugh, M. P. Klein, and D. E. Phelps, *J. Chem. Phys.* **48,** 3831 (1968).
60. R. Freeman and H. D. W. Hill, *J. Chem. Phys.* **51,** 3140 (1969).
61. R. Freeman and H. D. W. Hill, *J. Chem. Phys.* **53,** 4103 (1970).
62. R. Freeman and H. D. W. Hill, *J. Chem. Phys.* **54,** 3367 (1971).
63. J. L. Markley, W. J. Horsley, and M. P. Klein, *J. Chem. Phys.* **55,** 3604 (1971).
64. G. G. McDonald and J. S. Leigh, Jr., *J. Magn. Resonance* **9,** 358 (1973).
65. R. L. Vold, *J. Chem. Phys.* **56,** 3210 (1972).
66. A. C. McLaughlin, G. G. McDonald, and J. S. Leigh, Jr., *J. Magn. Resonance* **11.** 107 (1973).
67. R. Freeman and H. D. W. Hill, *J. Chem. Phys.* **55,** 1985 (1971).
68. D. C. Douglass and D. W. McCall, *J. Phys. Chem.* **62,** 1102 (1958).
69. D. E. Woessner, *Rev. Sci. Instrum.* **31,** 1146 (1960).
70. H. Y. Carr and E. M. Purcell, *Phys. Rev.* **94,** 630 (1954).
71. E. O. Stejskal, *J. Chem. Phys.* **43,** 3597 (1965).
72. E. O. Stejskal and J. E. Tanner, *J. Chem. Phys.* **42,** 288 (1965).
73. B. D. Boss and E. O. Stejskal, *J. Phys. Chem.* **71,** 1501 (1967).
74. J. E. Tanner, *J. Chem. Phys.* **52,** 2523 (1970).
75. L. J. Burnett and J. F. Harmon, *J. Chem. Phys.* **57,** 1293 (1972).
76. R. C. Wayne and R. M. Cotts, *Phys. Rev.* **151,** 264 (1966).
77. B. Robertson, *Phys. Rev.* **151,** 273 (1966).
78. J. S. Murday and R. M. Cotts, *J. Chem. Phys.* **48,** 4938 (1968).
79. T. L. James and G. G. McDonald, *J. Magn. Resonance* **11,** 58 (1973).
80. E. D. Ostroff and J. S. Waugh, *Phys. Rev. Lett.* **16,** 1097 (1966).
81. P. Mansfield and D. Ware, *Phys. Lett.* **22,** 133 (1966).
82. J. S. Waugh and L. M. Huber, *J. Chem. Phys.* **47,** 1862 (1967).
83. J. S. Waugh, C. H. Wang, L. M. Huber, and R. L. Vold, *J. Chem. Phys.* **48,** 662 (1968).
84. U. Haeberlen and J. S. Waugh, *Phys. Rev.* **175,** 453 (1968).
85. J. S. Waugh, L. M. Huber, and U. Haeberlen, *Phys. Rev. Lett.* **20,** 180 (1968).
86. W.-K. Rhim, D. D. Elleman, and R. W. Vaughan, *J. Chem. Phys.* **58,** 1772 (1973).
87. U. Haeberlen, J. D. Ellet, Jr., and J. S. Waugh, *J. Chem. Phys.* **55,** 53 (1971).
88. A. Pines and J. S. Waugh, *J. Magn. Resonance* **8,** 354 (1972).
89. M. Hanabusa, *J. Appl. Phys.* **42,** 1077 (1971).
90. J. D. Ellet, Jr., M. G. Gibby, U. Haeberlen, L. M. Huber, M. Mehring, A. Pines, and J. S. Waugh, *Advan. Magn. Resonance* **5,** 117 (1971).
91. J. A. Pople, W. G. Schneider, and H. J. Bernstein, "High-Resolution Nuclear Magnetic Resonance," p. 80. McGraw-Hill, New York, 1959.

CHAPTER 6

NMR STUDIES OF BIOMOLECULAR INTERACTIONS

Investigation of the structure, bonding, equilibria, and kinetics involved with complex formation and more subtle interactions of biomolecules furnishes one of the most powerful and exciting applications of nuclear magnetic resonance. The possible variety of interactions is considerable. Therefore, in this chapter, primarily the interaction of relatively small molecules or ions with macromolecules will be covered. The experimental tools that prove most valuable for these studies take advantage of NMR relaxation phenomena. Relaxation processes are very sensitive to variations in molecular environment. Various kinds of probes may be useful in providing information. In particular, paramagnetic ions (or free radicals) and quadrupolar ions (nuclei with a quadrupole moment) can be used to probe interactions. Nuclei on small molecules that bind to macromolecules may also be considered as probes of their molecular environment, as the nuclear relaxation times and chemical shifts change when the small molecule is bound to a macromolecule. Observation of an intermolecular nuclear Overhauser effect, which also depends on relaxation processes, gives direct and quite specific evidence of an interaction between a small molecule and a macromolecule.

The use of these relaxation techniques may provide knowledge about one or more of the following topics concerning interactions: (*1*) formation constants and stoichiometry; (*2*) accessibility of the binding site on a macromolecule to various molecules, ions, or solvent; (*3*) mobility at the binding site; (*4*) exchange kinetics; (*5*) identification of the site of binding on both the small molecule and macromolecule; (*6*) molecular environment of the "probe," including structure, configuration, and type of bonding; and (*7*) role of the interaction in biological processes. For the most part in this chapter, only the rationale, with illuminating examples, of the NMR methods for obtaining data are presented. For implications of the data in terms of the chemistry and biochemistry, the reader is referred to the original literature.

The extreme sensitivity of relaxation techniques to variations in molecular environment provides a means of overcoming the relatively low inherent signal-to-noise of NMR spectroscopy. For example, if the spin–lattice (T_1) relaxation time of a nucleus on a small molecule is drastically shortened when the small molecule (ligand) is bound to a macromoelcule, use of an excess of ligand will aid in detecting the nucleus' signal and, if there is rapid exchange between free and bound ligands, a T_1 value may be observed that is moderately shortened from the value of the free ligand in the absence of macromolecule. The high efficiency of the relaxation processes in the bound state furnishes an amplification factor enabling more concentrated solutions of the "probe" to be used, providing there is rapid exchange between free and bound ligands. If rapid exchange conditions do not prevail, it may be possible to gain information about the exchange kinetics.

6.1. Chemical Exchange

Many aspects of chemical exchange were presented in Section 2.6. Chemical exchange effects on linewidths and relaxation times of small molecules can provide useful information about the kinetics of complex formation between the small molecule and a macromolecule. For structural studies, moreover, it is generally necessary to ascertain that exchange effects are not influencing the measured relaxation times or linewidths and that there is rapid exchange of the ligand between free and bound sites. Two types of systems may be considered.

With the first type of system, the amount (population P_f) of the small ligand "free" in solution is approximately equal to the amount (population P_b) of ligand complexed by the macromolecule. A description of the relaxation times for this type of system under slow, intermediate, and rapid exchange conditions are given in Section 2.6. In the rapid exchange limit, a resonance peak is centered at frequency

$$\nu = P_f \nu_f + P_b \nu_b \tag{6-1}$$

with a linewidth $W_{1/2}$ $(= 1/\pi T_2)$ such that

$$\frac{1}{T_2} = \frac{P_f}{T_{2f}} + \frac{P_b}{T_{2b}} \tag{6-2}$$

where ν_f is the resonance frequency of the ligand nucleus in the free state, T_{2f} is the spin–spin relaxation time for the free ligand nucleus, and ν_b and T_{2b} are similarly defined for bound ligand. From Eqs. 2-71a and 2-71b, the spin–lattice relaxation time in the limit of rapid exchange,

$$\frac{1}{T_1} = \frac{P_f}{T_{1f}} + \frac{P_b}{T_{1b}} \tag{6-3}$$

can also be obtained. The rapid exchange limit may be operationally defined as the situation in which the nucleus exchanging between two sites behaves as though only one "average" environment were present; the experimental criteria are that (*a*) spin–lattice relaxation be described by a single exponential, (*b*) the lineshape be Lorentzian, and (*c*) the measured relaxation time vary continuously as either P_f or P_b is varied. It has been shown that the decay of the magnetization following an rf pulse generally depends on two (or more) exponentials in the case where exchange modulates the relaxation (1). Two exponentials will lead to a non-Lorentzian lineshape. In essence, the theoretical requirements and the criteria presented above are described by the following inequalities:

$$\tau_c \ll \tau_b \ll T_{1b}$$

$$\tau_b \ll T_{2b}$$

and

$$2\pi(\nu_f - \nu_b) \ll 1/\tau_b$$

where τ_c is the correlation time for the field fluctuations leading to relaxation in the bound state, and τ_b is the lifetime of the ligand in the bound state, i.e., the time a ligand molecule spends in the complex before the complex dissociates and the ligand is again free in solution. Less restrictive conditions for rapid exchange have recently been presented (2); however, the conditions described above provide a simple physical picture of rapid exchange.

With the second type of system, the amount of free ligand is much greater than the amount of complexed ligand ($P_f \gg P_b$). The NMR spectrum for the ligand nucleus is essentially a single peak with an angular frequency shift from the free ligand of (3)

$$\Delta\omega = 2\pi(\nu - \nu_f) = \frac{P_b \Delta\omega_b}{(1 + \tau_b/T_{2b})^2 + \tau_b^2(\Delta\omega_b)^2} \tag{6-4}$$

where $\Delta\omega_b = 2\pi(\nu_b - \nu_f)$. Unless $\Delta\omega_b$ results from a paramagnetic shift, $\Delta\omega$ will usually be small. Swift and Connick (3) also showed that the line-width and the spin–spin relaxation time could be expressed as

$$\frac{1}{T_2} = \frac{1}{T_{2f}} + \frac{P_b}{\tau_b}\left[\frac{1/T_{2b}^2 + 1/(T_{2b}\tau_b) + \Delta\omega_b^2}{(1/T_{2b} + 1/\tau_b)^2 + \Delta\omega_b^2}\right] \tag{6-5}$$

which is valid for all exchange rates. It was subsequently shown that the analogous equation for T_1 is (4, 5)

$$\frac{1}{T_1} = \frac{1}{T_{1f}} + \frac{P_b}{T_{1b} + \tau_b} \tag{6-6}$$

for the case where $P_f \ll P_b$. In the rapid exchange limit, the measured spin–lattice relaxation rate depends on T_{1b}. The observed spin–spin relaxation rate in the rapid exchange limit reveals two general relaxation mechanisms, one involving T_{2b} and the other involving the chemical shift difference between the bound and free states, $\Delta\omega_b$. Three limiting cases of Eq. 6-5 may be considered.

1. If the T_{2b} mechanism dominates ($1/T_{2b} \gg \Delta\omega_b$),

$$\frac{1}{T_2} = \frac{1}{T_{2f}} + \frac{P_b}{T_{2b} + \tau_b} \tag{6-7}$$

2. If the $\Delta\omega_b$ mechanism dominates the T_{2b} mechanism ($\Delta\omega_b \gg 1/T_{2b}$) and exchange is very rapid ($\tau_b^{-2} \gg \Delta\omega_b^2, T_{2b}^{-2}$),

$$\frac{1}{T_2} = \frac{1}{T_{2f}} + \frac{P_b}{T_{2b}} \tag{6-8}$$

3. If the $\Delta\omega_b$ mechanism dominates the T_{2b} mechanism ($\Delta\omega_b \gg 1/T_{2b}$) and exchange is not sufficiently rapid ($\Delta\omega_b^2 \gg \tau_b^{-2}$),

$$\frac{1}{T_2} = \frac{1}{T_{2f}} + \frac{P_b}{\tau_b} \tag{6-9}$$

Equation 6-7 will obviously serve regardless of the mechanism. The experimental question is to decide the relative weights of T_{2b} and τ_b in Eq. 6-7.

6.2. Paramagnetic Probes

The unpaired electron on paramagnetic species provides a very efficient relaxation mechanism (cf. Section 2.4.3), enabling the exploitation of paramagnetic probes in studies of biomolecular complex formation. Some of the many useful applications of contact shifts and pseudocontact shifts, which may occur in the presence of a paramagnetic species, were described in Section 3.3. Here, concern will be with the information that may be obtained by study of the relaxation phenomena engendered by paramagnetic probes. Most divalent transition metal ions and trivalent lanthanide ions are paramagnetic, as are stable free radicals. Some ions, e.g., Cu(II), Fe(III), and Mn(II), and free radicals are natural biological probes, being present as a constituent of the biomolecular system. In other cases, paramagnetic probes may be added to a system, e.g., replacing Mg(II) as an enzyme activator with Mn(II).

Paramagnetic transition metal ions were first used to probe the metal ion binding sites in nucleic acids (6) and enzymes (7) by measurements of the water proton relaxation rates. Many subsequent applications of paramagnetic probes in the study of enzyme mechanisms have been presented in recent reviews (8, 9).

6.2.1. Theory and Techniques

This section covers the rationale involved in the use of paramagnetic ions as probes of the binding of those ions and other molecules to macromolecules, as well as the interpretation of the NMR data for proper understanding of the dynamic and structural aspects of the binding. When a small molecule (e.g., water or an enzyme substrate) is bound to the paramagnetic metal ion (or near the ion) on the macromolecule, the relaxation times of nuclei on the small molecule will be decreased. The magnitude of the decrease depends on the distance of the nucleus from the paramagnetic center, the stoichiometry of the binding, the motional freedom of the ligand, and the time of the ligand's residence in the sphere of influence of the paramagnetic ion.

A. *Basic Equations*

The paramagnetic relaxation mechanism is generally quite efficient so the free ligand concentration is usually much greater than the bound ligand concentration, and the exchange equations pertinent to that condition are used. From Eq. 6-6 we obtain the paramagnetic contribution to the spin–

lattice relaxation rate

$$\frac{1}{T_{1p}} = \left(\frac{1}{T_1}\right)_{\text{obs}} - \left(\frac{1}{T_1}\right)_{\text{o}} = \frac{np}{T_{1M} + \tau_M} + \left(\frac{1}{T_1}\right)_{\text{OS}} \tag{6-10}$$

where $(1/T_1)_{\text{obs}}$ and $(1/T_1)_0$ are experimentally observed relaxation rates in analogous solutions in the presence and absence, respectively, of the paramagnetic metal ion. T_{1M} is the spin–lattice relaxation time for a nucleus in the coordination sphere of the paramagnetic ion, n is the number of coordinated ligands per paramagnetic ion, p is the concentration ratio of paramagnetic ion to total ligand in solution, and τ_M is the length of time the nucleus spends in the sphere of influence of the paramagnetic ion. $(1/T_1)_{\text{OS}}$ represents the outer sphere contribution to the relaxation rate caused by dipolar interactions of the paramagnetic ion with ligand nuclei not in the primary coordination sphere; $(1/T_1)_{\text{OS}}$ is generally small (and therefore is ignored) compared with the first term.

For several paramagnetic ions having long electron spin relaxation times (τ_s), e.g., Mn(II) or Gd(III), an equation for the paramagnetic contribution to the spin–spin relaxation rate, similar to Eq. 6-10, may be obtained from Eq. 6-7:

$$\frac{1}{T_{2p}} = \left(\frac{1}{T_2}\right)_{\text{obs}} - \left(\frac{1}{T_2}\right)_{\text{o}} = \frac{np}{T_{2M} + \tau_M} + \left(\frac{1}{T_2}\right)_{\text{OS}} \tag{6-11}$$

Paramagnetic ions having short τ_s values (cf. Table 6-1), e.g., Co(II) or Ni(II), can cause paramagnetic shifts so the $\Delta\omega_b$ terms in Eq. 6-5 must be included:

$$\frac{1}{T_{2p}} = \frac{np}{\tau_M}\left[\frac{1/T_{2M}^2 + 1/T_{2M}\tau_M + \Delta\omega_M{}^2}{(1/T_{2M} + 1/\tau_M)^2 + \Delta\omega_M{}^2}\right] \tag{6-12}$$

where $\Delta\omega_M (= 2\pi\Delta\nu_M)$ is the chemical shift difference between the free and bound states.

Hereafter, the discussion will generally be limited to the use of Mn(II) for cohesiveness in presentation and because Mn(II) is the most widely used paramagnetic probe. Allowances must be made for the properties of other paramagnetic probes if they are used. In particular, contact shifts must be recognized for those metal ions having short τ_s values.

There are several reasons Mn(II) is especially valuable as a paramagnetic probe. (*a*) Mn(II) is capable of activating most enzymes that normally require Mg(II) as the natural activator. Mn(II) (ionic radius 0.80 Å) and Mg(II) (ionic radius 0.65 Å) form similar complexes with many ligands. (*b*) Mn(II) has a labile hydration sphere, e.g., $\tau_M \approx 2.7 \times 10^{-8}$ sec

TABLE 6-1

ELECTRON SPIN QUANTUM NUMBERS AND APPROXIMATE VALUES OF THE ISOTROPIC g FACTOR FOR SEVERAL PARAMAGNETIC IONS

ION	SPIN S	g^a	τ_s (SEC)^b
V(IV)	1/2	2.00	$\sim 10^{-9}$
Cu(II)	1/2	2.0–2.5	$\sim 10^{-10}$
Fe(III), low spin	1/2	2.–6.	$\sim 10^{-12}$
Ni(II)	1	2.0–2.8	$\sim 10^{-13}$
Cr(III)	3/2	2.00	$\sim 10^{-10}$
Co(II)	3/2	2.1–2.7	$\sim 10^{-13}$
Fe(II)	2	2.1–2.3	$\sim 10^{-11}$
Mn(II)	5/2	2.00	$\sim 10^{-9}$
Fe(III), high spin	5/2	2.0	$\sim 10^{-10}$
Gd(III)	7/2	2.00	$\sim 10^{-10}$
Eu(II)	7/2	2.00	$\sim 10^{-10}$

[a] The g factor listed is a value that may be used in the Solomon-Bloembergen equations (eqs. 2-59 and 2-60). The Solomon-Bloembergen equations were derived assuming that a "spin-only" electron magnetic moment was appropriate. In fact, for ions toward the end of the first-row transition metal ions, and for the second- and third-row transition metal ions, orbital motion also contributes to the electron magnetic moments. For the lanthanide ions, it is necessary to use the total electron angular momentum J rather than simply the spin S(10). Therefore, the g factor listed in this table may not correspond to a true isotropic g factor.

[b] The electron spin relaxation time listed is only approximate, having some dependence on the temperature, frequency, and ligands coordinated to the metal ion.

for Mn(II) $(H_2O)_6$ at 27°C (3). These rapid exchange conditions permit access to structural information concerning the complexes. (*c*) Mn(II) has a large electron spin quantum number ($S = 5/2$). Because the relaxation rate varies as $S(S + 1)$, Mn(II) will be a more efficient "relaxer" than low-spin ions, and lower concentrations of Mn(II) may be used to effect observable relaxation. (*d*) Mn(II) has a long electron spin relaxation time (cf. Table 6-1) (10).

The Solomon-Bloembergen equations (Eqs. 2-59 and 2-60) describe the paramagnetic contribution to the spin–lattice and spin–spin relaxation rates of spin $\frac{1}{2}$ nuclei in the sphere of influence of the paramagnetic ion. Using constants compiled in Table 6-2 (11, 12), eqs. 2-59 and 2-60 may be

TABLE 6-2
VALUES OF CONSTANTS IN THE SOLOMON-BLOEMBERGEN EQUATIONS

SYMBOL	NAME	VALUE
S	Electron spin quantum number	5/2 [for Mn(II)]
γ	Gyromagnetic ratio	2.675×10^4 rad sec^{-1} Gauss^{-1} (for ^{1}H)
g	Electronic "g" factor	2.00 [for Mn(II)]
β	Bohr magneton	9.284×10^{-21} erg Gauss^{-1}
$\hbar$	Planck's constant/2π	1.055×10^{-27} erg sec
A	Hyperfine coupling constant	6.2×10^5 Hz[a]

[a] This is the isotropic hyperfine coupling constant for an aqueous Mn(II) solution (11). This value may be used as a first approximation of A for other solutions entailing Mn(II)–^{1}H interaction. In favorable cases, A may be estimated from the electron spin resonance spectrum (12).

written as

$$\frac{1}{T_{1M}} = \frac{2.878 \times 10^{-31}}{r^6} \times \left(\frac{3\tau_c}{1+\omega_I^2\tau_c^2} + \frac{7\tau_c}{1+\omega_s^2\tau_c^2}\right) + 5.84A^2\left(\frac{\tau_e}{1+\omega_s^2\tau_e^2}\right) \tag{6-13}$$

$$\frac{1}{T_{2M}} = \frac{1.439 \times 10^{-31}}{r^6} \times \left(4\tau_c + \frac{3\tau_c}{1+\omega_I^2\tau_c^2} + \frac{13\tau_c}{1+\omega_s^2\tau_c^2}\right) + 2.92A^2\left(\tau_e + \frac{\tau_e}{1+\omega_s^2\tau_e^2}\right) \tag{6-14}$$

for protons in the sphere of influence of Mn(II). The electron spin relaxation time for Mn(II) is long (cf. Table 6-1), so for reasonably large frequency values ($\lesssim 5$ MHz), $\omega_s^2\tau_e^2 \gg 1$, and the hyperfine term (i.e., the second term in Eq. 6-13) will not contribute to $1/T_{1M}$. By the same token, however, the hyperfine term may contribute to $1/T_{2M}$. The magnitude of the contribution of the hyperfine term depends on the hyperfine coupling constant A. As a first rough approximation, the Mn(II)–^{1}H value of A for aqueous Mn(II) solutions, 6.2×10^5 Hz (11), may be used for ligands in the first coordination sphere. For other nuclei, the appropriate gyromagnetic ratio γ must be used. For other paramagnetic metal ions, the appropriate electron spin quantum number S and g factor must be used. γ may be obtained from the data in Appendix 2. Values of S and g for several paramagnetic ions are given in Table 6-1.

For small molecules coordinated to Mn(II), e.g., water molecules in an aqueous Mn(II) solution, the rotational correlation time τ_r is the shortest correlation time and is therefore the dominant correlation time in Eq. 2-61, reproduced here as Eq. 6-15.

$$1/\tau_c = 1/\tau_r + 1/\tau_s + 1/\tau_M \tag{6-15}$$

When the ion is bound to a macromolecule, however, τ_r is several orders of magnitude larger, so τ_c may have contributions from τ_r, τ_s, or τ_M and the relaxation rates $1/T_{1M}$ and $1/T_{2M}$ may be much larger than in small complexes.

There will be two principal types of experiments we can perform that will utilize Eqs. 6-10 and 6-13 (or Eqs. 6-11 and 6-14 for T_2) to give structural information about macromolecular complexes. In one type of experiment, the water proton relaxation rates are measured. If a value for τ_c is estimated, T_{1M} may be obtained from Eq. 6-13, using a value for r of 2.86 Å determined from x-ray studies on several crystalline Mn(II) hydrates. If $T_{1M} \gg \tau_M$, a value for n, the number of water molecules in the first coordination sphere of Mn(II), may be estimated using Eq. 6-10. The value of n permits an estimate of the number of liganding groups donated to Mn(II) by the macromolecule, because Mn(II) has a strong propensity for octahedral coordination. Changes in n may be noted when additional ligands (e.g., enzyme substrates) are added. n is also an indicator of the accessibility of an enzyme's active site to the solvent water.

In a second type of experiment where the stoichiometry (n) of binding is known and, again, a value for τ_c may be estimated, a value for r may be calculated from Eq. 6-13. This experiment is valuable for determining distances of individual nuclei from Mn(II) by measuring the individual relaxation rates. For example, the distances from Mn(II) for the individual nuclei on a substrate may be determined for an enzyme–Mn(II)–substrate complex. For both types of experiment, relative values may be determined without knowledge of τ_c. However, if absolute values are desired, a knowledge of τ_c is required. The methods for estimating τ_c will now be covered.

B. τ_c *Estimation*

To determine molecular parameters, T_{1p} and T_{2p} may be measured as a function of frequency and temperature. In favorable cases, such studies will permit designation of the relaxation mechanism, determination of the relative magnitudes of the various dynamic processes, and evaluation of the pertinent correlation time.

T_{1M} and T_{2M} Dependence on τ_c. The correlation terms $f_1(\tau_c)$ and $f_2(\tau_c)$ for the dipolar contributions to T_{1M}^{-1} and T_{2M}^{-1} (the quantities in parentheses for the first terms of Eqs. 6-13 and 6-14) are given in Fig. 6-1 as a function of τ_c. For low values of τ_c, it is apparent from Fig. 6-1 (13) and Eqs. 6-13 and 6-14 that $1/T_{1M} = 1/T_{2M}$. For intermediate values of τ_c, i.e., $\omega_s^2\tau_c^2 \gg 1$

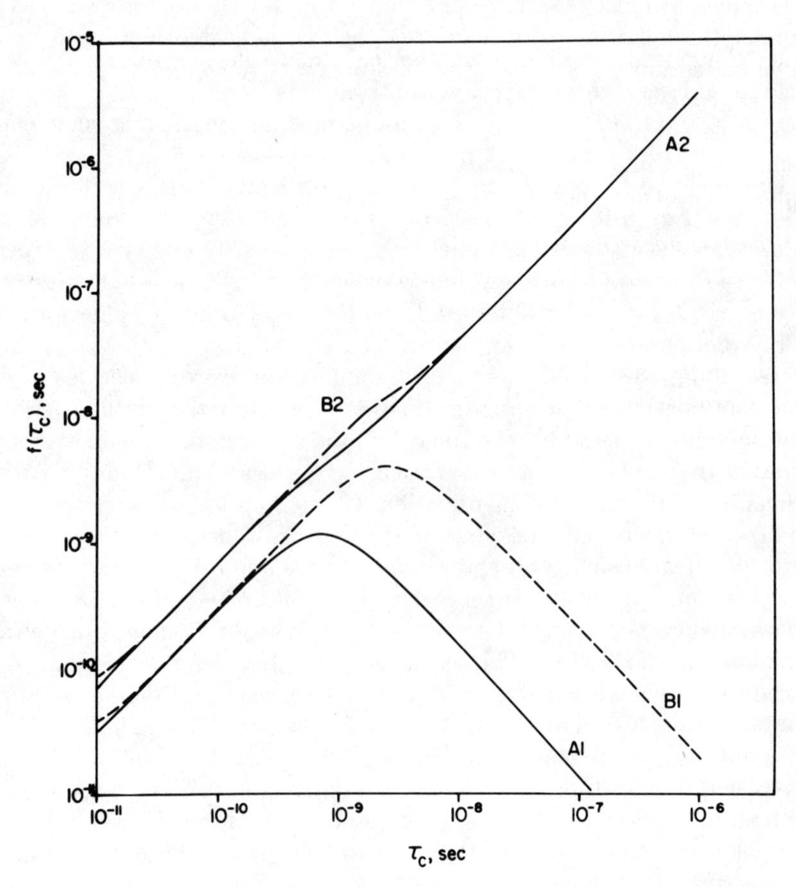

FIG. 6.-1. Theoretical dependence of the correlation terms $f_1(\tau_c)$ and $f_2(\tau_c)$ on the correlation time τ_c based on Solomon's equations. Curves A1 and B1 show $f_1(\tau_c)$ as a function of τ_c at 220 and 60 MHz, respectively. Curves A2 and B2 show $f_2(\tau_c)$ as a function of τ_c at 220 and 60 MHz, respectively (13).

but $\omega_I^2\tau_c^2 \ll 1$, $1/T_{1M} = 6/7T_{2M}$. For long values of τ_c, i.e., when $\omega_I^2 \gtrsim 1$, the ω_I frequency dispersion of $1/T_{1M}$ sets in with $1/T_{1M}$ proportional to $1/(\omega_I^2\tau_c^2)$, whereas $1/T_{2M}$ is proportional to τ_c. T_{1M} is therefore frequency dependent in this region, whereas T_{2M} is not.

It is evident from Fig. 6-1 that the relaxation rates may be frequency dependent. Changes in temperature are reflected in changes in τ_c, so the relaxation rates may also be temperature dependent. The dependence of the relaxation rates on these variables is contingent on the value of τ_c and the dominating relaxation process (if, in fact, one relaxation process dominates). Table 6-3 summarizes the relationship of the relaxation rates with these variables. It must be emphasized that Table 6-3 provides a guideline only under the following conditions: (*a*) the one indicated relaxation process dominates and (*b*) the hyperfine interaction term is negligible. As already noted, condition (*b*) is generally fulfilled for T_{1p} measurements in Mn(II) solutions but may not be fulfilled for T_{2p} measurements. The various aspects summarized in Table 6-3 are presented in the following discussion so that aberrations from these simple conditions may be understood.

Temperature Dependence. The individual correlation times may be described in terms of $(E_{\text{act}})_i$, activation energies, in an Arrhenius expression:

$$\tau_i = \tau_i^0 \exp[(E_{\text{act}})_i/RT] \tag{6-16}$$

The temperature dependence of the relaxation rates is related to the temperature dependence of the correlation time, which may also be given by an Eyring expression:

$$\tau_i = \frac{kT}{h} \exp\left[\frac{-\Delta H_i\ddagger}{RT} + \frac{\Delta S_i\ddagger}{R}\right] \tag{6-17}$$

where $\Delta H_i\ddagger$ and $\Delta S_i\ddagger$ are, respectively, the enthalpy and entropy of activation for the first-order rate process governing the correlation time, k is the Boltzmann constant, T is the absolute temperature, h is Planck's constant, and R is the gas constant.

As the temperature decreases, τ_r and τ_M increase but τ_s may either increase or decrease (14); i.e., it may have either an apparent "positive" or "negative" activation energy in Eq. 6-16. Generally τ_s increases with increasing temperature for small complexes (15, 16) and decreases with increasing temperature for macromolecular complexes (17), but it must be cautioned that this is not firmly established. The reason for the apparent "positive" or "negative" activation energy for τ_s processes is that τ_v (cf. Eq. 6-19) is actually the correlation time fitting Eq. 6-16. This aspect will be discussed in more detail below. The activation energies for the relaxation processes will usually fall in the order $(E_{\text{act}})_M > (E_{\text{act}})_r > (E_{\text{act}})_s$.

TABLE 6-3

DEPENDENCE OF $1/T_{1p}$ AND $1/T_{2p}$ ON TEMPERATURE AND FREQUENCY FOR VARIOUS RELAXATION PROCESSES[a]

		DOMINANT RELAXATION PROCESS						
		EXCHANGE	T_{1M}, T_{2M}			(T_1)os		
	REGION[b]	τ_M	τ_r	τ_s	τ_M	τ_r	τ_s	τ_d[c]
$\frac{d(T_{1p})^{-1}}{dT}$	X	+	−	±[d]	−	−	±[d]	−
	Y	+	−	±[d]	−	−	±[d]	−
	Z	+	+	±[d]	+	+	±[d]	+
$\frac{d(T_{2p})^{-1}}{dT}$	X	+[e]	−	±[f]	−	−	±[f]	−
	Y	+[e]	−	±[f]	−	−	±[f]	−
	Z	+[e]	−	±[f]	−	−	±[f]	−
E_{act} (kcal/mol)		6–25	2–4	<2.5	6–25	2–4	6–25	2–3
$\frac{d(T_{1p})^{-1}}{d\omega}$	X	0	−	$\overset{\pm}{0}$[g]	−	−	$\overset{\pm}{0}$[g]	−
	Y	0	~0	$\overset{\pm}{0}$[g]	~0	~0	$\overset{\pm}{0}$[g]	~0
	Z	0	−	±[g]	−	−	±[g]	−
$\frac{d(T_{2p})^{-1}}{d\omega}$	X	0	≤0	$\overset{\pm}{0}$[g]	≤0	≤0	$\overset{\pm}{0}$[g]	≤0
	Y	0	≤0	$\overset{\pm}{0}$[g]	≤0	≤0	$\overset{\pm}{0}$[g]	≤0
	Z	0	≤0	±[g]	≤0	≤0	±[g]	≤0
$\frac{T_{1p}}{T_{2p}}$	X	1[h]	1	1	1	1	1	1
	Y	1[h]	7/6	7/6	7/6	7/6	7/6	7/6
	Z	1[h]	>1	>1	>1	>1	>1	>1

[a] It is assumed that the hyperfine interaction term is negligible. For Mn(II) complexes, the assumption is good for T_{1p} but may not be so good for T_{2p}.

[b] Region X: $\omega_I^2\tau_c^2 < 1$, $\omega_s^2\tau_c^2 < 1$ (short τ_c); region Y: $\omega_I^2\tau_c^2 < 1$, $\omega_s^2\tau_c^2 > 1$ (intermediate τ_c); region Z: $\omega_I^2\tau_c^2 > 1$, $\omega_s^2\tau_c^2 > 1$ (long τ_c).

[c] Correlation time for diffusion past the Mn(II) complex.

[d] It is generally (+) for small complexes and (−) for macromolecular complexes in regions X and Y and has opposite signs in region Z.

[e] It is possible for T_{2p}^{-1} to be exchange limited, whereas T_{1p}^{-1} is not.

[f] It is generally (+) for small complexes and (−) for macromolecular complexes.

[g] τ_s may itself be frequency dependent, giving a positive frequency dependence for low frequencies.

[h] When chemical shift $\Delta\omega_M \ll T_{2M}^{-1}$, the T_{1p}/T_{2p} will equal 1, i.e., it is valid for Mn(II) complexes.

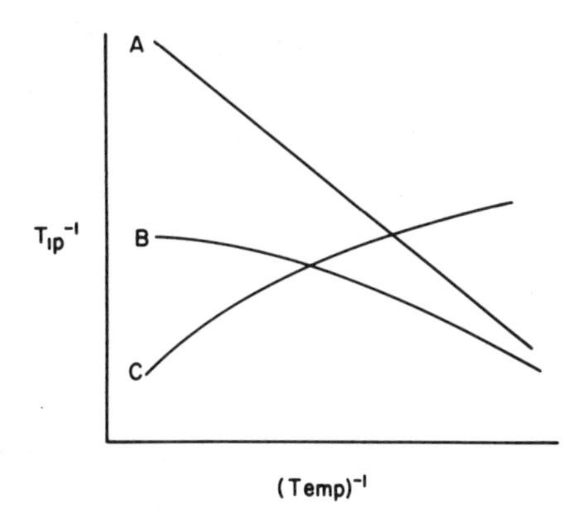

FIG. 6-2. Temperature dependence of $1/T_{1p}$. Case A: $\tau_m \gg T_{1M}$, i.e., exchange limited. Case B: $\tau_M \ll T_{1M}$ and $\omega_1^2\tau_c^2 > 1$. Case C: $\tau_M \ll T_{1M}$ and $\omega_I^2\tau_c^2 < 1$ (18.)

Figure 6-2 shows the three possible temperature dependences of T_{1p}^{-1} for macromolecular complexes (18). Curve A reflects the temperature dependence for the exchange-limited case: $\tau_M \gg T_{1M}$. The consequences of case A are seen in Eq. 6-10: $1/T_{1p}$ varies directly with $1/\tau_M$. For cases B and C, $T_{1M} \gg \tau_M$. Curve B reflects the temperature dependence of case B: $\omega_I^2\tau_c^2 > 1$. Curve B may be generated by varying τ_c at high values of τ_c, i.e., at values above the maximum in the $f_1(\tau_c)$ curve of Fig. 6-1. Curve C reflects the temperature dependence of case C: $\omega_I^2\tau_c^2 < 1$. Curve C may be generated by varying τ_c at low values of τ_c, i.e., at values below the maximum in the $f_1(\tau_c)$ curve of Fig. 6-1. τ_c is restricted to the region $\tau_c < 1/\omega_I$ (case C) by the negative temperature dependence of curve C. If a positive temperature dependence of $1/T_{1p}$ is noted, it may be necessary to investigate the temperature dependence of $1/T_{2p}$ to distinguish between cases A and B, the dependence being positive for case A and negative for case B (even if the hyperfine term contributes to $1/T_{2p}$.) This is not entirely unequivocal because $1/T_{2p}$ may be exchange limited with conditions for case A holding, whereas $1/T_{1p}$ is not exchange limited and case B holds. The energy of activation E_{act} may also provide a clue in favor of case A or B (cf. Table 6-3). A low activation energy (<5 kcal/mole) obtained over a reasonably large temperature range would militate against the exchange-limited case A.

If the temperature dependence of $1/T_{1p}$ passes through a maximum (i.e., temperature dependence slope = 0), case A may be ruled out and τ_c may be estimated from the maximum in the curve as $\tau_c = 1/\omega_I$ for the temperature producing the maximum. It should be verified that $1/T_{2p}$ does not also pass through a maximum. Observation of a maximum in the $1/T_{2p}$

curve would mean that τ_c is changing from τ_s dominance to τ_r or τ_M dominance. One aspect of the $1/T_{1p}$ dependence is illustrated in Fig. 6-3 for the water proton relaxation rate in a solution containing the ternary pyruvate kinase–Mn(II)–phosphoenol pyruvate complex in the presence of KCl. The negative temperature dependence at 8.13, 24.3, and 60 MHz permits the conclusion that case C prevails and $\tau_c < 1/\omega_I$. However, the data obtained at 220 MHz show no temperature dependence of $1/T_{1p}$, indicating that $\tau \approx 1/\omega_{220} = 7.2 \times 10^{-10}$ sec over the temperature range investigated. The activation energy calculated from the slope at the lower frequencies is less than 2 kcal/mole, implying that electron spin relaxation is the dominant relaxation process and $\tau_c = \tau_s$. The low activation energy also explains why the 220 MHz data show no temperature dependence over the range studied instead of exhibiting a maximum.

In certain situations, changes in temperature may lead to macromolecular conformation changes. An inflection in the temperature dependence of the water spin–lattice relaxation rate at about 15°C was interpreted as a change in the conformation of the Mn(II)–phosphorylase a complex (19). In certain instances, it may not be possible to easily make reliable T_2 measurements (e.g., with complex splitting patterns). In such situations, and for less ambiguous distinction between cases A and B in general, it is best to determine the frequency dependence of $1/T_{1p}$. $1/T_{1p}$ is frequency dependent for case B (cf. Eq. 6-13), but it is independent of frequency in the exchange-limited case A (cf. Eq. 6-10).

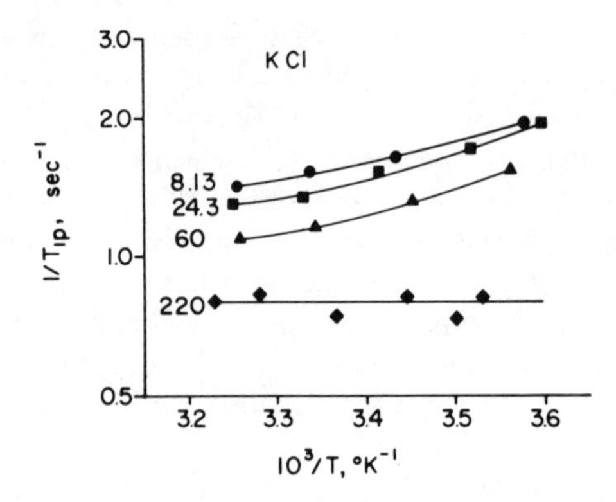

FIG. 6.3. The paramagnetic contribution to the water proton spin–lattice relaxation rate normalized to 0.10 m*M* Mn (II) as a function of the reciprocal of the absolute temperature for three frequencies (in MHz). Average of three samples containing 2.0 m*M* phosphoenol pyruvate, 100 m*M* KC1, between 0.136 and 0.195 m*M* pyruvate kinase sites, and between 0.10 and 0.15 m*M* $MnCl_2$ (18).

Frequency Dependence. Investigation of the dependence of relaxation rates on frequency will aid in determining whether or not a system is exchange limited and, if it is not, will enable τ_c to be calculated from the frequency dependence. T_{1p} will generally be more dependent on frequency than T_{2p} for values of τ_c usually encountered. Plots of $f_1(\tau_c)$, the quantity in parentheses in the dipolar term of Eq. 6-13, as a function of τ_c for various frequencies are shown in Fig. 6-4A (20). Because $\omega_s = 657\omega_I = 657(2\pi\nu)$, the frequency dispersion noted at $\sim 10^{-11}$ sec in Fig. 6-4A is caused by the ω_s term of $f_1(\tau_c)$. The much larger dispersion occurring in the region of $\sim 10^{-8}$ sec is caused by the ω_I term of $f_1(\tau_c)$. As shown in Fig. 6-4A and summarized in Table 6-3, $1/T_{1p}$ may or may not be a function of frequency, depending on the exact value of τ_c and the exact value of the frequencies used. Figure 6-4B presents the data in Fig. 6-4A in another manner, one that directly shows the frequency dependence of T_{1p} for fixed values of τ_c ranging from 5×10^{-11} to 5×10^{-7} sec. At sufficiently large values of τ_c, e.g., curves I and II, the ω_s dispersion may be neglected and the dipolar term of Eq. 6-13 rearranged to give (21)

$$T_{1M} = \frac{1}{3B}\left(\frac{1}{\tau_c} + \tau_c\omega_I^2\right) \qquad (6\text{-}18)$$

where B is a constant. τ_c may therefore be determined as the square root of the ratio of the slope to the intercept of T_{1M} vs ω_I^2 plots, such as those of curves I and II in Fig. 6-4B. A practical example of an ω_I^2 plot at three different temperatures is shown in Fig. 6-5 for the water proton relaxation time in an aqueous solution of the arginine kinase–Mn(II)–ADP–L-arginine–nitrate complex. The value of τ_c obtained from the slope-to-intercept ratio was found to be 1.1×10^{-8}, 8.2×10^{-9}, and 6.2×10^{-9} sec at 0°C, 11°C, and 22°C (22).

If only two frequencies are available, the correlation time may be estimated from the ratio of T_1 relaxation times. As shown in Fig. 6-6, however, the range of τ_c values for which this is a valid procedure is relatively narrow. Fortunately, for many macromolecular complexes of Mn(II), the value of τ_c falls in the appropriate range (13).

If measurements are performed at two frequencies $\omega_1 > \omega_2$, then Eq. 6-13 predicts that $1/T_{1M}(\omega_1) \leq 1/T_{1M}(\omega_2)$. A violation of this inequality is apparent with the 8.13 and 24.3 MHz data of Fig. 6-7. Such a violation may occur if the correlation time itself is frequency dependent.

The electron spin relaxation time τ_s may be frequency dependent and, therefore, τ_c may depend on frequency if τ_s is the dominant correlation time. According to Bloembergen and Morgan (15), the electron spin relaxation

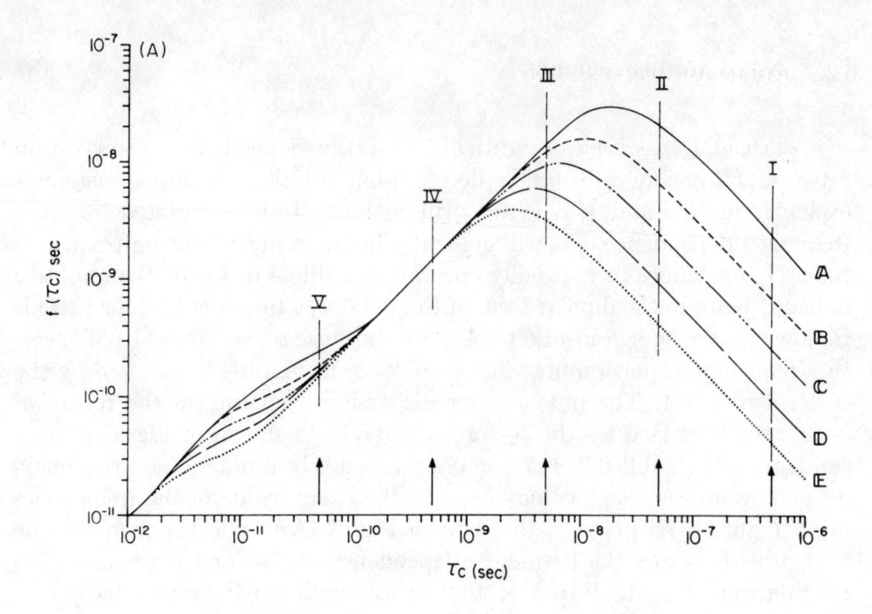

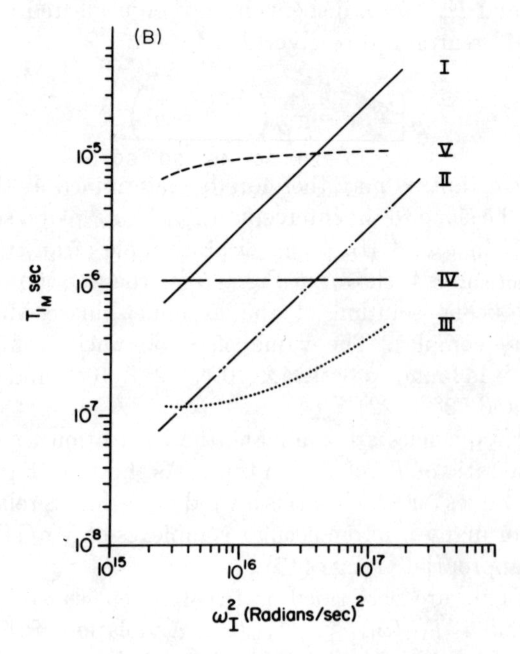

FIG. 6-4. Theoretical curves based on the Solomon-Bloembergen equation for the dependence of the correlation function $f_1(\tau_c)$ and the paramagnetic contribution to the relaxation time, T_{1M}, on the values of τ_c and the frequency under the assumption that the hyperfine contribution is negligible. (A) $f_1(\tau_c)$ as a function of τ_c: curve A, 8.13 MHz; curve B, 15.0 MHz; curve C, 24.3 MHz; curve D, 40.0 MHz; curve E, 60.0 MHz; (B) plot of T_{1M} of water caused by Mn(II) against ω_I^2 for five different values of τ_c ranging from 5×10^{-11} to 5×10^{-7} sec (Roman numerals I through V) (20).

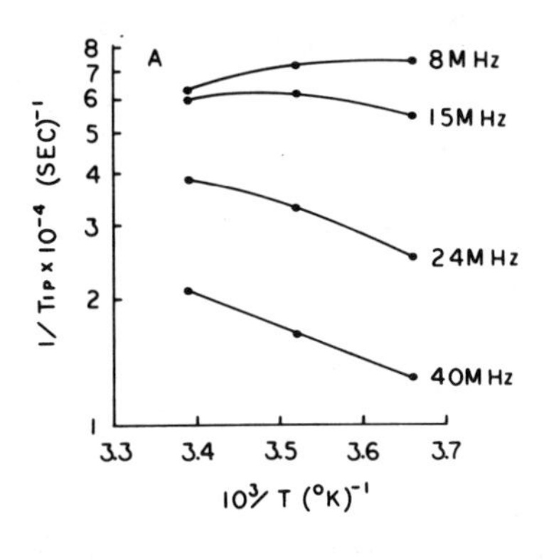

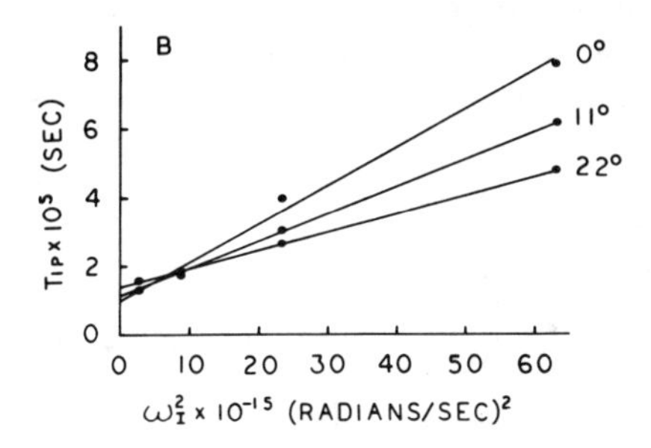

FIG. 6.5 Water proton relaxation rate as a function of temperature (A) and frequency (B) for an aqueous solution of the arginine kinase–Mn(II)–ADP–L-arginine–nitrate complex (22).

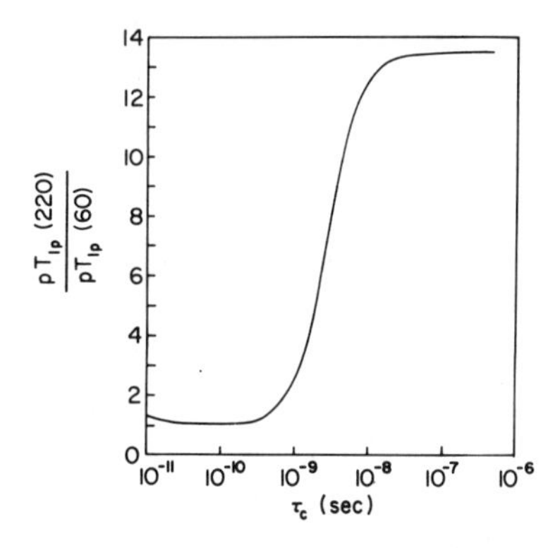

FIG. 6-6. The dependence of the ratio of paramagnetic contributions (cf. Eq. 6-13) to the spin–lattice relaxation times at 220 and 60 MHz on the correlation time τ_c (13).

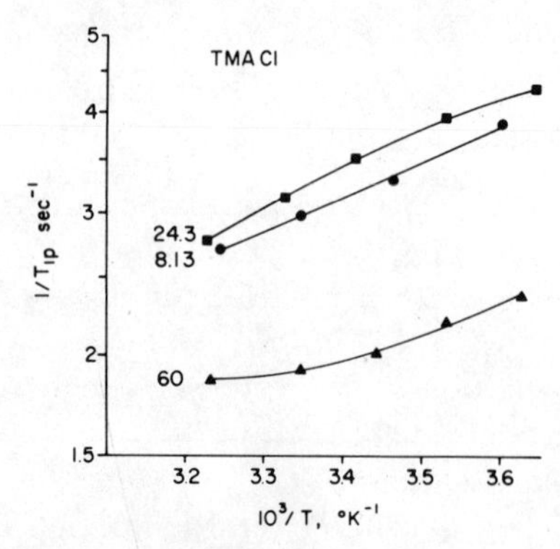

FIG. 6-7. The paramagnetic contribution to the water proton spin–lattice relaxation rate normalized to 0.10 mM Mn(II) as a function of the reciprocal of the absolute temperature for three frequencies (in MHz). Solutions contained 12.5 mM phosphoenol pyruvate, 62.5 mM $(CH_3)_4NCl$, 0.151 mM pyruvate kinase sites, and 0.125 mM $MnCl_2$ (18).

time may be expressed as

$$\frac{1}{\tau_s} = C\left(\frac{\tau_v}{1 + \omega_s^2\tau_v^2} + \frac{4\tau_v}{1 + 4\omega_s^2\tau_v^2}\right) \tag{6-19}$$

where C is a constant incorporating the spin S and the zero field splitting parameters, and τ_v is a correlation time that is a measure of the rate of modulation of the zero field splitting caused by solvent molecules colliding with the Mn(II) complex, i.e., fluctuating symmetry distortions of the complex. Equation 6-19 is based on several assumptions. The nature of these assumptions is not of direct interest here and will not be detailed.

There are two limiting cases for Eq. 6-19. In the first case, $\omega_s^2\tau_v^2 \ll 1$; consequently, $\tau_s = 1/(5C\tau_v)$ and is independent of frequency. This case usually holds for small Mn(II) complexes and, because

$$\tau_v = \tau_v^0 \exp[(E_{act})_v/RT] \tag{6-20}$$

τ_s will increase as the temperature increases. In the second limiting case, $\omega_s^2\tau_v^2 \gg 1$; consequently,

$$\tau_s = \omega_s^2\tau_v/2C \tag{6-21}$$

and is frequency dependent. This case appears to hold for macromolecular Mn(II) complexes (17) and, as a consequence of Eq. 6-20, τ_s will decrease as the temperature increases. Substituting Eq. 6-21 in Eq. 6-18 with $\tau_s = \tau_c$, we have an expression for the frequency dependence of T_{1M}:

$$T_{1M} = \frac{1}{3B}\left(\frac{2C}{657^2\omega_I^2\tau_v} + \frac{657^2\omega_I^4\tau_v}{2C}\right) \tag{6-22}$$

This dependence is illustrated in Fig. 6-8 (23), which shows the deviation from linearity at low ω_I^2 values characteristic of a frequency-dependent correlation time, i.e., a system in which $\tau_c = \tau_s$.

Other Means of Estimating τ_c. The estimation of τ_c from a maximum in a temperature dependence plot of $1/T_{1p}$ and the determination of τ_c from the frequency dependence of $1/T_{1p}$ have already been mentioned. It is apparent from the curves in Fig. 6-4A that two values of τ_c will satisfy one value of $f_1(\tau_c)$, so the analysis must be sufficiently thorough that the correct range for τ_c is ascertained before a value for τ_c is specified. If the conditions place τ_c in the range such that $\omega_I^2\tau_c^2 \lesssim 1$, τ_c may be estimated from the T_{1p}/T_{2p} ratio if the hyperfine contact term is deemed negligible (cf. Fig. 6-9):

$$\frac{T_{1p}}{T_{2p}} = \frac{7 + 4\omega_I^2\tau_c^2}{6} \tag{6-23}$$

The hyperfine contact term will, of course, be negligible if there is no unpaired electron spin density at the nucleus. Navon (24) has considered the contribution of the hyperfine terms for water proton relaxation rates in aqueous Mn(II) complexes. The hyperfine term in $1/T_{2p}$ amounts to about 2% of the dipolar term when $\tau_c = \tau_e$. Inclusion of the hyperfine contribution for water ligands gives (24)

$$\frac{T_{1p}}{T_{2p}} = \frac{1}{2} + \left(\frac{2}{3} + 0.022\,\frac{\tau_e}{\tau_c}\right)(1 + \omega_I^2\tau_c^2) \tag{6-24}$$

The Mn(II) electron paramagnetic resonance (EPR) spectrum for some Mn(II) complexes may be used to provide a lower limit for τ_s from the EPR linewidth {$\tau_s \lesssim [\sqrt{3}\pi$ (peak-to-peak separation in derivative sig-

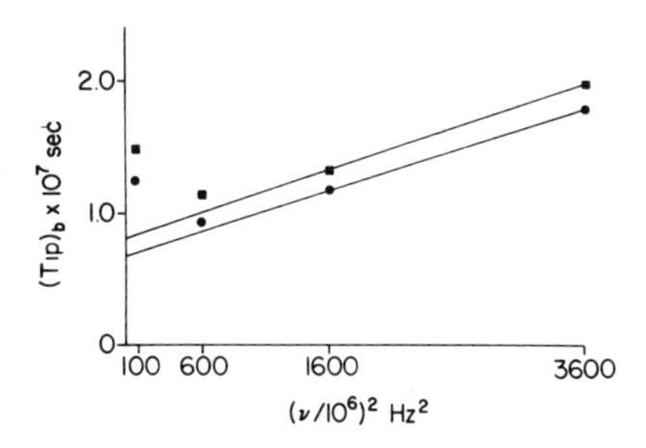

FIG. 6-8. The proton relaxation time $(T_{1p})_b$ of the manganese–pyruvate kinase complex plotted against the square of the resonance frequency at 17.5°C(●) and 35.4°C(■). The curves are linear extrapolations from the points at the two highest frequencies (23).

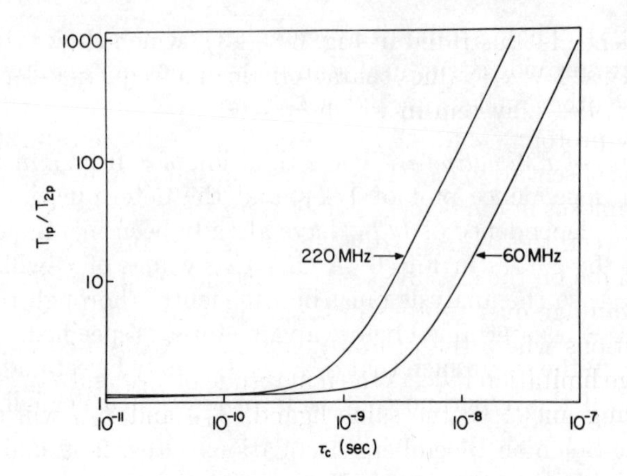

FIG. 6-9. Dependence of T_{1p}/T_{2p} at 60 MHz and at 220 MHz on the correlation time assuming the hyperfine contribution is negligible (cf. Eq. 6-23) (13).

nal)]$^{-1}$}. This may serve to eliminate τ_s as the dominant correlation time if the value of τ_c calculated from NMR results is smaller or, it may support τ_s as the relevant correlation time if the electron spin relaxation time calculated from EPR is approximately equal to τ_c.

It is usually more difficult to calculate τ_c for a nucleus on a ligand other than the solvent because T_1 values are not as readily determined and the capability for frequency-dependent measurements is generally limited by available instrumentation. With those hardships, some workers have resorted to the assumption that the value of τ_c determined by temperature and frequency dependence of the water proton relaxation rates for a particular complex is also the value of τ_c pertinent to the relaxation of nuclei on other ligands. Such an assumption is certainly valid if τ_s is the relevant correlation time for both ligands. However, that is not known with certainty *a priori*. At the very least, a temperature dependence study of the nuclear relaxation rate for the other ligand should be made. If it has been established that $\tau_c = \tau_s$ for the water protons, and the temperature dependence study yields an E_{act} value of less than 2.5 kcal/mole for the ligand nuclear resonance relaxation rate, the assumption of $\tau_c = \tau_s$ for the ligand nucleus is reasonable. For most other cases, the assumption can be misleading, generally giving values of τ_c that are too low.

C. *Variation of Nuclei and Ions*

Proton NMR studies are quite common. However, it may be advantageous to perform NMR relaxation rate studies with other nuclei, e.g.,

^{13}C or ^{31}P (25). These studies may complement proton NMR studies or may, in fact, provide structural information not obtainable by proton NMR. These other nuclei may be part of molecules or ions that cannot be examined by proton NMR, e.g., CO_3^{2-} or PO_4^{3-}. An interesting example of this use is the observation of ^{205}Tl NMR relaxation rates in pyruvate kinase–Mn(II) complexes in the presence of TlCl for the purpose of estimating the distance between the requisite divalent metal ion and the requisite monovalent metal ion on the enzyme (26). Other nuclei on the same ligand may have an advantage over protons in that they may not be exchange limited under conditions where the proton relaxation rates are exchange limited. The exchange limitation reflects the magnitude of T_{1M} and T_{2M} compared to τ_M. In varying nuclei on the same ligand, T_{1M} and T_{2M} will change; the factor in the Solomon-Bloembergen equations (Eqs. 2-59 and 2-60) that changes is γ^2/r^6. Comparison of ^1H and ^{13}C nuclei at the same distance from Mn(II) reveals that the exchange rate in the case of ^{13}C may be approximately 16 times slower before ^{13}C will also be exchange limited.

The use of paramagnetic ions other than Mn(II) may also provide valuable information. In certain cases with Mn(II), it may be difficult to unambiguously unravel the temperature or frequency dependence of T_{1p} and T_{2p}. The use of ions, e.g., Co(II) and Ni(II), possessing small electron relaxation times provides another parameter, the paramagnetic shift, to aid in deciphering the mechanism. In particular, the chemical shift is a function of the exchange rate (cf. Eq. 6-4), so chemical shift measurements may be used in conjunction with linewidth (or T_1) measurements to obtain values for τ_M and T_{2M} (or T_{1M}) (27). Other transition metal ions are not as efficient as Mn(II) in promoting relaxation (smaller S and smaller τ_c). If τ_M is comparable or even smaller when Mn(II) is replaced by another ion, the dominant paramagnetic relaxation method may change from an exchange-limited process to a T_{1M} process, permitting structural information to be obtained. Varying the ion may likewise produce the T_{1M} (or T_{2M}) and τ_M conditions necessary for investigating ligand exchange kinetics.

In a prelude to studies of macromolecular complexes, Barry *et al.* (28) examined the proton resonance line broadening and chemical shifts of adenylic acid and thymidylic acid in the presence of different lanthanide ions. With certain assumptions, perhaps the major one being that the lanthanide ions each coordinate the nucleotides in exactly the same manner, it was possible to deduce the nucleotide complex conformations in solution from the angular and distance information calculated from pseudocontact shifts using shift-promoting lanthanides and from line broadening using relaxation-promoting lanthanides. The results are in qualitative agreement with the conclusion that the bases prefer the anti conformation on grounds of chemical shift considerations (29–31). Quantitative conformational studies using nuclear Overhauser effect (NOE) measurements have also been made for adenine and thymine nucleosides (32, 33). It will be interesting to compare the conclusions of the two methods when NOE studies on the nucleotides have been made.

D. *Assumptions and Limitations of the Technique*

Even if the relaxation data are correctly analyzed, there are certain assumptions inherent in the paramagnetic probes method that should be understood for a proper perspective of the significance of the results obtained from analysis of the data. The purpose of this section is to point out briefly some of the complications that are generally ignored.

It was noted in Table 6-1 that the Solomon-Bloembergen equations were derived under the assumption that a field caused by a "spin-only" electron magnetic moment gave rise to nuclear relaxation. In fact, for many ions (second- and third-row transition metal ions, in particular), the orbital angular moment, which is sensitive to the particular ligands in the complex, couples with the electron spin moment (the spin–orbital coupling), leading to variations in g. The Solomon-Bloembergen equations also assume that g is isotropic, i.e., the electric field around the ion is symmetric. For most complexes of interest, the electric field produced by the ligands will not be cubically symmetric about the ion. The ligand field will then be anisotropic (dependent on orientation) and the g value will also be anisotropic. The anisotropy of g is usually expressed as a second-rank tensor with principle values g_x, g_y, and g_z. If the ligand field is symmetric, $g_x = g_y = g_z$, and a single isotropic g factor obtains. In complexes possessing axial symmetry, i.e., those having a threefold or higher axis of symmetry, it is common to define a g component parallel to the axis, $g_{\|} = g_z$, and a g component perpendicular to the axis, $g_{\perp} = g_x = g_z$. The existence of anisotropy will cause deviations in $g_{\text{isotropic}}$ assumed for the Solomon-Bloembergen equations, and the extent of the deviations will depend on the metal ion, the nature of the ligands, the stoichiometry of complexation, and the symmetry of the complex. Metal ions possessing half-filled electron shells, such as Mn(II) ($3d^5$), have little ligand field stabilization energy and, consequently, will be little affected by changes in the orientation or the nature of the ligands in the complex. The g factor for Mn(II) may exhibit a small degree of anisotropy in macromolecular complexes, but the anisotropy will be much less than for other transition metal ions. The variation in g value with ligand for some ions is reflected in the range of g values given for those ions in Table 6-1.

The effect of anisotropy in the g factor is to introduce additional frequency dispersion terms ($\omega\tau_c$ terms) to the Solomon-Bloembergen equations. These dispersion effects are most easily observed at low magnetic fields. Indeed, Koenig *et al.* (34) have found that the Solomon-Bloembergen equations give a poor description of the water proton relaxation rates for solutions of Mn(II)–carboxypeptidase A at frequencies less than 1 MHz. Such low fields are rarely used in most types of studies; however, to safely minimize possible problems with low fields, it is recommended that studies be generally carried out at fields in excess of 10 MHz for protons.

It was noted in a footnote in Section 2.4.3 that a rigorous treatment necessitates a distinction between the longitudinal (τ_{s_1}) and transverse (τ_{s_2}) electron spin relaxation times in cases where τ_s is the dominant correlation time. The correlation terms $f_1(\tau_c)$ and $f_2(\tau_c)$ (cf. Eqs. 6-13 and 6-14) are then

$$f_1(\tau_c) = \frac{3\tau_{s1}}{1 + \omega_I^2\tau_{s_1}} + \frac{7\tau_{s2}}{1 + \omega_s^2\tau_{s2}} \quad (6\text{-}25)$$

and

$$f_2(\tau_c) = 4\tau_{s_1} + \frac{3\tau_{s1}}{1 + \omega_I^2\tau_{s1}} + \frac{13\tau_{s2}}{1 + \omega_s^2\tau_{s2}} \quad (6\text{-}26)$$

It was also pointed out in the footnote that for real systems the distinction between τ_{s_1} and τ_{s_2} need rarely be considered.

Rubenstein *et al.* (35) considered relaxation in solutions of transition metal ions with spin $S = 3/2$ and $5/2$ and noted that, for example, a spin 5/2 ion, such as Mn(II), will actually have three different τ_s values if the ligand field around the ion is assymetric. Unless detailed information about the electronic environment of the ion is desired, the effective τ_s (a weighted average of the three individual τ_s values) will generally suffice.

In certain cases, τ_r may be the dominant correlation time. It has been assumed that the rotational motion corresponding to τ_r is isotropic. Anisotropy in the rotational motion has been studied for small molecules (e.g., 36, 37). In general, it is expected that anisotropy in the rotational motion also exists for macromolecular complexes, leading to more than one correlation time. In a study of water proton relaxation rates in Mn(II)–tRNA solutions, Peacocke *et al.* (21) noted that the Mn(II)–^{1}H dipolar interaction energy may be decreased by as much as 3/5 if the Mn(II) ion is bound to the macromolecule by only a single bond, and there is rapid rotation of the aquo complex Mn(II)$(H_2O)_5$ about the Mn(II)–phosphate bond. A reduction in the dipolar energy means that the dipolar terms of the Solomon-Bloembergen equations (i.e., the first terms in Eqs. 6-13 and 6-14) should be effectively multiplied by a factor between 2/5 and 1, depending on the efficiency of the reduction. Slower motions with a characteristic τ_r modulate the residual dipolar interaction, leading to the observed relaxation. This problem of anisotropic motion could have a particularly deleterious effect on attempts to determine n, the number of water molecules in the coordination sphere of the metal ion, for Mn(II) complexes with nucleic acids. It might be presumed, however, that complexation of Mn(II) with most other macromolecules (principally enzymes) involves more than one bond between Mn(II) and the macromolecule, thus vitiating a dipolar coupling reduction caused by rapid rotation about a single Mn(II)–macromolecule bond.

The use of paramagnetic probes in NMR studies has been primarily limited to investigations utilizing paramagnetic metal ions. The Solomon-Bloembergen equations should also be applicable to studies involving nitroxide spin labels. This has been illustrated with a spin-labeled nicotinamide adenine dinucleotide analog used as a probe to perturb protons on substrates and inhibitors of alcohol dehydrogenase (38) and by the effect of spin-labeled ATP bound to DNA polymerase (*Escherichia coli*) on AMP (39). The use of "natural" free radicals, such as flavin adenine dinucleotide or some of the vitamin semiquinones, could also yield useful information. The major problem with most "natural" free radicals is that the unpaired electron is delocalized over several atoms. In such cases it would be virtually impossible to obtain quantitative distance data. The investigations employing artificial nitroxide spin labels have the benefit that a substantial fraction of the unpaired electron is localized on the nitrogen atom as well as the advantage of application to systems not naturally having an unpaired electron. The disadvantage varies in severity with the system, being dependent on the degree of distortion from the natural system caused by the presence of the spin label.

Mn(II) is very efficient in promoting nuclear relaxation. Other paramagnetic probes, however, may not be as efficient, and it may be necessary to use a probe concentration within an order of magnitude of the relaxing nucleus' concentration to obtain a measurable change in relaxation rate. In such cases, where p is not much less than 1, Eqs. 6-10, 6-11, and 6-12 will not be valid. It will then be necessary to employ the equations in Sections 2.6 or 6.1 appropriate to the exchange conditions: slow, intermediate, or rapid (e.g., equations 2-83 and 6-3 for fast exchange).

We have assumed that all the paramagnetic ions in solution are in the macromolecular complex. Knowledge of the concentrations of all species containing the paramagnetic ion is essential for distance determinations, hydration number determinations, and studies of ligand exchange kinetics. In the following section, methods for determination of the dissociation constants using water proton relaxation rate measurements are presented.

6.2.2. Water Proton Relaxation Rates

Measurements of the spin–lattice relaxation rate of water protons were among the first NMR studies of biochemical systems (6, 7). Water proton relaxation rate (PRR) investigations have continued to provide useful information about macromolecular complexes containing paramagnetic metal ions. Water PRR measurements have no great sensitivity requirements, are readily amenable to temperature and frequency dependence studies, and require only relatively inexpensive pulsed NMR equipment.

Although $1/T_{1p}$ is equally meaningful, the enhancement of the water PRR was defined early and has seen continued usage (6):

$$\epsilon = \frac{1/T_{1p}}{(1/T_{1p})_0} \tag{6-27}$$

$1/T_{1p}$ is the increment in the relaxation rate caused by the paramagnetic ion [$T_{1p}^{-1} = T_1^{-1} - (T_1^{-1})^0$ with $(T_1^{-1})^0$ being the relaxation rate in the absence of paramagnetic ions] in the macromolecule-containing solution and $(T_{1p}^{-1})_0$ is similarly defined for a solution containing no macromolecule or other nonwater ligand. Therefore, the enhancement of freely aquated paramagnetic ion is unity by definition. In the following discussion the enhancement ϵ is often used. However, the numerator of Eq. 6-27, $1/T_{1p}$, can be interchanged with ϵ (provided all ϵ values are exchanged with $1/T_{1p}$ values) in any equation; the denominator of Eq. 6-27 has no intrinsic significance.

A. *Dissociation Constants and Stoichiometry*

The basis for determination of dissociation constants is that the water PRR in solutions containing paramagnetic ions generally varies with addition of macromolecules or ligands. If, for example, Mn(II) is bound to a macromolecule, the binary complex will have a characteristic value of the enhancement, ϵ_b, which differs from a value of 1 characteristic of free Mn(II) in water. Addition of another ligand to form a ternary complex will yield a value ϵ_t characteristic of that particular ternary complex. In general, $\epsilon_t \neq \epsilon_b$. The value of ϵ_b or ϵ_t is characteristic of the environment around the metal ion and therefore contains structural information. In addition to their use as a parameter for obtaining dissociation constants, changes in ϵ are interpreted in terms of alteration in the structure of the metal ion complex.

In the following discussion, it will be assumed that the macromolecule is an enzyme and the paramagnetic ion is Mn(II). Modification for other macromolecules or ions is straightforward.

Binary Complexes. The dissociation constant for a metal ion from a site on an enzyme is

$$K_D = [\mathrm{E}][\mathrm{M}]/[\mathrm{EM}] \tag{6-28}$$

where [E], [M], and [EM] represent the concentrations of enzyme binding site, metal ion, and enzyme–metal complex, respectively. The total or analytical concentration of metal ion is

$$C_{\mathrm{M}} = [\mathrm{M}] + [\mathrm{EM}] \tag{6-29}$$

The enhancement observed for any metal ion–macromolecule solution will be a weighted average of the enhancements for water at two different sites, ϵ_b for water around the metal ion in the binary complex and 1 for water around the freely aquated metal ion:

$$\epsilon = \frac{[\mathrm{M}]}{C_{\mathrm{M}}} + \frac{[\mathrm{EM}]}{C_{\mathrm{M}}}\,\epsilon_b \tag{6-30}$$

By suitable combination of Eqs. 6-28, 6-29, and 6-30, we obtain the expression for K_D:

$$K_D = \frac{[mC_{\mathrm{E}}(\epsilon_b - 1) - C_{\mathrm{M}}(\epsilon - 1)](\epsilon_b - \epsilon)}{(\epsilon - 1)(\epsilon_b - 1)} \tag{6-31}$$

where C_{E} is the total enzyme concentration with m binding sites per molecule. To determine K_D, it will be necessary to know ϵ_b and m. Mildvan and Cohn (40) have presented three methods for determining ϵ_b: (1) titration of metal ion with protein and extrapolation of measured values of ϵ to infinite protein concentration, using a double reciprocal plot; (2) titration of protein with metal ion and extrapolation of measured values of ϵ to zero metal ion concentration, yielding a lower limit to ϵ_b; and (3) measurement of ϵ for a single concentration of metal ion and protein in conjunction with determination of the amount of free metal ion from the EPR spectrum, giving ϵ_b from Eq. 6-30. A value for m was obtained using a Hughes-Klotz plot (i.e., $C_{\mathrm{M}}/[\mathrm{EM}]$ vs $1/[\mathrm{M}]$). An example of a graphical method is shown by Fig. 6-10, where the value $\epsilon_b = 11.5$ is obtained by extrapolation to infinite bovine serum albumin (BSA) concentration and, assuming formation of a 1:1 complex, a value of $K_D = 31\ \mu M$ is obtained from the BSA concentration at an enhancement halfway between 1 and ϵ_b (40).

The problems associated with measurement of K_D are simply those inherent in basic binding theory. Appropriate methods of graphical analysis have recently been discussed by Deranleau (41, 42), and by Klotz and Hunston (43). In light of these analyses, it appears that use of a Scatchard plot in conjunction with a protein titration entailing PRR measurements

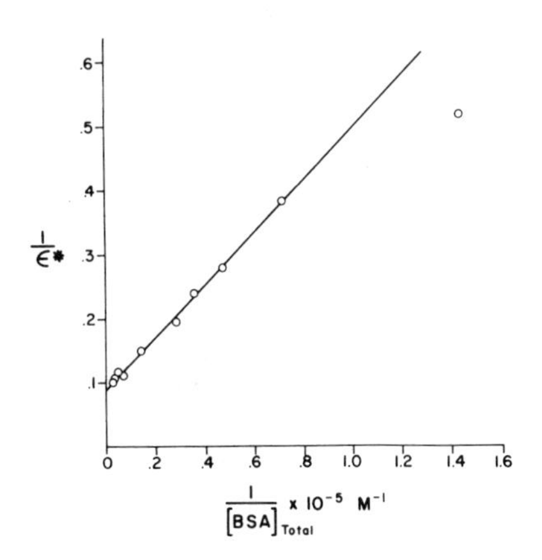

FIG. 6-10. Double reciprocal plot for the titration of Mn(II) with bovine serum albumin (BSA). The solutions contained 0.05 mM $MnCl_2$, 0.5 M $(CH_3)_4NCl$, and pH 7.5, 50 mM Tris buffer (40).

is the preferred procedure. The data analyzed should cover the range from 20–80% of completion of the titration. The Schatchard equation,

$$[EM]/C_E[M] = m/K_D - [EM]/C_EK_D \tag{6-32}$$

may be obtained directly from eq. 6-31:

$$\frac{\epsilon - 1}{C_E(\epsilon_b - \epsilon)} = \frac{m}{K_D} - \frac{C_M(\epsilon - 1)}{C_EK_D(\epsilon_b - 1)} \tag{6-33}$$

A value for ϵ_b is obtained independently via a double reciprocal plot as explained above. Plotting $(\epsilon - 1)/C_E\ (\epsilon_b - \epsilon)$ vs $C_M(\epsilon - 1)/C_E(\epsilon_b - 1)$ gives a straight line with an intercept on the abscissa of m and an intercept on the ordinate of m/K_D with a slope $-1/K_D$. A variation of this procedure with a Mn(II) titration using the Scatchard equation (Eq. 6-32) and Mn(II) EPR spectra to determine the free Mn(II) concentration (44) is shown in Fig. 6-11.

Ternary Complexes. We will designate ternary complexes as those composed of a paramagnetic ion, a macromolecule, and an additional ligand. Again we consider the macromolecule to be an enxyme and the additional ligand to be a substrate. In solutions of Mn(II) containing enzyme and substrate there may be four manganese-containing species with characteristic enhancements: free Mn(II) (1), enzyme–Mn(II) (ϵ_b), substrate–Mn(II) (ϵ_a), and enzyme–Mn(II)–substrate (ϵ_t). To simplify the equilibria as much as possible, it is presumed that the pH is adjusted so that various protonated complexes need not be taken into consideration.

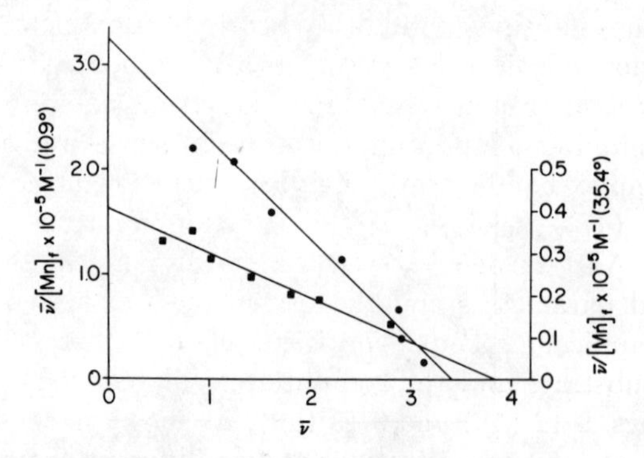

FIG. 6-11. Scatchard plots for the binding of Mn(II) to pyruvate kinase. The concentrations are: 0.0429 mM enzyme, 0.1 M $(CH_3)_4NCl$, 40–335 μM $MnCl_2$. (●), 10.9°C (left scale) and (■), 35.4°C (right scale). $\bar{\nu} = [EM]/C_E$ (23).

The measured water PRR enhancement will therefore be a weighted average of all four species,

$$\epsilon = ([M] + [MS]\epsilon_a + [EM]\epsilon_b + [EMS]\epsilon_t)/C_M \tag{6-34}$$

Concentration conservation relationships are

$$C_M = [M] + [MS] + [EM] + [EMS] \tag{6-35}$$

$$C_E = [E] + [EM] + [ES] + [EMS] \tag{6-36}$$

$$C_S = [S] + [MS] + [ES] + [EMS] \tag{6-37}$$

The pertinent dissociation constants, valid for a particular specified pH, are:

$$K_1 = [M][S]/[MS] \tag{6-38}$$

$$K_s = [E][S]/[ES] \tag{6-39}$$

$$K_D = [E][M]/[EM] \tag{6-40}$$

$$K_2 = [E][MS]/[EMS] \tag{6-41}$$

$$K_3 = [EM][S]/[EMS] = K_1K_2/K_D \tag{6-42}$$

$$K_A = [ES][M]/[EMS] = K_1K_2/K_S \tag{6-43}$$

where [E] represents the concentration of free enzyme binding sites. The values of the parameters for the binary EM (ϵ_b, K_D), MS (ϵ_a, K_1), and perhaps ES (K_s) complexes are determined in separate experiments.

Graphical methods, varying somewhat depending on whether $\epsilon_t < \epsilon_b$ or $\epsilon_t > \epsilon_b$, for estimating the enhancement and binding parameters of the ternary complex have been presented (45, 46). The numerous complications arising from the multiple equilibria dictate a nonlinear least-squares analysis to accurately fit the parameters to the observed data without making *a priori* assumptions concerning linearity. One approach has been to employ a computer adaptation of the usual successive approximation method for complex equilibria to analyze a titration of enzyme–Mn(II) with substrate (47). Analysis of the titration data for the adenylate kinase–Mn(II)–ADP system showed a large disparity between the computer-calculated parameters and the parameters calculated by the more approximate graphical methods (47). Examples of the titration of enzyme–Mn(II) with substrate and the computer-calculated values of ϵ_t and K_2 are shown in Figs. 6-12 (48) and 6-13 (49). As shown in those figures, it is necessary to do at least two titrations at two different enzyme concentrations to obtain a unique fit of the parameters to the titration data. It will be noted in both Figs. 6-12 and 6-13 that the measured enhancement initially increases as ligand is added. This is characteristic for a case where $\epsilon_t > \epsilon_b$. The observed decrease in ϵ with addition of either ADP or ATP in excess of the enzyme binding site concentration is caused by competition between the nucleotide and binary Mn(II)–nucleotide for the enzyme binding site. The decrease in ϵ results as the free ligand pushes Mn(II)–nucleotide off the ternary enzyme complex.

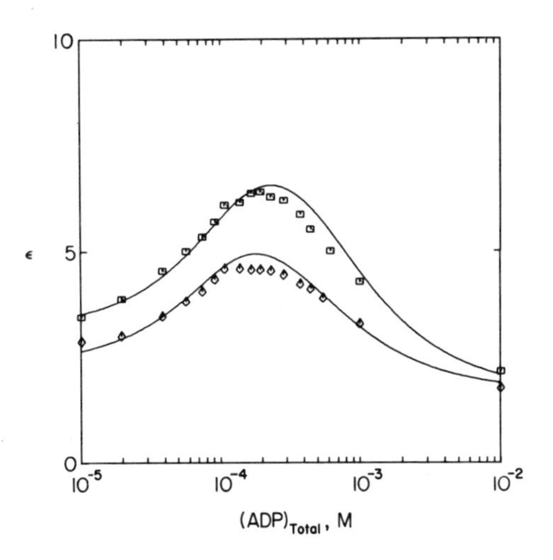

FIG. 6-12. Water PRR (24.3 MHz) titration of dansylated creatine kinase (0.30 mM sites, upper curve and 0.15 mM, lower curve) and manganese acetate (0.10 mM) at 24°C. There are two sites per molecule of enzyme. The theoretical curves represent the parameters $K_1 = 30.0\ \mu M$, $K_s = 0.112$ mM, $K_D = 0.40$ mM, $K_2 = 70.0\ \mu M$, $\epsilon_a = 1.7$, $\epsilon_b = 6.5$, and $\epsilon_t = 10.0$ (48).

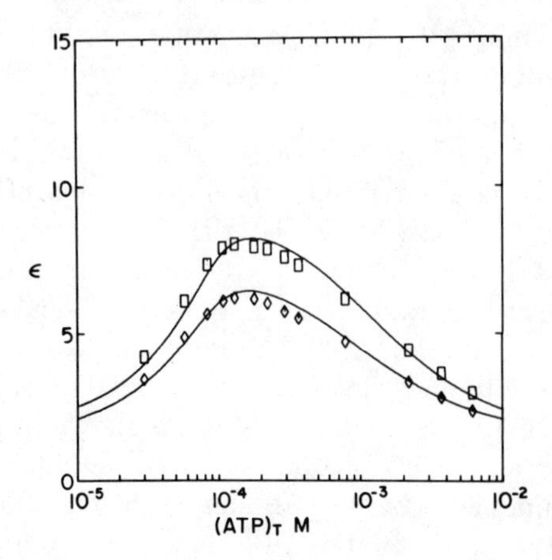

FIG. 6-13. Water PRR (24.3 MHz) titration of formyltetrahydrofolate synthetase (0.196 mM sites, upper curve; and 0.118 mM sites, lower curve) and $MnCl_2$ (0.10 mM) with ATP at 25°C. Four sites per molecule of enzyme are assumed. The theoretical curves shown represent the parameters $K_1 = 10.0\ \mu M$, $K_s = 0.316$ mM, $K_D = 2.3$ mM, $K_2 = 79.4\ \mu M$, $\epsilon_a = 1.6$, $\epsilon_b = 11.2$, and $\epsilon_t = 13.5$. The percent standard deviation of the ϵ_t values for the analysis is 5.3% (49).

Use of the method of Marquardt (50), which combines the Gauss-Newton (Taylor series) method and the method of steepest descent, for nonlinear least-squares analysis offers an improvement over the standard successive approximation method for calculating parameters from titration data.

Quarternary and Higher Complexes. In the case of many bimolecular reactions catalyzed by a metal-activated enzyme, a quaternary complex incorporating two substrates must be considered. The PRR enhancement for a solution containing the quarternary complex will, for the general case, entail weighting the individual relaxation enhancements 1, ϵ_a, ϵ_a', ϵ_b, ϵ_t, ϵ_t' and ϵ_q by the relative concentrations of the metal ion-containing species: M, MS, MS′, EM, EMS, EMS′, and EMSS′. In principle, the method of nonlinear least squares can be applied to the multiple equilibria involved with a quaternary complex. It should be kept in mind, however, that less exact values of the parameters are obtained as the number of unknown parameters is increased. A compromise may be obtained for some particular system in which the concentrations may be adjusted such that certain species and certain dissociation constants can be neglected, thus simplifying the calculation. For example, if the two substrate concentrations are sufficient to saturate the enzyme, both ternary complexes and the binary complex may be neglected.

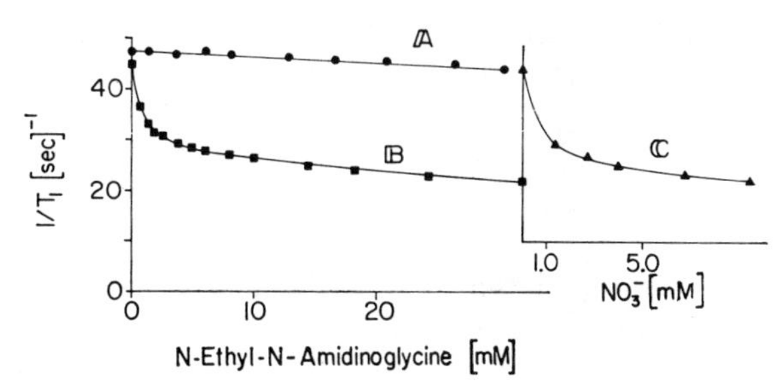

FIG. 6-14. PRR of MnADP–creatine kinase as a function of N-ethyl-N-amidinoglycine concentration in the presence and absence of nitrate. Curve A: manganese acetate, 0.50 mM; ADP, 0.55 mM; creatine kinase, 0.61 mM in sites; potassium HEPES, 50 mM; potassium acetate, 50 mM; pH 8.0; T = 1°C. Curve B, manganese acetate, 0.48 mM; KNO_3, 9.6 mM; potassium HEPES, 50 mM; potassium acetate, 50 mM; pH 8.0; T = 1°C. The inset (Curve C) shows a titration with KNO_3 of MnADP–creatine kinase in the presence of N-ethyl-N-amidinoglycine. Solutions contained manganese acetate, 0.50 mM; ADP, 0.55 mM; creatine kinase, 0.61 mM in sites; N-ethyl-N-amidinoglycine, 32 mM; potassium HEPES, 50 mM; potassium acetate, 50 mM; pH 8.0; T = 1°C (51).

Approximate graphical methods for calculating the dissociation constant of a substrate from a quaternary complex have been employed (46, 51). One such determination for the dissociation of the weak substrate N-ethyl-N-amidinoglycine from the quaternary complex formed in the presence of nitrate with creatine kinase–Mn(II)–ADP is illustrated in Figs. 6-14 and 6-15 (51). The titration data are shown in curve B of Fig.

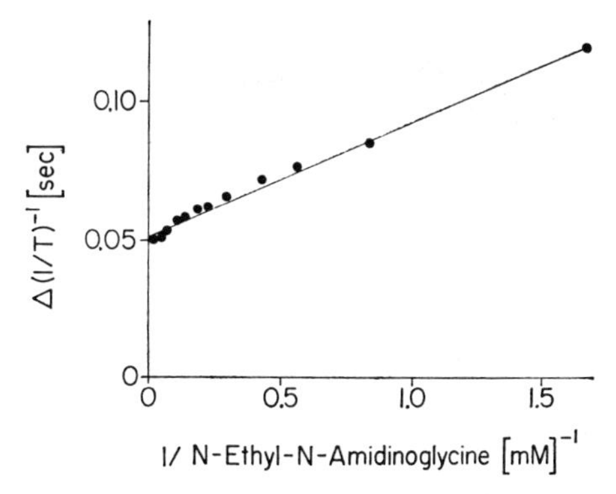

FIG. 6-15. Effect of N-ethyl-N-amidinoglycine on the PRR of the MnADP–creatine kinase complex in the presence of nitrate. A double reciprocal plot of the difference between the observed relaxation time and that of the ternary complex, $\Delta(1/T)$, versus substrate concentration; points were derived from data presented in Fig. 6-14, curve B, corrected as described in the text (51).

6-14. At the high ligand concentrations employed, the water PRR tended to continually decrease with increasing ligand concentration as indicated in curve A of Fig. 6-14. Correction for this "drift" was made by subtracting curve A from curve B and using the difference in a double reciprocal plot, as shown in Fig. 6-15, to determine the approximate dissociation constant.

In some other enzyme systems, higher complexes are conceivable. The analysis of these complexes is difficult indeed.

One unanswered question yet remains. In work with enzymes, it is not unusual for studies to be carried out at activities of, e.g., 60–80% maximum. For these dissociation constant determinations, is it appropriate to use the entire enzyme (protein) concentration, or should the concentration used in the calculations be weighted by the activity? This question has not been answered; however, general practice has been to use the entire protein concentration.

B. Hydration Numbers

It was mentioned above that it is possible to calculate n, the number of water molecules in the first coordination sphere of the metal ion, using Eqs. 6-10 and 6-13 if it is established that rapid exchange prevails and if a value for τ_c may be estimated. In the case of Mn(II), a value for r, the ion–proton internuclear distance, of 2.86 Å is generally used. It is possible to set limits of $2.815\ \text{Å} < r < 2.923\ \text{Å}$ from crystallographic data on octahedrally coordinated Mn(II) hydrates (52–55). Mn(II) is capable of forming tetrahedral complexes in nonaqueous environments (56) (the interior of a macromolecule may be such an environment). If Mn(II) is in a tetrahedral complex, the value for $r = 2.86\ \text{Å}$ is no longer correct. It would appear, however, that even in the case of macromolecule complexes of Mn(II), the metal ion coordination is probably octahedral. For example, the optical spectra of pyruvate kinase–Co(II) complexes reveal the metal ion site to be octahedral (57) and Mn(II) has much less propensity for tetrahedral coordination than does Co(II) (56).

The value of n may be used as an indication of the accessibility of the enzyme active site to solvent. By subtracting n from the octahedral coordination number of 6, it is possible to estimate the number of liganding groups on the macromolecule that are coordinated to Mn(II). Changes in n on addition of another ligand to the macromolecule–metal ion complex may, with such other information as τ_c values for the binary and ternary complexes, indicate whether the additional ligand is coordinated to the metal ion or to another group on the macromolecule. The change in n may provide evidence for a substrate-induced conformational change in the case of ternary and quarternary enzyme complexes (18).

The number of exchangeable water molecules in the Mn(II) coordination sphere has been estimated to be 1 for carbonic anhydrase–Co(II) (58), 10 for RNA-Co(II) (59), 1 for Mn(II)–carboxypeptidase A (34), 3 for pyruvate kinase–Mn(II) (23), and 0.2–0.5 for pyruvate kinase–Mn(II)–phosphoenol pyruvate (18). Although integral values of n might be expected, there are several plausible reasons for obtaining a fractional value for n (18).

6.2.3. Distance Determinations

The spatial relationship of ligands with respect to the paramagnetic ion in ternary or higher macromolecular complexes may be examined using Eqs. 6-10 and 6-13 or Eqs. 6-11 and 6-14. Of course, small binary complexes can also be examined. If it is established that the relaxation rate for a nucleus on the ligand is not exchange limited and if a suitable value for τ_c is obtained through proper analysis of the temperature and frequency dependence of the relaxation rate, a value for r, the distance from the nucleus to the paramagnetic ion, may be calculated. Use of the spin–lattice relaxation rate with Eqs. 6-10 and 6-13 is recommended, although the spin–spin relaxation rate with Eqs. 6-11, or 6-12, and 6-14 may be used with some caution. $1/T_{2p}$ is more likely to be exchange limited and there is a greater probability that the hyperfine interaction term will contribute to $1/T_{2p}$, whereas the hyperfine term can be neglected for $1/T_{1p}$ in Mn(II) complexes. If $1/T_{2p}$ is used and the hyperfine term is neglected, the calculated value of r should be considered a lower limit. For both $1/T_{1p}$ and $1/T_{2p}$, if there is some contribution from exchange to the relaxation rate, the calculated value of r, neglecting exchange, will be an upper limit. This loss in structural information may be compensated by the gain in kinetic information obtainable.

If a knowledge of τ_c is lacking, relative distances may still be obtained for different nuclei in the same complex. Some researchers have used this approach to compare relative distances in different complexes, e.g., comparison of an EMS (enzyme–metal ion–substrate) complex with an EMSS′ complex (having a second substrate in the complex). This can be misleading because τ_c may be different in the two complexes. The comparison should generally be avoided unless evidence is available showing that τ_c is the same in the two complexes.

An example showing the calculated distances from Mn(II) to the different protons of α-(dihydroxyphosphinylmethyl)acrylate, a weak competitive inhibitor of the pyruvate kinase-catalyzed phosphoryl transfer reaction, in the ternary complex with pyruvate kinase–Mn(II) is given in Fig. 6-16 (13). A correlation time of 1×10^{-9} sec was calculated from the

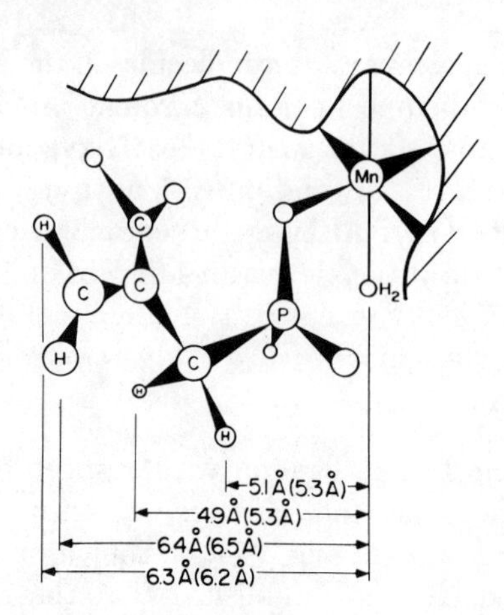

FIG. 6-16. Structure of the pyruvate kinase–manganese(II)–α-(dihydroxyphosphinylmethyl)acrylate complex. Distances given are measured from the molecular model with experimentally determined values given in parentheses (13).

frequency variation of $1/T_{1p}$ between 220 and 60 MHz for each of the resonances and also from the T_{1p}/T_{2p} ratios of the vinyl proton resonances. A molecular model was constructed using the experimental distances as a guideline. The distances given in Fig. 6-16 were measured from the molecular model, with the experimentally determined values reported in parentheses. A more sophisticated, although not necessarily superior, approach is to use the experimental distances in a computer program that matches sterically possible conformations with the experimental data and provides a display of the most probable conformation.

The facility for obtaining a fairly accurate "picture" of macromolecular complexes has been advanced considerably by the recent advent of Fourier transform NMR methods for selective measurement of relaxation times (cf. Section 5.7.3). That the "pictures" of macromolecular complexes obtained using less accurate CW relaxation time measurements can be considered at all is because of the r^{-6} dependence in the Solomon-Bloembergen equations. For example, if a 100% error is made in the estimation of τ_c, i.e., the true τ_c is twice as large as the estimated τ_c, the maximum error in the r calculation is 12%. If τ_c possesses a value near the maximum in the $f_1(\tau_c)$ curve (cf. Fig. 6-1), the error in r may be much less than 12%.

One of the virtues of the NMR method for calculating distances is that the complexes studied are in solution and therefore have direct biological significance, in contrast to studies of solid-state complexes using x-ray techniques. Nevertheless, the x-ray diffraction results might be expected to corroborate the NMR measurements in a majority of cases. Table

6-4 (8, 60-63) shows a comparison of the distances calculated from the NMR method with the distances obtained from x-ray diffraction for those few cases where a comparison can be made. The agreement between the NMR results and the x-ray diffraction results is fairly good.

For reliable distance determinations in macromolecular complexes, one feature is essential. The solutions used for the NMR studies must contain the paramagnetic ion almost exclusively in the macromolecular complex. In general, this means that knowledge of the various dissociation constants must be available. For example, in the determination of metal ion-to-substrate nuclei internuclear distances in the ternary EMS complex, the presence of a significant amount (a few perent) of MS complex in solution may be very detrimental to reliable distance determinations in the EMS complex. However presence of a few percent of EM complex will have little effect on the distance measurements. The degree to which presence of an MS complex perverts distance measurements for the EMS complex depends on the relative distances in the MS and EMS complexes and on the frequency at which measurements are made.

TABLE 6-4

COMPARISON OF DISTANCES FROM THE PARAMAGNETIC CENTER TO A LIGAND NUCLEUS USING THE NMR RELAXATION METHOD (FOR SOLUTIONS) AND X-RAY DIFFRACTION (FOR SOLIDS)

COMPLEX		DISTANCE (Å)		
		NMR	X-RAY	REF.
Gd(III)-lysozyme-β-MeNAG	Gd(III)-$H_{acetamido}$	6.5 ± 0.3	7.3 ± 0.8	60
	Gd(III)-$H_{glycosidic}$	5.4 ± 0.3	4.6 ± 0.6	60
Lysozyme (spin labeled)-diNAG	Spin label near (Tryp 123)	>15	18	61
	To acetamidomethyl protons (subsites B and C)	>15	28	61
Mn(F)	Mn-F	2.1 ± 0.2	2.1	62
Mn(FPO_3)	Mn-F	5.0 ± 0.8	4.4 ± 0.4	62
Mn (urocanate)	Mn-imidazole C2	3.4 ± 0.2	3.27	8
	Mn-imidazole C5	3.5 ± 0.2	3.24	8
Mn (imidazole)	Mn-imidazole C2	≥ 3.1	3.27	8
	Mn-imidazole C5	3.4 ± 0.2	3.24	8
MbFe(III) (F)	Fe-H	2.9 ± 0.1	2.9-3.4	63
	Fe-F	< 6.1	1.92	63
MbFe(III) (N_3H)	Fe-H	3.1 ± 0.1	2.7-3.6	63

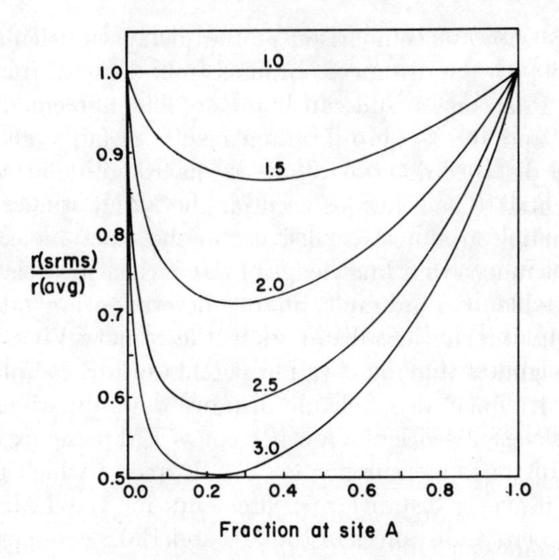

FIG. 6-17. Effect of the fraction of nuclei at sites A and B on the ratio of the average radius calculated according to the Solomon-Bloembergen equation (sixth root mean sixth) to the weighted average radius, where the distance from the paramagnetic center to a nucleus at site A or site B is r_A and r_B, respectively. The five curves differ for $r_B/r_A =$ 1.0, 1.5, 2.0, 2.5, and 3.0. Equal correlation times are assumed for the two sites.

For purposes of illustration the question may be posed: What is the distance calculated via the Solomon-Bloembergen equation for a solution containing 95% of the paramagnetic metal ion in the EMS complex, with the substrate nucleus having a correlation time of 3.0×10^{-9} sec and being 6.0 Å from the metal ion and containing 5% in the MS complex with the substrate nucleus having a correlation time of 3.0×10^{-11} sec and an internuclear distance of 2.0 Å. A simple weighted average would give 5.8 Å. However, the distances will actually enter as a weighted sum with terms $Bf_1(\tau_c)/r^6$ giving a calculated distance of 5.4 Å for data at 60 MHz and 5.5 Å at 220 MHz. The calculated distance would have been smaller had it not been for the difference in τ_c for the two complexes. The case for equal correlation times at both site A and site B is illustrated in Fig. 6-17, which shows the ratio of the distance calculated via the Solomon-Bloembergen equation (sixth root mean sixth*) to the weighted average distance as a function of the fraction of metal ions in complex A and the

* I.e., the distance calculated is a mean of the sixth roots obtained from the Solomon–Bloembergen equation for all complexes where the distance for any particular complex enters the Solomon–Bloembergen equation for that complex and is raised to the sixth power.

ratio r_B/r_A. An example of this case is the existence of two ternary complexes with about the same τ_c but possessing two different ion–nucleus distances, $r_A \leq r_B$. If only 3% of the metal ion is in complex A with r_A = 3.0 Å and 97% is in complex B with r_B = 7.5 Å, the weighted average distance is 7.4 Å. However, according to Fig. 6-14, the distance calculated from the NMR data will be 5.2 Å. The r^{-6} factor gives considerable leverage to the complex with a smaller internuclear distance. This averaging should be kept in mind particularly for systems in which the metal ion has no specific biochemical role and, therefore, could conceivably have more than one complex [cf. Morallee *et al.* 60)], or for systems containing a small excess of paramagnetic species [cf. Wien *et al.* (61)].

6.2.4. Ligand Exchange Kinetics

In those cases where the relaxation rate is exchange limited, kinetic information may be obtained in lieu of structural information. Because $1/T_{2p}$ will become exchange limited before $1/T_{1p}$, linewidth or chemical shift measurements are generally most suitable for exchange rate studies. For ions possessing long τ_s values, e.g., Mn(II), a value for the lifetime of the complex τ_M may be obtained using Eq. 6-11, with the assumption that the outer sphere contribution is negligible. Equation 6-12 is used for a complex containing an ion with a short τ_s value, e.g., Co(II). Proper analysis generally requires temperature- and frequency-dependent measurements of the relaxation rates (or linewidths) and chemical shifts to establish that τ_M indeed does dominate T_{2M} or T_{1M}, and the measured relaxation rate is determined by the exchange rate. The analysis would show, for ions such as Co(II), which of Eqs. 6-7, 6-8, and 6-9 is applicable. In certain careful studies, it may be possible to extract both structural and kinetic information from the relaxation rate data. If the relaxation rate is not entirely dominated by chemical exchange or in the absence of a thorough analysis, the observed relaxation time places an upper limit on τ_M and, consequently, a lower limit on the exchange rate.

For illustrative purposes we may consider the dissociation of a ternary enzyme complex:

$$\mathrm{EMS} \underset{k_{\mathrm{ass}}}{\overset{k_{\mathrm{diss}}}{\rightleftharpoons}} \mathrm{EM} + \mathrm{S}$$

If nuclei on the substrate S are monitored, it may be possible to determine τ_M which, in this case, can be identified with $k_{\mathrm{diss}} = 1/\tau_M$. The dissociation constant

$$K_3 = k_{\mathrm{diss}}/k_{\mathrm{ass}} = 1/(\tau_M k_{\mathrm{ass}}) \tag{6-44}$$

K_3 may be determined independently; therefore, a value for k_{ass} may also be obtained. It should be noted that for NMR the exchange rates are determined at equilibrium, so Eq. 6-44 is strictly valid, unlike in other techniques, which deal with nonequilibrium exchange. In general, the exchange rate $1/\tau_M$ must be examined as a function of concentration. Concentration dependence may determine whether the ligand exchange is a uni- or bimolecular process and will enable a rate constant to be calculated. Details of the mechanism, including establishing which end of a multidentate ligand dissociates first, may be elucidated by study of different nuclei on the same ligand.

Reviews of NMR applications in the investigation of ligand exchange rates in small complexes have been presented (64, 65). A list of estimated values of $1/\tau_M$ for several macromolecular complexes containing paramagnetic probes has also been tabulated (8). The investigation of various ternary inhibitor complexes with Co(II)–carbonic anhydrase C, performed by Taylor *et al.* (27), may serve as a model for future studies. The rate constants for exchange of the inhibitors, given in Table 6-5, were calculated following temperature, frequency, and concentration dependence studies of the linewidths and chemical shifts of nuclei on the inhibitors.

Gupta and Redfield (66) have taken advantage of NMR double resonance to calculate the association and dissociation rate constants for azide ions associating with ferricytochrome c. The lifetime of the complex was determined from T_1 and signal intensity measurements when the proton resonance of one of the porphyrin ring methyl groups of azidoferricytochrome c was monitored in the absence and presence of a second strong irradiating field at the frequency corresponding to the methyl group in ferricytochrome c. The degree of saturation transfer between ferricyto-

TABLE 6-5

ESTIMATED ASSOCIATION AND DISSOCIATION RATE CONSTANTS FOR VARIOUS INHIBITOR COMPLEXES OF Co(II)–CARBONIC ANHYDRASE C AT 25°C[a]

LIGAND	K_{diss} (M)	k_{ass} (M^{-1} SEC^{-1})[b]	k_{diss} (SEC^{-1})[c]
Formate	1.5×10^{-4}	3.9×10^{8}	6.0×10^{4}
Monofluoroacetate	9.1×10^{-4}	2.2×10^{8}	2.0×10^{5}
Difluoroacetate	5.3×10^{-4}	1.9×10^{8}	1.0×10^{5}
Trifluoroacetate	7.7×10^{-4}	2.0×10^{8}	1.5×10^{5}

[a] From Taylor *et al.* (27).

[b] Calculated from k_{diss}/K_{diss}.

[c] Determined from lifetime calculated from NMR measurements and extrapolation to 25°C.

chrome c and azidoferricytochrome c is a function of the lifetime of the two states and is therefore a function of the azide exchange rate.

6.3. Quadrupolar Nuclei

A nucleus with a nuclear spin $>\frac{1}{2}$ possesses a quadrupole moment eQ. Interaction of the quadrupole moment with the fluctuating electric field gradient at the nucleus generally furnishes the dominant relaxation process for quadrupolar nuclei as discussed in Section 2.4.4. Examples of quadrupolar ions, each having a spin of 3/2, are $^{23}Na^+$, $^{39}K^+$, $^{35}Cl^-$, and $^{81}Br^-$. The interaction of these quadrupolar ions with other ions or molecules may be investigated by examining relaxation rate changes. Equation 2-63 of Section 2.4.4 describes the relaxation rate, which for a spin 3/2 nucleus, assuming an axially symmetric electric field gradient, is

$$\frac{1}{T_1} = \frac{1}{T_2} = \frac{2\pi^2}{5}\left(\frac{e^2qQ}{h}\right)^2 \tau_r \tag{6-45}$$

Relaxation rate increases caused by complexation of the quadrupolar ions can arise from changes in the electric field gradient at the nucleus, eq. This will be the predominant effect noticed in the formation of small complexes. In the case of macromolecular complexes, an additional increase in relaxation rate may be noted because the rotational correlation time τ_r generally becomes larger as the macromolecule becomes larger (cf. Section 2.3.1).

Stengle and Baldeschwieler (67) introduced the technique of using $^{35}Cl^-$ as an indirect chemical probe of the interactions of mercury and organomercury compounds with the sulfhydryl groups of hemoglobin by utilizing ^{35}Cl NMR linewidth measurements. James and Noggle (68) subsequently showed that the relatively weak interactions of certain quadrupolar ions with macromolecules could be investigated, illustrating the use with ^{23}Na NMR relaxation time measurements. NMR studies of quadrupolar nuclei in cellular systems will be discussed in Chapter 8.

The basis for use of the indirect $^{35}Cl^-$ probe technique is that Cl^- forms complexes with Zn(II) and Hg(II). For example, addition of NaCl to an $HgCl_2$ solution results in formation of an $HgCl_4^{2-}$ complex. Hg(II) and Zn(II) are also capable of accepting Cl^- as a ligand, even in the presence of macromolecular liganding groups. It therefore appears that $^{35}Cl^-$ may be used as an innocuous probe of the site where Zn(II) or Hg(II) is bound to the macromolecule. It is generally observed that (*a*) the ^{35}Cl linewidth of NaCl solutions is about 10 Hz (there is actually a small concentration and temperature dependence), (b) the linewidth increases considerably in the presence of Hg(II) or Zn(II) and is strongly dependent on concentration, and (*c*) the linewidth increases even more when a protein with free sulfhydryl groups is added to a solution containing Hg(II) and Cl^-. The linewidth for $^{35}Cl^-$ bound to the macromolecule is usually several thousand Hertz. To bring the observed linewidth into a reasonable range (<100 Hz), an excess of chloride ion is used and the effect of the macromolecule is realized by rapid exchange (cf. Eq. 6-2) of freely solvated chloride and chloride bound to the macromolecule. The method has been used to study the rotational motion of an organomercury inhibitor of α-chymotrypsin (69), the denaturation of bovine mercaptalbumin (70), and the sulfhydryl groups on hemoglobin (71).

It has also been shown that the chloride probe technique is not limited to Hg(II) and chloride binding to sulfhydryl groups. Chlorine-35 linewidths were used to study the helix–coil transition of poly-L-glutamate (72), zinc adenosine diphosphate complexes (73), and hapten–antibody interactions (74).

The ^{35}Cl NMR results for the interaction of a mercury-labeled hapten with an antibody are illustrated in Fig. 6-18. The ^{35}Cl linewidth obviously increases as the ratio of the 2,4-dinitro-4′-(chloromercuri)diphenylamine hapten to antidinitrophenyl antibody increases, reaching a maximum at a stoichiometry of two haptens per antibody. The linewidth increase in the presence of antibody is caused by an increase both in e^2qQ/h and τ_r. Most of the DNP-mercurial may be displaced by DNP–lysine according to ^{35}Cl linewidth measurements. It is interesting to note how this technique overcomes the usual sensitivity problems of NMR. It was noted (74) that

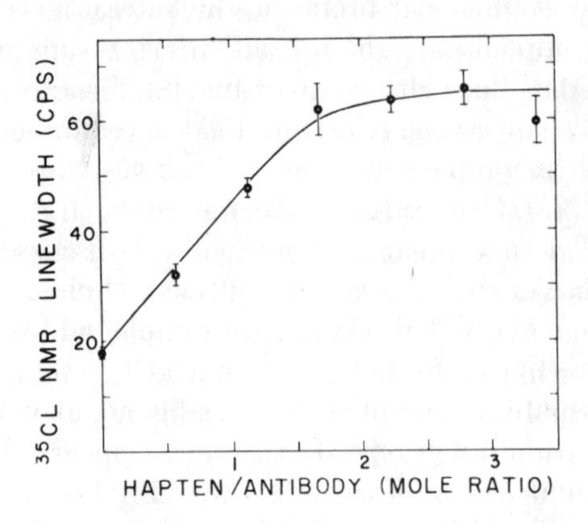

FIG. 6-18. Chlorine-35 NMR titration of the antidinitrophenyl antibody with a mercurial dinitrophenyl hapten showing changes in the observed ^{35}Cl linewidths (74).

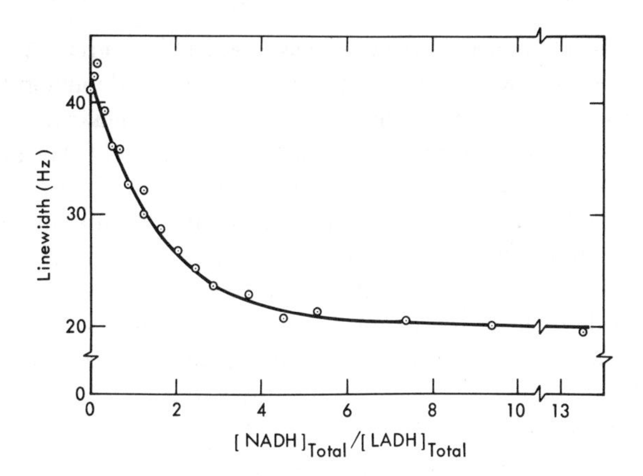

FIG. 6-19. Elimination of Zn(II)–chloride interaction on liver alcohol dehydrogenase (LADH) by addition of NADH as monitored by the ^{35}Cl linewidth. Conditions are 0.13 m*M* LADH and 0.5 *M* NaCl at pH 7.5 (76).

linewidth variations in 1 *M* NaCl could be observed with as little as 6×10^{-7} *M* antibody.

The addition of extraneous Hg(II) or Zn(II) may not be necessary for investigating certain systems. Ellis *et al.* (75) observed the interaction of chloride ion with horseradish peroxidase using ^{35}Cl NMR in the absence of Hg(II). In another ^{35}Cl NMR investigation, the Zn(II), required for enzymic activity at the active site of liver alcohol dehydrogenase was studied with $^{35}Cl^-$ as a probe (76). The titration curve in Fig. 6-19 shows that the coenzyme NADH binds to the enzyme near the Zn(II) [or perhaps directly complexes Zn(II)], preventing interaction of Zn(II) and chloride ion. Adenosine diphosphate ribose also eliminates the Zn(II)–chloride interaction, but the competitive inhibitor orthophenanthroline does not. These results were interpreted as suggesting octahedral coordination of Zn(II) on the enzyme.

The lifetime of the chloride ion bound to Hg(II) [or Zn(II)] may provide additional ^{35}Cl NMR line broadening in those situations in which rapid exchange is not attained. Ward and Cull (77) have found the lifetime of chloride bound to Zn(II) on human carbonic anhydrase B to be the dominating factor in the ^{35}Cl linewidth from an examination of the temperature dependence. In the limit of rapid exchange, it is expected that the rotational correlation time τ_r will increase (and therefore the linewidth will increase) with a decrease in temperature (cf. Eqs. 6-45 and 6-7 with $T_{2b} \gg \tau_b$). In the limit of slow exchange, the lifetime in the bound state τ_b will be long enough to dominate T_{2b} in Eq. 6-7 and will increase with a decrease in temperature, causing a decrease in linewidth. A positive dependence of ^{35}Cl linewidth on temperature for human carbonic anhydrase B suggests that chloride exchange is the dominant relaxation mechanism. The rate constant k_{off} calculated at 25°C from Eq. 6-7 was 1×10^6 sec^{-1} at pH 6.46 and 7.3×10^5 sec^{-1} at pH 8.59 (77).

A recent ^{35}Cl NMR study has shown that chloride binds to human oxy- and carbon monoxyhemoglobin with a lifetime between 10^{-8} sec and 10^{-6} sec for the complex (78). Bromine-79 and ^{81}Br T_2 measurements have also been used to determine the bromide ion probe exchange rates and correlation time estimates for bromide binding to mercury on the F9(93) β-sulfhydryl groups of horse methemoglobin at pH 7 and pH 10 (79).

Sodium ions and potassium ions generally have little capacity for complex formation. Nevertheless, weak complex formation can occur and may be important in nerve transmission, active transport across biological membranes, and maintenance of biopolymer conformations. Examination of ^{23}Na and ^{39}K NMR relaxation times as a function of pH, metal ion concentration, and ligand concentration provides a method for investigating weak complex formation.

Observation of changes in ^{23}Na spin–lattice relaxation rates as a function of pH in sodium ion solutions containing various small ligands has been used to provide a qualitative indication of the strength of sodium ion interactions, to demonstrate a chelate effect, and to identify important coordinating groups (80–82). The pH dependence of the ^{23}Na relaxation rate for a 1:1 solution of Na^+ with N'-(2-hydroxyethyl)ethylenedinitrilo-N,N,N'-triacetic acid (HEEDTA) is depicted in Fig. 6-20. The small difference in relaxation rate over the pH range from -0.3 to 6.5 can be compared with the large increase that occurs over the range from pH 8.5 to 11.5 where the last proton is lost by a nitrogen of HEEDTA. This illustrates one manifestation of the formation of a chelate with the HEEDTA trianion utilizing some sodium–nitrogen bonding (80). In contrast to the large $1/T_1$ change in Fig. 6-20, the changes in ^{23}Na relaxation rate for 1:1 solutions of sodium ion with natural amino acids are small over the entire pH range (81).

Changes in the ^{23}Na relaxation rate in an aqueous phosphatidylserine dispersion over a pH range limited at low pH by coagulation and at high pH by phosphate ester hydrolysis are shown in Fig. 6-21. The pH dependence is similar to that for phosphoserine, which may serve as a water-soluble model of phosphatidylserine; a relatively larger relaxation rate for phosphatidylserine may be ascribed to a larger correlation time for the vesicles as compared to the mobile phosphoserine molecule.

Equilibria may be investigated via the dependence of the ^{23}Na relaxation

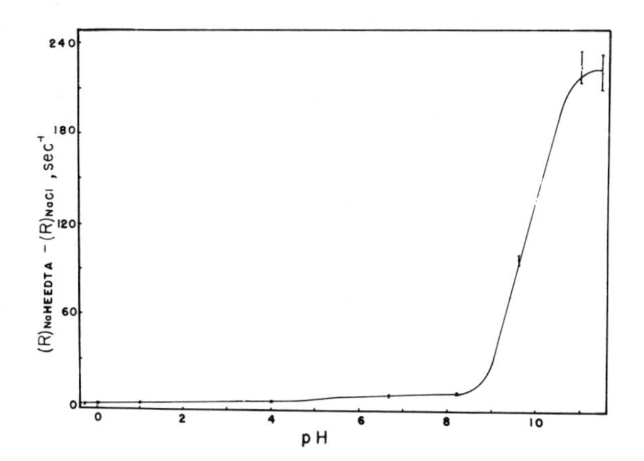

FIG. 6-20. pH dependence of $^{23}NaT_1^{-1}$ for 0.100 M NaCl–0.095 M HEEDTA solution. The pH was adjusted with concentrated HCl or $(CH_3)_4NCl$ (80).

rate on sodium ion and ligand concentration. The observed relaxation rate R $(= 1/T_1)$ is a weighted composite of the relaxation rates of the sodium ion in the "free" state R_F and in the bound state R_B if rapid equilibrium exists:

$$R = R_F X_F + R_B X_B = R_F + (R_B - R_F) X_B \tag{6-46}$$

where X_F and X_B are the respective mole fractions of "free" and bound ions.

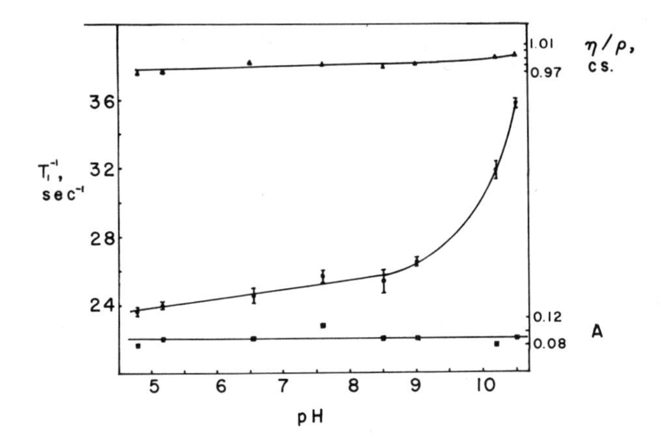

FIG. 6-21. pH dependence of the viscosity, absorbance and ^{23}Na relaxation rate of a 5.29 mg/ml phosphatidylserine–0.10 M NaCl dispersion. The pH was varied by addition of concentrated HCl or $(CH_3)_4NCl$ (82).

Equation 6-46 is simply Eq. 6-3 with a change in notation. If we assume that only one complex is formed:

$$\mathrm{MS} \overset{K_d}{\rightleftharpoons} \mathrm{M} + \mathrm{S} \tag{6-47}$$

and the dissociation constant

$$K_d = [\mathrm{M}][\mathrm{S}]/[\mathrm{MS}] \tag{6-48}$$

where [MS], [M], and [S] are the respective concentrations of complex, "free" ion, and ligand. In the case of a macromolecule, [S] is the binding site concentration (all sites assumed to be identical) and K_d is an intrinsic dissociation constant.

Defining C_M and C_S as the total concentrations of metal ion and ligand, respectively, it may be established that

$$X_\mathrm{B} = \frac{C_\mathrm{S}}{K_d + C_\mathrm{S} + C_\mathrm{M} - C_\mathrm{M}X_\mathrm{B}} \tag{6-49}$$

Equation 6-49 is a quadratic equation that may be solved iteratively.

The relaxation rate R_F can be measured in the absence of ligand. For weak complexes, however, it is usually not possible to measure R_B directly. Equations 6-46 and 6-49 must be considered to contain two undetermined parameters, K_d and R_B, with two independent variables C_M and C_S. In the case of small molecules, measurements of R ($= 1/T_1$) as a function of C_S and C_M are fitted to the mathematical model represented by Eqs. 6-46 and 6-49 using a nonlinear least-squares computer program (80). The ^{23}Na relaxation rate as a function of NaCl and HEEDTA^{-3} concentrations is shown in Fig. 6-22. The curves shown in Fig. 6-22 represent the mathematical model with "best fit" parameters $R_\mathrm{B} = 610 \pm 30$ sec^{-1} and $K_d = 0.115 \pm 0.01$ *M* at an ionic strength of about 1.0. The dependence of ^{23}Na T_1^{-1} for 0.100 *M* NaCl over a wide range of HEEDTA^{-3} concentration reveals agreement of the experimental data with the theoretical curve calculated using the parameters given above and thus corroborates the two-parameter model and the assumed 1:1 stoichiometry.

Macromolecular complexes may also be treated in the same manner. However, it has been found convenient to use a graphical method of analysis for macromolecules (68). Equations 6-46 and 6-49 may be combined to give

$$R - R_\mathrm{F} = \frac{C_\mathrm{S}(R_\mathrm{B} - R_\mathrm{F})}{K_d + C_\mathrm{M} + C_\mathrm{S} - [\mathrm{MS}]} \tag{6-50}$$

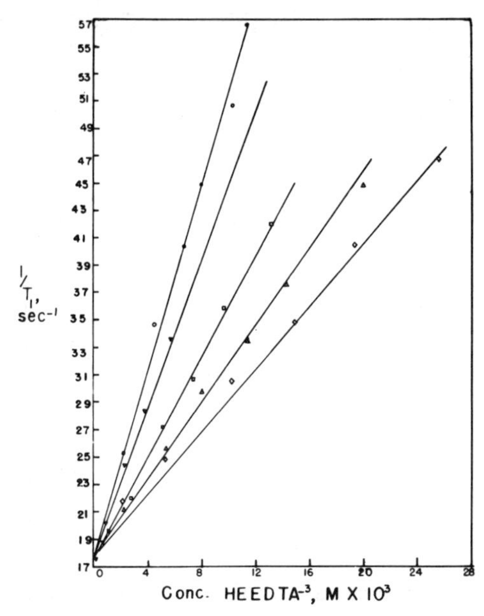

FIG. 6-22. Sodium-23 relaxation rate as a function of NaCl and $HEEDTA^{-3}$ concentration at pH 11.5. The curves are drawn using Eqs. 6-46 and 6-49, with $K_d = 0.115\ M$ and $R_B = 610\ sec^{-1}$ as calculated from the data. NaCl concentrations are 0.050 (○), 0.100 (▽), 0.200 (□), 0.294 (△), and 0.392 *M* (◊). The concentration of $(CH_3)_4NCl$ is roughly 0.1 to 0.4 *M* (80).

In the limit where $C_S \ll C_M$, Eq. 6-50 reduces to

$$(R - R_F)^{-1} = \frac{K_d}{C_S(R_B - R_F)} + \frac{C_M}{C_S(R_B - R_F)} \tag{6-51}$$

Measurements of R for constant C_S and varying C_M, with $C_S \ll C_M$, will permit K_d to be determined from the intercept-to-slope ratio of a plot of $(R - R_F)^{-1}$ versus C_M. The only assumption, $C_S \ll C_M$, is usually a practical consequence anyway because NMR sensitivity places a lower limit on C_M and solubility or availability of macromolecules places an upper limit on C_S; the case $C_S \ll C_M$ is therefore often unavoidable. It is also interesting that K_d may be determined without a knowledge of C_S. If C_S can be estimated, R_B may be calculated from the slope.

This NMR relaxation method for ion–macromolecule interactions is illustrated in Fig. 6-23 for the binding of sodium ion by yeast sRNA (68). The deviation from the linearity predicted by Eq. 6-51 at higher NaCl concentrations could be caused by an increase in either R_B or C_S. Other ex-

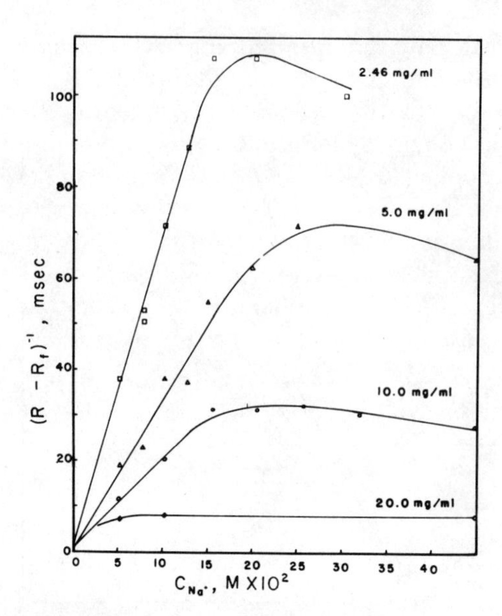

FIG. 6-23. ^{23}Na spin–lattice relaxation rate for aqueous solutions of yeast sRNA as a function of NaCl concentration at pH 7.0. $R_F = 17.5$ sec^{-1} (68).

perimental evidence (83) obtained from proton NMR spectra has indicated that the sRNA forms aggregates at higher NaCl concentrations. Formation of aggregates could produce the observed deviation by an increase in the rotational correlation time of bound sodium ion caused by the overall decreased rotational motion of the aggregate, or by a decrease in the motional freedom of localized regions of the complex caused by aggregation. From Eq. 6-45, it is apparent that an increase in τ_r would result in a larger R_B, producing the nonlinear behavior. Using the linear part of the curves, a value of $\log K_d = -2.8 \pm 0.3$ was calculated. Assuming that C_S is equal to the concentration of phosphate groups in the sRNA, a value of 222 ± 19 sec^{-1} was obtained for R_B.

This technique was subsequently used to examine the binding of sodium ion to a partially purified beef brain (Na–K)-stimulated transport ATPase (84). A value of 1.5 mM was obtained for K_d at 9°C. It was also found that the binding site is specific for Na^+, addition of K^+ having no effect on the ^{23}Na relaxation rate.

Haynes *et al.* (85) have examined the complexing of sodium ion by several ionophores (antibiotics) that are capable of acting as mobile carriers of alkali cations in membranes. Sodium-23 NMR linewidths of 1:1 ionophore–sodium ion complexes in methanol solutions were observed. Using Eq. 2-84, it was possible to calculate an upper limit for the lifetime and, consequently, a lower limit for the dissociation rate constant of $k_{off} > 100$ sec^{-1} for all ionophores studied, with the exception of monensin. Using the Debye-Stokes relationship (Eq. 2-40) for estimating τ_r, it was possible to calculate the quadrupole coupling constant (e^2qQ/h) from the linewidth of $^{23}Na^+$ in the 1:1 complex. The values for the ionophore complexes are compared with those of sodium ion in other environments in Table 6-6 (68, 80, 85–87). The complexes generally do have greater coupling constants than the ionic crystals. Nevertheless, the bonding in the complexes must still be

TABLE 6-6

^{23}Na QUADRUPOLE COUPLING CONSTANTS IN VARIOUS COMPLEXES AND CRYSTALS

SUBSTANCE	ENVIRONMENT	τ_r (BASIS) (SEC)	e^2qQ/h (MHZ)
$Na^+/HEEDTA^{-3}$ [a]	H_2O	1.3×10^{-10} (Debye-Stokes)[b]	1.1
		5.7×10^{-11} (microvisc.)[b]	1.7
		1.0×10^{-11} (tumbling)[b]	4.0
Na^+/NTA^{-3} [a]	H_2O	1.3×10^{-10} (Debye-Stokes)[b]	1.0
		5.7×10^{-11} (microvisc.)[b]	1.6
		1.0×10^{-11} (tumbling)[b]	3.7
Na^+/ionophores[c]	CH_3OH	$0.3–4.0 \times 10^{-10}$ (Debye-Stokes)[b]	0.47–1.64
$NaBrO_3$[d]	Crystal	—	0.842
$NaClO_3$[d]	Crystal	—	0.779
$NaNO_3$[e]	Crystal	—	0.334
Na^+/sRNA[f]	H_2O	2×10^{-8} to 1.4×10^{-7} (Debye-Stokes)	0.049–0.130

[a] James and Noggle (80). HEEDTA is *N'*-(2-hydroxyethyl)ethylenedinitrilo-*N*,*N*,*N'*-triacetic acid and NTA is nitrilotriacetic acid.

[b] Correlation times were calculated using the Debye-Stokes theory (Eq. 2-40), microviscosity theory (Eqs. 2-41 and 2-42), or the value of τ_r for tumbling of water as a lower limit for τ_r.

[c] Haynes *et al.* (85).

[d] Gutowsky and Williams (86). Using pure quadrupole resonance spectroscopy.

[e] Bernheim and Gutowsky (87). Using pure quadrupole resonance spectroscopy.

[f] James and Noggle (68).

described as very ionic by comparison with the values of ^{35}Cl coupling constants ($Q_{Na} \sim Q_{Cl}$) in covalent carbon–chlorine bonds, which range from 40 to 90 MHz.

Bryant has shown that, in spite of the inherent low sensitivity, it may still be possible to use ^{43}Ca (88), ^{39}K (89), and ^{25}Mg (90) NMR linewidth measurements to study the interaction of those ions with biologically interesting compounds, using sufficiently high concentrations ($\lesssim 1\ M$) and, in the case of ^{43}Ca, a high isotopic enrichment.

6.4. Binding of Small Diamagnetic Molecules to Macromolecules

The interaction of a small molecule (usually present in excess) with a macromolecule will generally result in chemical shifts and broadening of the NMR resonances of the small molecule. The chemical shifts arise from a change in the chemical environment for the bound ligand molecule, and the broadening is a result of binding to a macromolecule with consequent increase in the rotational correlation time. The dipole–dipole relaxation process is generally the dominant relaxation mechanism, so an increase in τ_r will result in broader lines (cf. Section 2.4.1). An upper limit on τ_r for any of the resonances is given by the rotational reorientation time of the macromolecule (which may be calculated from the Debye-Stokes theory with knowledge of the macromolecule's dimensions or diffusion coefficient, as discussed in Section 2.3.1). The various nuclei on the bound ligand may have smaller correlation times than the macromolecule and, in fact, the correlation times may differ for various nuclei on the same bound ligand. It is expected that τ_r will be larger for those nuclei in functional groups that interact directly with the macromolecule, and τ_r will not be as large for nuclei in groups that do not interact because these groups may have greater motional freedom. It is therefore possible to learn which groups of the ligand are directly bound to the macromolecule by observing the selective broadening of the ligand resonances. In general, the resonances for those nuclei in (or near) interacting functional groups will also be shifted the most when the ligand is bound, although this is subject to more uncertainty, being very dependent on the shielding contributions of nearby groups in the macromolecule.

In many cases, it is also possible to study the rate of exchange of the ligand between two sites, on the macromolecule and free in solution, by examining the chemical shifts or relaxation times of ligand nuclei if the exchange rate is not in the limit of very rapid or very slow exchange.

NMR has proved very useful in studies of substrate– or inhibitor–enzyme binding, drug–receptor interactions, mutagen–DNA interactions, and hapten–antibody interactions. Perhaps the earliest evidence of the capability of NMR for providing detailed information about interactions was given by Jardetzky *et al.* (91), who noted the disappearance of the pyridine proton resonances of nicotinamide adenine dinucleotide in solutions with yeast alcohol dehydrogenase and suggested a specific role for the nicotinamide moiety of NAD in binding to ADH. This initial study was pursued in greater detail by Hollis (92), who, using active enzyme, came to the opposite conclusion, namely that the selective broadening of the NAD and NADH proton resonances indicated specific binding of the adenine moiety to yeast ADH. In contrast, it is interesting that Sarma and Kaplan (93) concluded that both the adenine and nicotinamide moieties of NADH were directly bound to chicken lactate dehydrogenase.

The possibilities of specific drug binding to proteins were originally explored for the binding of penicillin G (94) and some sulfonamides (95) to bovine serum albumin (BSA). Figure 6-24 shows the effect of added BSA on the relaxation rate of the phenyl protons and methyl protons of sulfacetamide,

$$H_2N-C_6H_4-SO_2NHCOCH_3$$

For a few solutions, T_1 was measured by the direct saturation method and found to be equal to T_2 within experimental error. The selective broadening evident in Figure 6-24 is taken as being indicative of the specific interaction of the *p*-aminobenzene moiety with the protein. From the calculated dissociation constant it was determined that the ratio $T_2(\text{free})/T_2(\text{bound})$ is 2870 for the phenyl protons and 610 for the methyl protons. An increase in viscosity with added BSA should not exhibit this selectivity. It was also noted that the relaxation rate is a function of the sulfonamide-to-BSA ratio, decreasing with an increase in sulfonamide concentration for a fixed BSA concentration. If the binding is nonspecific, it would be expected that the relaxation should increase with increasing sulfonamide concentration.

Actinomycin D inhibits DNA-dependent RNA synthesis and binds double-stranded DNA. Arison and Hoogsteen (96) applied NMR to the problem of how actinomycin D (see Fig. 6-25) binds to DNA by initially examining the proton chemical shifts of actinomycin D in solutions with 5′-deoxyguanylic acid. The observed shifts are recorded in Table 6-7. It is apparent that the ring protons and methyl protons of the three-ring system in actinomycin D are all shifted upfield in the presence of dGMP, whereas the threonyl methyl protons are deshielded. These observations imply that the three-ring system is stacked parallel to the base in dGMP according to the arguments presented in Section 3.1.2,E.

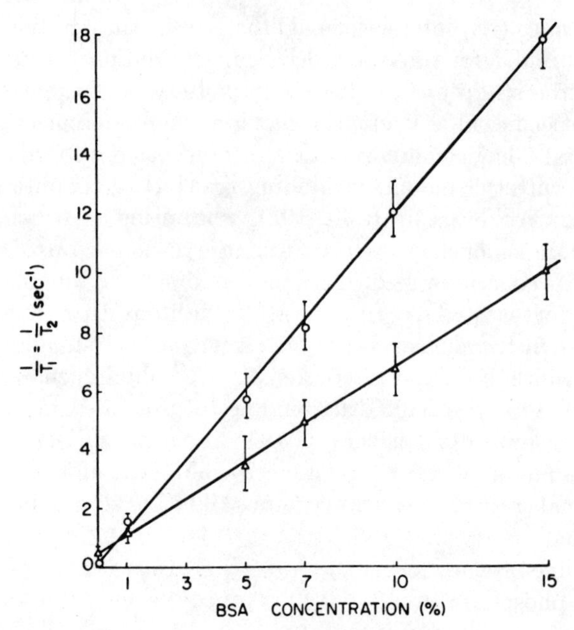

FIG. 6-24. Relaxation rate of the *p*-aminobenzene sulfonamide protons (○) and methyl protons (△) of 0.1 *M* sulfacetamide as a function of bovine serum albumin concentration in D_2O at pH 8.8 (95).

Acridines have been shown to possess mutagenic and carcinogenic properties, presumably associated with their ability to bind DNA. Blears and Danyluk (97) examined the NMR spectrum of acridine,

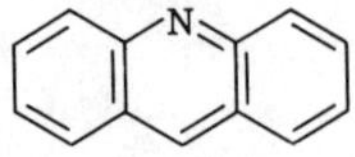

in the presence of DNA in 1:1 D_2O–CH_3OH solutions as a function of acridine and DNA concentration, temperature, salt content, and paramagnetic ion content. Resonance peaks of the acridine protons experienced both line broadening and signal intensity diminution on addition of DNA to the solution. The intensity loss was interpreted as strong binding with the acridine motion very restricted (large τ_r) at an acridine-to-phosphate ratio <0.2. The observed linewidth increase at higher acridine-to-phosphate ratios was interpreted as signifying the existence of weak binding with less restrictive motion. The acridine proton linewidths were also observed to decrease when the temperature was raised above 50°C and DNA passed through a helix–coil transition.

McMurray *et al.* (98) have continued the NMR study of the binding of substrate analogs to aspartate transcarbamylase (ATCase) initiated earlier (99, 100). In the study by McMurray *et al.* (98), ATCase was modified by reacting a specific sulfhydryl residue at the active site with the chromophoric mercurial 2-chloromercuri-4-nitrophenol, causing loss of enzyme activity. Results with this modified ATCase were compared with those of the native enzyme. For example, the proton resonance linewidth of acetyl phosphate, an analog of the substrate carbamyl phosphate, as a function of acetyl phosphate concentration in the presence of native or modified ATCase is depicted in Fig. 6-26. The parameters listed in the caption of Fig. 6-26 were calculated by nonlinear least-squares analysis of the experimental data. The linewidth of acetyl phosphate in the modified enzyme is virtually identical to that in the native enzyme, indicating the same rotational mobility of acetyl phosphate in both enzymes, but the dissociation constant is larger by a factor of approximately three. The results for succinate, which is an analog of the substrate L-aspartate, are shown in Table 6-8. Equilibrium dialysis studies showed that succinate did not bind to modified ATCase. The line broadening for succinate in solution with native ATCase and carbamyl phosphate, noted in Table 6-8, arises from slow exchange of

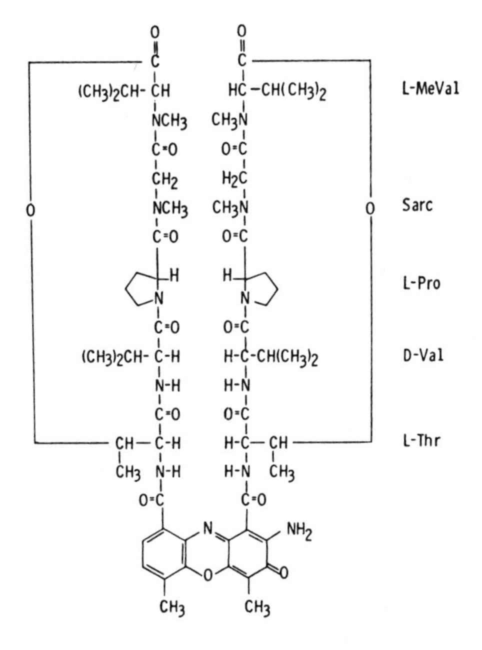

FIG. 6-25. Actinomycin D.

TABLE 6-7A

THE EFFECT OF 5′-dGMP ON THE PROTON NMR CHEMICAL SHIFTS OF ACTINOMYCIN D[a,b]

PROTON	ACT. D (PPM)[c]	ACT. D + DGMP (PPM)[c]	DIFFERENCE (PPM)
Ar (C-7, C-8)	7.51	7.32	−0.19
MeVal–α-CH	6.17	∼6.22	+0.05 ± 0.02
NCH_3 (6-H)	3.01	3.02	+0.01
NCH_3 (6-H)	2.91	2.90	−0.01
6-CH_3	2.55	2.31	−0.24
4-CH_3	2.16	1.95	−0.21
Thr–CH_3	1.36	1.47	+0.11
D-Val–CH_3[d]	1.09	1.14	+0.05
MeVal–CH_3[d]	0.98	0.99	+0.01
D-Val–CH_3[d]	0.87	0.90	+0.03
MeVal–CH_3[d]	0.81	0.83	+0.02

[a] Arison and Hoogsteen (96).
[b] The molar ratio of actinomycin D to 5′-dGMP is 1:2.
[c] Measured in ppm from internal DSS.
[d] Tentative assignment.

TABLE 6-7B

THE EFFECT OF ACTINOMYCIN D ON THE PROTON NMR CHEMICAL SHIFTS OF 5′-dGMP[a,b]

PROTON	5′-DGMP (PPM)[c]	5′-DGMP + ACT. D (PPM)[c]	DIFFERENCE (PPM)
C-8	8.11	8.01	−0.10
C-1′	6.27	6.11	−0.16
C-2′	∼1.91, 1.70	NO[d]	
C-3′	4.69	NO[d]	
C-4′	4.23	4.09	−0.14
C-5′	3.96	∼3.80	∼−0.16

[a] Arison and Hoogsteen (96).
[b] The molar ratio of actinomycin D to 5′-dGMP is 1:2.
[c] Measured in ppm from internal DSS.
[d] Not observed.

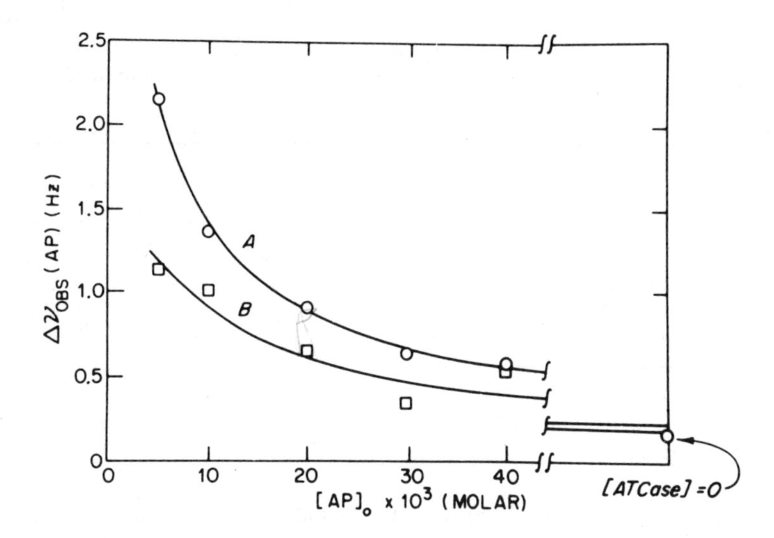

FIG. 6-26. The proton linewidth of the individual resonances of the acetyl phosphate doublet as a function of initial AP concentration. (○) In the presence of 10 mg/ml ATCase at pH 7.4. The solid line was calculated with K_d = 2.5 mM, $\Delta\nu_{EI}$ = 72 Hz, and $\Delta\nu_I$ = 0.23 Hz. (□) In the presence of 10 mg/ml modified ATCase at pH 7.4. The solid line was calculated with K_d = 9.3 mM, $\Delta\nu_{EI}$ = 74 Hz, and $\Delta\nu_I$ = 0.15 Hz. The linewidths recorded are differences between the observed linewidths and the linewidth of an acetone internal standard (98).

succinate between the free solution and the ternary complex as shown by temperature dependence studies (100). The dissociation rate constant k_{-1} $(= 1/\tau_b)$, calculated using Eq. 6-9, was determined to be 170 $\sec^{-1}$ (100). The NMR study utilizing the modified ATCase shows that the mercurial modifier blocks succinate binding but affects the extent of acetyl phosphate binding only slightly.

The first NMR studies of inhibitor binding to enzymes were made by Thomas (101, 102) who observed the chemical shifts of *N*-acetyl-D-glucosamine (NAG) on binding to lysozyme and noted that the single acetamido proton resonance of NAG is shifted upfield and split into two components on addition of lysozyme. These components were identified with the α and β anomers of NAG. Thomas found that the α- and β-methyl glycosides of NAG (which do not interconvert in solution) qualitatively produced the same results. Subsequent studies by Dahlquist and Raftery (103) revealed that the upfield peak, ascribed to the α anomer, was broadened more than the β-anomer peak. Detailed studies of the linewidths enabled the exchange rate of β-NAG with lysozyme to be examined as a function of temperature and pH (104). Studebaker *et al.* (105) have also

TABLE 6-8

LINEWIDTH OF SUCCINATE (RELATIVE TO LINEWIDTH OF INTERNAL STANDARD ACETONE) AT pH 7.4, 0.1 *M* IMIDAZOLE·HCL BUFFER, D_2O, 1 m*M* EDTA[a]

SOLUTION	$W_{1/2}$ (SUCC.) *Hz*
a. 15 m*M* succinate + 25 m*M* AP[b] + 10 mg/ml ATCase	0.1 ± 0.2
b. 15 m*M* succinate + 25 m*M* AP + 5 m*M* CAP[c] + 10 mg/ml ATCase	2.9 ± 0.2
c. 15 m*M* succinate + 5 m*M* CAP + 10 mg/ml modified ATCase	0.1 ± 0.2
d. 15 m*M* succinate + 5 m*M* CAP + 10 mg/ml ATCase	3.2 ± 0.2

[a] McMurray *et al.* (98).
[b] AP = acetyl phosphate.
[c] CAP = carbamyl phosphate.

determined the effect of pH and dimerization on the rate of exchange of α-NAG with lysozyme. Butchard *et al.* (106) have continued the study of lysozyme–inhibitor complexes using ^{19}F NMR.

Chemical exchange studies of inhibitor binding to ribonuclease using ^{31}P NMR (107) and to α-chymotrypsin using ^{1}H NMR (108) and ^{19}F NMR (109) further illustrate the applications of NMR to enzyme binding studies. A ^{19}F NMR study also provided information about the exchange kinetics involved in binding of some aromatic hydrophobes to β-lactoglobulin (110).

Joffe (111) showed that the same techniques could be used to study hapten–antibody interactions, in particular, the binding of *p*-nitrophenyl-β-D-galactoside to antigalactosyl antibody. Burgen *et al.* (112) examined the binding of choline compounds to antiphenoxycholine antibody. It was determined for the *N*-methyl proton resonance of acetamidophenylcholine ether that exchange between the free solution and the binding site of the antibody was in the intermediate exchange region below 30°C.

6.5. Application of the Intermolecular Nuclear Overhauser Effect

Very few applications of nuclear Overhauser effect (NOE) measurements to the investigation of intermolecular interactions have been made to date. It is expected, however, that the detailed information obtainable via NOE measurements will prove quite useful in many areas of study.

Kaiser (113) was the first to observe an intermolecular NOE. Saturating the cyclohexane proton resonance in a solution composed of chloroform, cyclohexane, and TMS in the mole ratios of 1:4.5:0.5 resulted in a 31%

increase in the chloroform proton signal intensity. More recently, a perturbation of the CS_2 ^{13}C signal intensity was noted when the proton resonance of $CHCl_3$ was subjected to a strong irradiating field in a $CHCl_3$–CS_2 solution (114). The negative NOE observed (approximately 30% decrease in intensity) was tentatively ascribed to a time-dependent intermolecular scalar coupling (114).

The base stacking interactions of purines in nucleotides and nucleic acids were discussed in Section 3.1.2,E with regard to induced chemical shifts. Chan and Kreishman (115) have cleverly applied NOE measurements to further elucidate the base stacking interactions by examining the intermolecular NOE between purine,

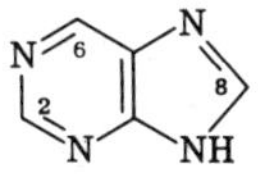

and 3′,5′-polyuridylic acid. The NOE's observed for the purine protons on irradiation of the ribose protons of poly U are summarized in Table 6-9. The tabulated results were interpreted as implying that the purine stacks with the uridine, the poly U having either the H6 or H8 protons preferentially oriented toward the H5′ and H3′ of the ribose moiety.

As discussed in Section 2.5.2, it can be seen from the equation of Balaram *et al.* (116) (Eq. 2-69) that a negative NOE may be obtained in macromolecular systems. A negative NOE was indeed demonstrated for the proton resonances of some peptides when proton resonances of the protein bovine neurophysin II (NP-II) were saturated (117). The results are sum-

TABLE 6-9

INTERMOLECULAR NUCLEAR OVERHAUSER EFFECT BETWEEN PURINE PROTONS AND POLYURIDYLIC ACID PROTONS IN A SOLUTION OF 0.4 *M* PURINE AND POLYURIDYLIC ACID EQUIVALENT TO 0.1 *M* URIDYLIC ACID[a]

PURINE PROTON OBSERVED	RIBOSE PROTON IRRADIATED			
	H5′	H2′, H4′	H3′	H1′
H6	11 ± 3%	0	0	0
H2	0	0	0	0
H8	0	0	6 ± 3%	0

[a] Chan and Kreishman (115).

marized in Table 6-10. It is apparent that the aromatic residue at position 2 of the tripeptide hormone analogs interacts with NP-II, as evidenced by the NOE's. Phenylalanine at position 3 did not exhibit an NOE. Although it was not possible to assign the $\delta 1.9$ and $\delta 3.1$ resonances to specific alkyl groups of the protein, the $\delta 6.86$ resonance was attributed to the sole tyrosine residue of NP-II.

Many NOE studies of the interactions of small ligands with macromolecules will entail solutions containing an excess of ligand molecules. If the exchange rate is sufficiently rapid, i.e., $>1/T_{1b}$, an averaged NOE value will be observed. The NOE will depend on the ligand-to-macromolecule ratio. This dependence is illustrated in Fig. 6-27 for the formate proton in a quaternary creatine kinase complex (118). Formate is hypothesized to occupy the site of the migrating phosphoryl group, thus transforming the enzyme–Mg(II)–creatine–ADP complex into an analog of the transition state of the active complex. The dependence in Fig. 6-27 is linear at high formate–enzyme ratios, reaching a maximum negative NOE at a formate-to-active site ratio $\lesssim 15$. The leveling off at the lower ratios is probably

TABLE 6-10

PERCENT DECREASE IN PEPTIDE PROTON SIGNAL AMPLITUDE ON SATURATION OF NEUROPHYSIN II PROTON RESONANCES[a,b]

PEPTIDE	RESONANCE OBSERVED	RESONANCE SATURATED[c]		
		$\delta 1.9$	$\delta 3.1$	$\delta 6.86$
Ala-Tyr-PheNH$_2$[d]	Tyr ortho	38	36	*h*
	Tyr meta	14	24	*h*
(S-Me)Cys-Tyr-PheNH$_2$[e]	Tyr ortho	21	19	*h*
	Tyr meta	3	9	*h*
Met-Tyr-PheNH$_2$[f]	Tyr ortho	33	35	*h*
	Tyr meta	20	29	*h*
(S-Me)Cys-Phe-IleNH$_2$[g]	Phe 2, 3, 4	20	18	22
	Phe 1, 5	17	15	10

[a] Balaram *et al.* (117).
[b] The fact that the NOE's are concentration dependent is ignored.
[c] δ ppm from DSS.
[d] 4.5 mM peptide, 0.3 mM NP-II, pH 6.5.
[e] 5.4 mM peptide, 0.3 mM NP-II, pH 3.5.
[f] 2.0 mM peptide, 0.3 mM NP-II, pH 6.6.
[g] 4.9 mM peptide, 0.23 mM NP-II, pH 6.5.
[h] Observing and irradiating frequencies overlap.

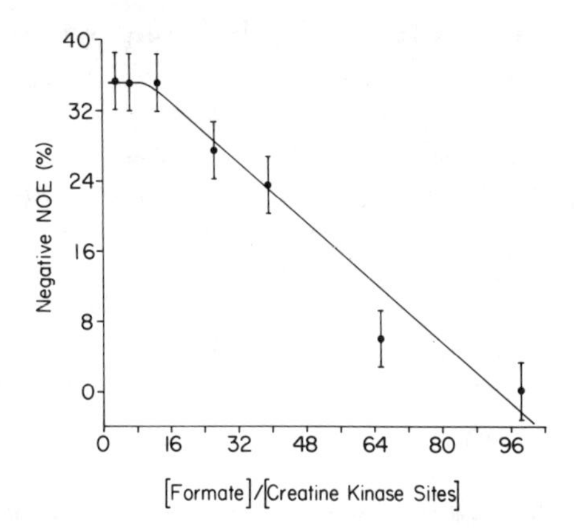

FIG. 6-27. Formate NOE dependence on the ratio of formate concentration to creatine kinase binding site concentration. Solution conditions are: 61 mg/ml creatine kinase, 12.4 m*M* magnesium acetate, 10 m*M* creatine, and 2.5 m*M* ADP in D_2O at pD 8.2 (118).

because the extent of relaxation in the free state for any particular formate molecule is no longer significant before the formate molecule is bound again and the proton is subjected to the polarizing effects of a proton on the enzyme. James and Cohn (118) have shown that the ϵ-CH_2 protons of a lysyl residue at the active site are responsible for the formate NOE. The assignment was based on the resonance frequency of the protons effecting the formate NOE, previous studies implicating a lysyl residue at the active site, and lack of an observable NOE with creatine kinase modified only by chemical reaction of the active site lysyl NH_3^+ group with dansyl chloride. The implication is that formate, and therefore presumably the transferable phosphoryl group, interacts with the lysyl NH_3^+ group at the active site. Observation of the formate resonance revealed little or no NOE for binary or ternary creatine kinase complexes with Mg(II), ADP, and creatine, implying that the proper conformation of the transition state complex is achieved only in the presence of the requisite divalent metal ion activator and both substrates (118).

Further advances in the application of the intermolecular NOE may enable the extraction of internuclear distances because the basis of the nuclear Overhauser effect is dipolar relaxation. It may also be possible to utilize NOE measurements in determining chemical exchange rates in biomolecular complex formation as it has been used in intramolecular exchange studies (119).

6.6. Application of Chemically Induced Dynamic Nuclear Polarization

The technique of chemically induced dynamic nuclear polarization (CIDNP) holds promise for investigation of photosynthesis or other biological processes entailing free radical intermediates. CIDNP involves unpaired electron spins interacting with nuclear spins to modify the intensity of NMR absorption peaks and to produce NMR emission peaks. The phenomenon was first observed independently by Ward and Lawler (120) and by Bargon *et al.* (121) in 1967. Ward (122) and Lawler (123) have also reviewed CIDNP.

The changes in NMR peak intensity reflect the nonequilibrium populations of nuclear energy states caused by coupling of an unpaired electron with the nuclear spin. The unpaired electrons are generated during the course of a chemical reaction. It is also necessary for observation of CIDNP that the unpaired electrons be present in radical pair cages that decompose to give either products or free unpaired radicals (124–126).

CIDNP experiments may provide mechanistic information about (*a*) the existence of short-lived radical pairs in a reaction, (*b*) identification of the spin state (triplet or singlet) of the product precursors by the characteristics of the CIDNP spectrum, (*c*) reaction sequences, and (*d*) an indication that a reaction is occurring in the case of exchange reactions where reactant and product are identical.

Tomkiewicz and Klein (127) have illustrated the potential of CIDNP by applying CIDNP to the investigation of the photoxidation of chlorophyll b by 1,4-benzoquinone and 2,6-dimethyl-1,4-benzoquinone. Irradiation of a mixture of chlorophyll b and 1,4-benzoquinone in CD_3OD with visible light resulted in an enhanced absorption peak for the quinone. Irradiation of a mixture of chlorophyll b and 2,6-dimethyl-1,4-benzoquinone resulted in an enhanced absorption peak for the aromatic protons and an emission peak for the methyl protons of the quinone. No polarization of the quinones was detected in the absence of chlorophyll b. The results were consistent with the following proposed mechanism:

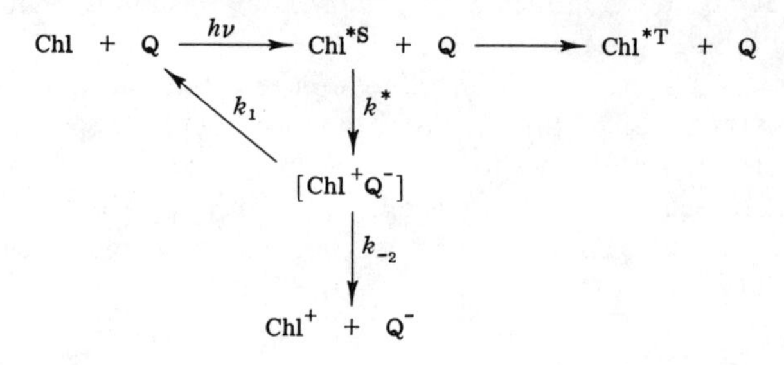

The photon excites chlorophyll to its excited singlet state, which can go to the triplet state via intersystem crossing or to the radical pair via quenching of the excited singlet by quinone. Annihilation of the radical pair gives the starting materials. Alternatively, the radical pair can dissociate, forming the isolated radicals. A theoretical spectrum matching the observed CIDNP spectrum was calculated using the above mechanism.

References

1. D. E. Woessner, *J. Chem. Phys.* **35,** 41 (1961).
2. A. C. McLaughlin and J. S. Leigh, Jr., *J. Magn. Resonance* **9,** 296 (1973).
3. T. J. Swift and R. E. Connick, *J. Chem. Phys.* **37,** 307 (1962).
4. Z. Luz and S. Meiboom, *J. Chem. Phys.* **40,** 2686 (1964).
5. D. E. O'Reilly and C. P. Poole, Jr., *J. Phys. Chem.* **67,** 1762 (1963).
6. J. Eisinger, R. G. Shulman, and B. M. Szymanski, *J. Chem. Phys.* **36,** 1721 (1962).
7. M. Cohn and J. S. Leigh, Jr., *Nature* (*London*) **193,** 1037 (1962).
8. A. S. Mildvan and M. Cohn, *Advan. Enzymol.* **33,** 1 (1970).
9. M. Cohn, *Quart. Rev. Biophys.* **3,** 61 (1970).
10. G. Herzberg, "Atomic Spectra and Atomic Structure," pp. 109 and 209. Dover, New York, 1944.
11. B. B. Wayland and W. L. Rice, *Inorg. Chem.* **5,** 54 (1966).
12. H. M. Swartz, J. R. Bolton, and D. C. Borg, "Biological Applications of Electron Spin Resonance," p. 36ff. Wiley (Interscience), New York, 1972.
13. T. L. James and M. Cohn, *J. Biol. Chem.* **249,** 3519 (1974).
14. B. B. Garrett and L. O. Morgan, *J. Chem. Phys.* **44,** 890 (1966).
15. N. Bloembergen and L. O. Morgan, *J. Chem. Phys.* **34,** 842 (1961).
16. G. H. Reed, J. S. Leigh, Jr., and J. E. Pearson, *J. Chem. Phys.* **55,** 3311 (1971).
17. G. H. Reed and W. J. Ray, Jr., *Biochemistry* **10,** 3190 (1971).
18. T. L. James, J. Reuben, and M. Cohn, *J. Biol. Chem.* **248,** 6443 (1973).
19. R. A. Dwek, G. K. Radda, R. E. Richards, and A. G. Salmon, *Eur. J. Biochem.* **29,** 509 (1972).
20. G. H. Reed, H. Diefenbach, and M. Cohn, *J. Biol. Chem.* **247,** 3066 (1972).
21. A. R. Peacocke, R. E. Richards, and B. Sheard, *Mol. Phys.* **16,** 177 (1969).
22. D. Buttlaire and M. Cohn, *J. Biol. Chem.* **249,** 5733 (1974).
23. J. Reuben and M. Cohn, *J. Biol. Chem.* **245,** 6539 (1970).
24. G. Navon, *Chem. Phys. Lett.* **7,** 390 (1970).
25. C. H. Fung, A. S. Mildvan, A. Allerhand, R. Komoroski, and M. C. Scrutton, *Biochemistry* **12,** 620 (1973).
26. J. Reuben and F. J. Kayne, *J. Biol. Chem.* **246,** 6227 (1971).
27. P. W. Taylor, J. Feeney, and A. S. V. Burgen, *Biochemistry* **10,** 3866 (1971).
28. C. D. Barry, A. C. T. North, J. A. Glasel, R. J. P. Williams, and A. V. Xavier, *Nature* (*London*) **232,** 236 (1971).
29. I. C. P. Smith, B. J. Blackburn, and T. Yamane, *Can. J. Chem.* **47,** 513 (1969).
30. P. O. P. Ts'o, N. S. Kondo, M. P. Schweitzer, and D. P. Hollis, *Biochemistry* **8,** 997 (1969).
31. J. H. Prestegard and S. I. Chan, *J. Amer. Chem. Soc.* **91,** 2843 (1969).
32. P. A. Hart and J. P. Davis, *J. Amer. Chem. Soc.* **91,** 512 (1969).

33. P. A. Hart and J. P. Davis, *Biochem. Biophys. Res. Commun.* **34,** 733 (1969).
34. S. H. Koenig, R. D. Brown, and J. Studebaker, *Cold Spring Harbor Symp. Quant. Biol.* **36,** 551 (1971).
35. M. Rubenstein, A. Baram, and Z. Luz, *Mol. Phys.* **20,** 67 (1971).
36. K. T. Gillen, *J. Chem. Phys.* **56,** 1573 (1972).
37. D. E. Woessner, *J. Chem. Phys.* **36,** 1 (1962).
38. A. S. Mildvan and H. Weiner, *J. Biol. Chem.* **244,** 2465 (1969).
39. T. R. Krugh, *Biochemistry* **10,** 2594 (1971).
40. A. S. Mildvan and M. Cohn, *Biochemistry* **2,** 910 (1963).
41. D. A. Deranleau, *J. Amer. Chem. Soc.* **91,** 4044 (1969).
42. D. A. Deranleau, *J. Amer. Chem. Soc.* **91,** 4050 (1969).
43. I. M. Klotz and D. L. Hunston, *Biochemistry* **10,** 3065 (1971).
44. M. Cohn and J. Townsend, *Nature* (*London*) **173,** 1090 (1954).
45. A. S. Mildvan and M. Cohn, *J. Biol. Chem.* **241,** 1178 (1965).
46. W. J. O'Sullivan and M. Cohn, *J. Biol. Chem.* **243,** 2737 (1968).
47. G. H. Reed, M. Cohn, and W. J. O'Sullivan, *J. Biol. Chem.* **245,** 6547 (1970).
48. T. L. James, unpublished results.
49. D. Buttlaire and G. H. Reed, *J. Biol. Chem.*, to be published.
50. D. L. Marquardt, *J. Soc. Ind. Appl. Math.* **2,** 443 (1963).
51. A. C. McLaughlin, M. Cohn, and G. L. Kenyon, *J. Biol. Chem.* **247,** 4382 (1972).
52. H. Montgomery, R. V. Chastain, and E. C. Lingafelter, *Acta Crystallogr.* **20,** 731 (1966).
53. B. Morosin and E. J. Groeber, *J. Chem. Phys.* **42,** 898 (1965).
54. A. Zalkin, J. D. Forrester, and D. H. Templeton, *Inorg. Chem.* **3,** 529 (1964).
55. R. Chidambaram, *J. Chem. Phys.* **36,** 2361 (1961).
56. F. A. Cotton and G. Wilkinson, "Advanced Inorganic Chemistry," p. 701ff. Wiley (Interscience), New York, 1962.
57. A. S. Mildvan and M. Cohn, *J. Biol. Chem.* **240,** 238 (1965).
58. M. E. (Reipe) Fabry, S. H. Koenig, and W. E. Schillinger, *J. Biol. Chem.* **245,** 4256 (1970).
59. Z. Luz and R. G. Shulman, *J. Chem. Phys.* **43,** 3750 (1965).
60. K. G. Morallee, E. Niebohr, F. J. C. Rossotti, R. J. P. Williams, A. V. Xavier, and R. A. Dwek, *Chem. Commun.* p. 1132 (1970).
61. R. W. Wien, J. D. Morrisett, and H. M. McConnell, *Biochemistry* **11,** 3707 (1972).
62. A. S. Mildvan, J. S. Leigh, and M. Cohn, *Biochemistry* **6,** 1805 (1967).
63. A. S. Mildvan, N. M. Rumen, and B. Chance, *in* "Probes of Structure and Function of Macromolecules and Membranes" (B. Chance, T. Yonetani, and A. S. Mildvan, eds.), Vol. 2, p. 205ff. Academic Press, New York, 1971.
64. R. G. Pearson and M. M. Anderson, *Angew. Chem., Int. Ed. Engl.* **4,** 281 (1965).
65. T. R. Stengle and C. H. Langford, *Coord. Chem. Rev.* **2,** 349 (1967).
66. R. K. Gupta and A. G. Redfield, *Biochem. Biophys. Res. Commun.* **41,** 273 (1970).
67. T. R. Stengle and J. D. Baldeschwieler, *Proc. Nat. Acad. Sci. U.S.* **55,** 1020 (1966).
68. T. L. James and J. H. Noggle, *Proc. Nat. Acad. Sci. U.S.* **62,** 644 (1969).
69. A. G. Marshall, *Biochemistry* **7,** 2450 (1968).
70. R. G. Bryant, *J. Amer. Chem. Soc.* **91,** 976 (1969).
71. T. R. Stengle and J. D. Baldeschwieler, *J. Amer. Chem. Soc.* **89,** 3045 (1967).
72. R. G. Bryant, *J. Amer. Chem. Soc.* **89,** 2496 (1967).
73. R. L. Ward and J. A. Happe, *Biochem. Biophys. Res. Commun.* **28,** 785 (1967).

74. R. P. Haugland, L. Stryer, T. R. Stengle, and J. D. Baldeschwieler, *Biochemistry* **6**, 498 (1967).
75. W. D. Ellis, H. B. Dunford, and J. S. Martin, *Can. J. Chem.* **47**, 157 (1969).
76. R. L. Ward and J. A. Happe, *Biochem. Biophys. Res. Commun.* **45**, 1444 (1971).
77. R. L. Ward and M. D. Cull, *Arch. Biochem. Biophys.* **150**, 436 (1972).
78. T. E. Bull, J. Andrasko, E. Chiacone, and S. Forsén, *J. Mol. Biol.* **73**, 251 (1973).
79. T. R. Collins, Z. Starčuk, A. H. Burr, and E. J. Wells, *J. Amer. Chem. Soc.* **95**, 1649 (1973).
80. T. L. James and J. H. Noggle, *J. Amer. Chem. Soc.* **91**, 3424 (1969).
81. T. L. James and J. H. Noggle, *Bioinorg. Chem.* **2**, 69 (1972).
82. T. L. James and J. H. Noggle, *Anal. Biochem.* **49**, 208 (1972).
83. I. C. P. Smith, T. Yamane, and R. G. Shulman, *Science* **159**, 1361 (1968).
84. F. Ostroy, T. L. James, J. H. Noggle, A. Sarrif, and L. E. Hokin, *Fed. Proc., Fed. Amer. Soc. Exp. Biol.* **31**, 1207 (1972); *Arch. Biochem. Biophys.* **162**, 421 (1974).
85. D. H. Haynes, B. C. Pressman, and A. Kowalsky, *Biochemistry* **10**, 852 (1971).
86. H. S. Gutowsky and G. A. Williams, *Phys. Rev.* **105**, 464 (1957).
87. R. A. Bernheim and H. S. Gutowsky, *J. Chem. Phys.* **32**, 1072 (1960).
88. R. G. Bryant, *J. Amer. Chem. Soc.* **91**, 1870 (1969).
89. R. G. Bryant, *Biochem. Biophys. Res. Commun.* **40**, 1162 (1970).
90. R. G. Bryant, *J. Magn. Resonance* **6**, 159 (1972).
91. O. Jardetzky, N. G. Wade, and J. J. Fischer, *Nature (London)* **197**, 183 (1963).
92. D. P. Hollis, *Biochemistry* **6**, 2080 (1967).
93. R. H. Sarma and N. O. Kaplan, *Proc. Nat. Acad. Sci. U.S.* **67**, 1375 (1970).
94. J. J. Fischer and O. Jardetzky, *J. Amer. Chem. Soc.* **87**, 3237 (1965).
95. O. Jardetzky and N. G. Wade-Jardetzky, *Mol. Pharmacol.* **1**, 214 (1965).
96. B. H. Arison and K. Hoogsteen, *Biochemistry* **9**, 3976 (1970).
97. D. J. Blears and S. S. Danyluk, *Biopolymers* **5**, 535 (1967).
98. C. H. McMurray, D. R. Evans, and B. D. Sykes, *Biochem. Biophys. Res. Commun.* **48**, 572 (1972).
99. P. G. Schmidt, G. R. Stark, and J. D. Baldeschwieler, *J. Biol. Chem.* **244**, 1860 (1969).
100. B. D. Sykes, P. G. Schmidt, and G. R. Stark, *J. Biol. Chem.* **245**, 1180 (1970).
101. E. W. Thomas, *Biochem. Biophys. Res. Commun.* **24**, 611 (1966).
102. E. W. Thomas, *Biochem. Biophys. Res. Commun.* **29**, 628 (1967).
103. F. W. Dahlquist and M. A. Raftery, *Biochemistry* **7**, 3269 and 3277 (1968).
104. F. W. Dahlquist and M. A. Raftery, *Biochemistry* **8**, 713 (1969).
105. J. F. Studebaker, B. D. Sykes, and R. Wien, *J. Amer. Chem. Soc.* **93**, 4579 (1971).
106. C. G. Butchard, R. A. Dwek, P. W. Kent, R. J. P. Williams, and A. V. Xavier, *Eur. J. Biochem.* **27**, 548 (1972).
107. G. C. Y. Lee and S. I. Chan, *Biochem. Biophys. Res. Commun.* **43**, 142 (1971).
108. J. T. Gerig and J. D. Reinheimer, *J. Amer. Chem. Soc.* **92**, 3146 (1970).
109. B. D. Sykes, *J. Amer. Chem. Soc.* **91**, 949 (1969).
110. K. A. Robillard, Jr. and A. Wishnia, *Biochemistry* **11**, 3841 (1972).
111. S. Joffe, *Mol. Pharmacol.* **3**, 399 (1967).
112. A. S. V. Burgen, O. Jardetzky, J. C. Metcalfe, and N. G. Wade-Jardetzky, *Proc. Nat. Acad. Sci. U.S.* **58**, 447 (1967).
113. R. Kaiser, *J. Chem. Phys.* **42**, 1838 (1965).
114. D. P. Miller, B. Ternai, and G. E. Maciel, *J. Amer. Chem. Soc.* **95**, 1336 (1973).
115. S. I. Chan and G. P. Kreishman, *J. Amer. Chem. Soc.* **92**, 1102 (1970).

116. P. Balaram, A. A. Bothner-By, and J. Dadok, *J. Amer. Chem. Soc.* **94**, 4015 (1972).
117. P. Balaram, A. A. Bothner-By, and E. Breslow, *J. Amer. Chem. Soc.* **94**, 4017 (1972).
118. T. L. James and M. Cohn, *J. Biol. Chem.* **249**, 2599 (1974).
119. J. H. Noggle and R. E. Schrimer, "The Nuclear Overhauser Effect: Chemical Applications," p. 134ff. Academic Press, New York, 1971.
120. H. R. Ward and R. G. Lawler, *J. Amer. Chem. Soc.* **89**, 5518 (1967).
121. J. Bargon, H. Fischer, and U. Johnson, *Z. Naturforsch. A* **22**, 1551 (1967).
122. H. R. Ward, *Accounts Chem. Res.* **5**, 18 (1972).
123. R. G. Lawler, *Accounts Chem. Res.* **5**, 25 (1972).
124. R. Kaptein and L. Oosterhoff, *Chem. Phys. Lett.* **4**, 195 (1969).
125. R. Kaptein and L. Oosterhoff, *Chem. Phys. Lett.* **4**, 214 (1969).
126. G. Gloss and A. Trifunac, *J. Amer. Chem. Soc.* **92**, 2183 (1970).
127. M. Tomkiewicz and M. P. Klein, *Proc. Nat. Acad. Sci. U.S.* **70**, 143 (1973).

CHAPTER 7

NMR SPECTRA OF BIOPOLYMERS

The high degree of specificity in most biochemical processes, protein biosynthesis, the immune response, enzyme catalysis, etc., is a result of specific biopolymer structures. The nuclear magnetic resonance spectrum of a biopolymer inherently contains much structural information. The general problem is to extract structural information from an overabundance of data. Chiefly for reasons of sensitivity, the discussion will generally be limited to proton magnetic resonance spectra of macromolecules (although some ^{13}C and ^{31}P examples will be cited). Even a small protein or nucleic acid will have on the order of a thousand protons with individual resonance lines characteristic of the chemical environment of each proton. The problems of sensitivity and broad overlapping lines are exemplified in the early 60 MHz proton magnetic resonance spectrum of ribonuclease obtained by Kowalsky (1) and reproduced in Fig. 7-1. Changes in a broad envelope of lines will reflect gross conformational alterations, such as the differences evident in Fig. 7-1 on denaturation of ribonuclease.

Determination of more subtle structural characteristics, however, usually requires the identification and study of individual resonance lines. This has been aided considerably by the introduction of superconducting

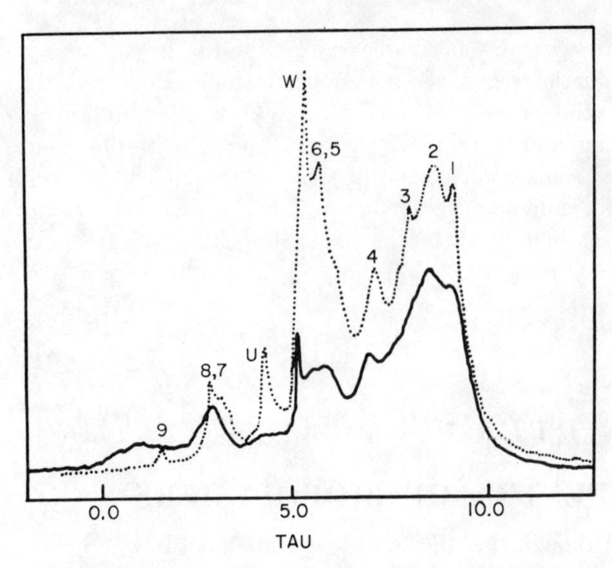

Fig. 7-1. 60 MHz proton magnetic resonance spectrum of 20% native ribonuclease (solid line) and ribonuclease in 8 *M* urea (dotted line) using D_2O as solvent. W represents the water line from residual protons in the D_2O and U represents the line from urea protons (1).

magnets, permitting proton NMR studies at frequencies of 220–360 MHz with consequent improvement in resolution and sensitivity.

There are three general approaches by which resonances may be identified. The first is to examine only those resonances that are shifted from the main envelope of resonances by ring current effects, paramagnetic shifts, or some other effect. The second approach is to use selective isotope labeling, i.e., ^{13}C, ^{2}H, or ^{19}F, with examination of a few nuclear resonances in the absence of other complicating lines. A third approach, that of NMR difference spectroscopy, has recently been explored (2, 3). NMR difference spectroscopy entails subtracting one NMR spectrum from another that has been slightly perturbed (e.g., pH change) in some known manner. NMR signals will be obtained only at those frequencies corresponding to the resonance lines that have been perturbed (shifted or broadened).

The sensitivity problem encountered with NMR studies of biopolymers was abated to a certain extent with the introduction of a computer of average transients (CAT), permitting the detection of a single proton (~1 m*M* concentration) on a small macromolecule in concentrated solutions. The recent advent of Fourier transform NMR and superconducting magnets yielded further sensitivity improvements such that proton concentrations on the order of 0.1 m*M* can be detected in favorable cases.

There are several conceivable causes for the broad lines observed in proton magnetic resonance spectra of macromolecules (4). Ignoring, for a moment, the overlap of lines arising from different kinds of protons (e.g., tyrosine and tryptophan protons), the linewidth for a single proton resonance of a macromolecule may still be expected to be larger than the linewidth for a monomer. The sum of the possible contributions to the linewidth (at half height) of a single resonance will give the total linewidth (4):

$$(W_{1/2})_t = (W_{1/2})_{\text{monomer}} + (W_{1/2})_{\text{CS}} + (W_{1/2})_{\text{DD}} + (W_{1/2})_{ex} + (W_{1/2})_{\text{paramag}} + (W_{1/2})_{\text{other}} \quad (7\text{-}1)$$

These various factors will now be considered in turn.

The linewidth for the resonance in the monomeric unit is represented by $(W_{1/2})_{\text{monomer}}$. This term includes the sources of T_2 ($W_{1/2} = 1/\pi T_2$) relaxation in the monomer, magnetic field inhomogeneity, and spin–spin splitting. Although the splitting may be resolved in the monomer, resolution of the multiplet will generally not be observed in the biopolymer. Therefore, the components of the multiplet will simply contribute to a broader line.

A biopolymer will usually contain more than one unit of a particular monomer, e.g., ribonuclease contains six tyrosine residues. There will generally be slight differences in chemical environment for a particular kind of proton in the various like monomeric units of the macromolecule and, consequently, slight variations in the resonance frequency. $(W_{1/2})_{\text{CS}}$ represents the contribution from these chemical shift variations. In several cases presented in this chapter, the chemical shifts are sufficiently large that some of the individual resonances may be resolved.

As in the case of the monomer, the dipole–dipole relaxation mechanism will generally be dominant. $(W_{1/2})_{\text{DD}}$ is the linewidth contribution from dipolar relaxation in excess of that present in the monomer. The additional contribution may be caused by the proximity of other nuclear dipole moments near the nucleus in the macromolecule but is often the result of an increased rotational correlation time in the macromolecule. An upper limit to the rotational correlation time may be approximated from the dimensions of the macromolecule by using the Debye-Stokes equation (Eq. 2-40). The dipolar contribution to the linewidth is given by Eq. 2-53. It is evident that a larger correlation time will lead to increased linewidths. It is also apparent from Eq. 2-53 that the dipolar contribution to the linewidth will depend on frequency if the exchange-narrowing limit ($\omega^2\tau_c^2 \ll 1$) does not obtain. The exchange-narrowing limit often is not applicable to nuclei on biopolymers. The pertinent correlation times are usually in the range 10^{-10}–10^{-7} sec. The frequency dependence of the linewidth provides another incentive for using the large fields generated by superconducting magnets. For a correlation time of 10^{-9} sec, the dipolar contribution to the linewidth at 300 MHz will be less than half that at 60 MHz.

As discussed in Section 2.6, chemical exchange effects can produce line broadening. The amount of broadening caused by exchange of the proton between different sites is given by $(W_{1/2})_{ex}$. If a paramagnetic species is present either in the biopolymer or in solution, the line may be broadened by an amount $(W_{1/2})_{\mathrm{paramag}}$ according to the discussion in Sections 2.4.3 and 3.3. The flavoproteins are examples of biopolymers containing a group that may be paramagnetic. In other cases, a paramagnetic transition metal ion may be added to a solution containing a biopolymer. The last term in Eq. 7-1, $(W_{1/2})_{\mathrm{other}}$, represents any contributions to the linewidth that may be from other relaxation mechanisms (cf. Section 2.4). It is expected that these contributions will be negligible.

The linewidth of any particular resonance in a biopolymer will therefore be greater than that in a monomer. The situation is complicated by overlapping of neighboring lines and by some rather large chemical shifts of individual nuclear resonances relative to their positions in the spectrum of the monomer. These complications often must be dealt with when a detailed analysis of a spectrum is desired.

7.1. Proteins and Polypeptides

The most fruitful NMR studies on biopolymers have been carried out with small proteins ($\lesssim$25,000 molecular weight). Roberts and Jardetzky (5) have reviewed some of the proton NMR studies of the enzymes bovine pancreatic ribonuclease, lysozyme, and staphylococcal nuclease. Wuthrich (6) has reviewed some proton NMR studies of heme proteins. A more extensive discussion of the proton NMR spectra of proteins and polypeptides has been presented by Bovey (6a).

As a first approximation, the NMR spectrum of a protein may be constructed from the nuclear resonances of the constituent amino acids using broader lines. From a consideration of the chemical shifts of the proton resonances of amino acids and peptides, McDonald and Phillips (7) listed the expected resonance frequencies for the proton resonances of the amino acid residues in a protein and calculated protein spectra from the amino acid composition, assuming a triangular lineshape for the individual resonances. Table 7-1 is derived largely from the compilation of McDonald and Phillips (7). As shown in Fig. 7-2 for lysozyme, the calculated spectrum

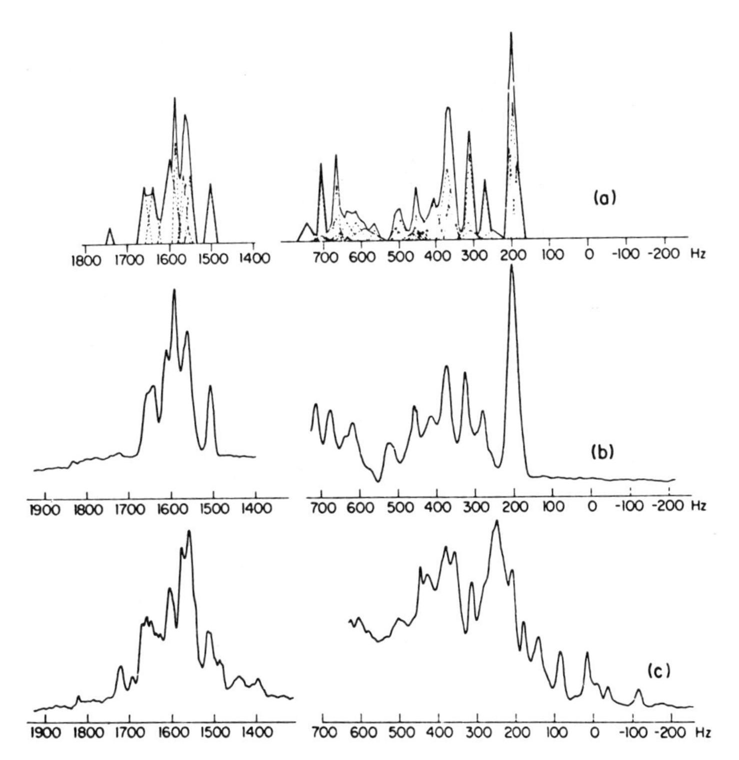

FIG. 7-2. Comparison of the 220 MHz proton NMR spectra of hen egg white lysozyme in D_2O: (a) spectrum calculated on the basis of triangular lineshapes using the amino acid composition of lysozyme and the chemical shifts of the constituent amino acids; (b) observed spectrum at 80°C (random coil); (c) observed spectrum at 65°C (native). The chemical shifts are expressed in Hertz relative to the methyl resonance of DSS (sodium 2,2-dimethyl-2-silapentane-5-sulfonate). The NH protons have been deuterium exchanged by preheating in D_2O (8).

provides a good representation for a random-coil protein but exhibits many differences from the spectrum of the native protein (8). Evidently, the folding of the protein chain in the native enzyme results in many chemical shift variations because of the influence of proximate groups in the native state. Simulated spectra for ribonucleuse and *Clostridium* MP flavodoxin, calculated from their amino acid compositions assuming Lorentzian lineshapes, are given in Fig. 7-3. Comparison with the native enzyme spectra presented later in this chapter indicates that although there are certain major features characteristic of protein spectra in general, the spectrum of each of these small proteins also has individualistic features.

The plausibility of obtaining natural abundance ^{13}C NMR spectra of proteins has now been demonstrated (9–13). Glushko *et al.* (14) have recently compiled a table (reproduced here as Table 7-2) of ^{13}C chemical shifts of amino acids and peptides, using much of the data originally presented by Horsley *et al.* (15), which is found to be in agreement with later studies by Freedman *et al.* (16). The ^{13}C spectrum of native hen egg white lysozyme is compared in Fig. 7-4 with the random-coil spectrum and a "stick" spectrum prepared on the basis of the number and chemical shift of constituent amino acids (13). As in the case of proton NMR, it is obvious that the calculated spectrum corresponds closely to the random-coil spectrum. The ^{13}C spectral lines of bovine pancreatic ribonuclease have also been tentatively assigned on the basis of the amino acid and peptide chemical shifts (14).

The following discussion will cover several of the applications of ^{1}H and ^{13}C NMR spectra of small proteins and polypeptides, touching on observa-

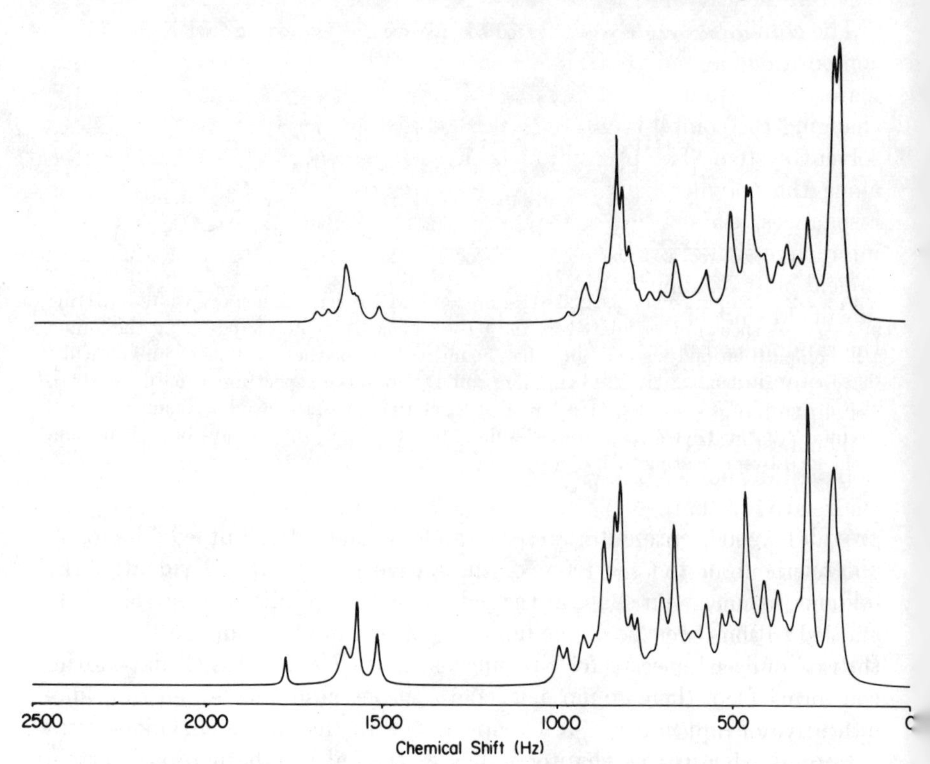

FIG. 7-3. Computed simulation of the spectra of *Clostridium* MP flavodoxin (top) and ribonuclease (bottom) employing the chemical shifts of amino acids and model peptides. The simulations were made using the amino acid compositions of the proteins and assuming Lorentzian lineshapes.

tion of denaturation, conformational equilibria, assignment of spectral lines, and studies of protein binding sites. The use of selective isotope labeling for simplification of spectra is also considered.

7.1.1. Helix–Coil Transitions in Polypeptides

Polypeptides have been investigated as protein models suitable for examining secondary structure characteristics without the complications arising from tertiary structure. It was mentioned above that the NMR spectrum of a random-coil protein or polypeptide is simply a composite of the spectra of its constituent amino acids. The transition of a polypeptide from a random-coil structure to a helical structure is marked by deviations from the simple random-coil spectrum. These deviations have permitted study of the equilibria and kinetics involved in the reversible transition between helical and random-coil forms.

The equilibria and kinetics of the helix–coil transition depend on solvent composition, amino acid side-chain composition, and temperature. The majority of studies have involved monitoring the transition effected by changing the composition of the $CDCl_3$–trifluoroacetic acid (TFA) mixed solvent system. The investigations have involved observation of protons along the polymer backbone, the NH (peptide bond) and α-CH proton resonances, these being, of course, the most sensitive to the helix–coil interconversion. Formation of a helical structure generally results in an upfield shift of about 0.3–0.6 ppm for the α-CH proton resonances and a downfield shift of about 0.2 ppm for the NH proton resonances relative to the random-coil polypeptide (18–22). Other methods (e.g., optical properties) of monitoring the amount of helical form generally parallel the chemical shift change.

The first NMR observation of the random-coil and helix forms of a polypeptide, poly-γ-benzyl-L-glutamate, by Bovey *et al.* (17) revealed that the relatively sharp NH and α-CH lines present with the polypeptide in 10% TFA–90% trichloroethylene were broadened beyond detection in 100% trichloroethylene. Subsequent studies, using TFA–$CDCl_3$ solvent systems, confirmed the early observations on poly-γ-benzyl-L-glutamate and led to the suggestion that the line broadening is a result of changes in the rate of exchange of the individual protons between helical and random-coil forms (18). The sharp lines present in the mixed solvent system are indicative of rapid exchange. It was presumed that the observed broadening occurs when the rate of interconversion is comparable to the chemical shift difference ($\sim$25 Hz) between the two forms. However, this calculated rate of exchange is several orders of magnitude slower than the value of 10^7–10^8 sec^{-1} for the interconversion exchange rate measured by temper-

TABLE 7-1

PROTON MAGNETIC RESONANCE POSITIONS OF AMINO ACID RESIDUES FOR COMPUTING THE SPECTRA OF RANDOM-COIL PROTEINS[a]

PROTON TYPE	EQUIV. PROTONS/RESIDUE	RESONANCE POSITION, (PPM)[b]	$W_{1/2}$ (HZ)
Leucine CH_3	6	0.89	15
β-CH_2 + γ-CH	3	1.64	20
Isoleucine CH_3	6	0.83	20
CH_2	1	1.14	30
CH_2	1	1.41	30
β-CH	1	1.93	25
Valine CH_3	6	0.93	17
β-CH	1	2.25	25
Alanine CH_3	3	1.41	18
Threonine CH_3	3	1.23	16
Lysine γ-CH_3	2	1.43	30
δ-CH_2 + β-CH_3	4	1.68	30
ϵ-CH_2	2	3.02	22
Arginine γ-CH_2	2	1.66	28
β-CH_2	2	1.84	24
δ-CH_2	2	3.20	14
Proline γ-CH_2	2	2.02	21
β-CH_2	2	2.11	25
δ-CH_2	2	3.30	30
Glutamic acid β-CH_2	2	1.98	20
γ-CH_2	2	2.27	20
Glutamine β-CH_2	2	2.07	20
γ-CH_2	2	2.32	20
Aspartic acid β-CH_2	2	2.68	55
Asparagine β-CH_2	1	2.79	30
β-CH_2	1	2.90	30
Methionine CH_3	3	2.06	10
β-CH_2	2	2.06	22
γ-CH_2	2	2.57	16
Cysteine β-CH_2	2	3.02	12
Histidine β-CH_2	2	3.18	28
Imidazole C-4	1	7.07	10
Imidazole C-2	1	7.91	10

TABLE 7-1—(Continued)

PROTON TYPE	EQUIV. PROTONS/RESIDUE	RESONANCE POSITION, (PPM)[b]	$W_{1/2}$ (HZ)
Tyrosine β-CH_2	2	2.98	30
Aromatic ortho to OH	2	6.82	17
Aromatic meta to OH	2	7.09	17
Phenylalanine β-CH_2	1	2.95	30
β-CH_2	1	3.18	30
Aromatic	5	7.26	30
Tryptophan β-CH_2	2	3.39	27
Indole C-2	1	7.20	10
Indole C-5, C-6	1, 1	7.04, 7.12	15, 15
Indole C-4, C-7	1, 1	7.45, 7.54	18, 18

[a] McDonald and Phillips (7).
[b] From internal DSS.

ature jump, dielectric relaxation, viscoelastic relaxation, and ultrasonics (23–28). Possible explanations for that discrepancy will be discussed.

The large discrepancy between the rate constants determined by non-NMR methods and the rate constants derived from interpretation of the NMR results was explained by Ferretti *et al.* (29), using a statistical model, to be from observation of different phenomena, the NMR calculated rates being associated with the rate of nucleation of ordered structure and the non-NMR observations being related to the rate of helix propagation. In contrast, Bradbury *et al.* (30) have explained the discrepancy by suggesting that the random-coil–helix transition is rapid (observed by the non-NMR techniques), but that protonation of the random-coil moiety of the polypeptide (presumably observed by NMR) is relatively slow. The interconversion therefore involves a protonated random-coil structure and an unprotonated helix structure. This interpretation is supported by the observation of two α-CH peaks in poly-DL-alanine, which should not be able to assume a helix form (31). The additional peak was attributed to protonated species.

Several studies have revealed double peaks for the α-CH and NH resonances (19, 21, 31–35) in several polypeptides, e.g., poly-L-alanine (21, 31), poly-L-phenylalanine (33), and poly-γ-benzyl-L-glutamate (19, 34). The separate signals were ascribed to the coil and helix forms. From the peak separations, values of the exchange lifetimes (τ) were estimated to be on the order of 10^{-3}–10^{-1} sec (19, 21, 31, 32, 34).

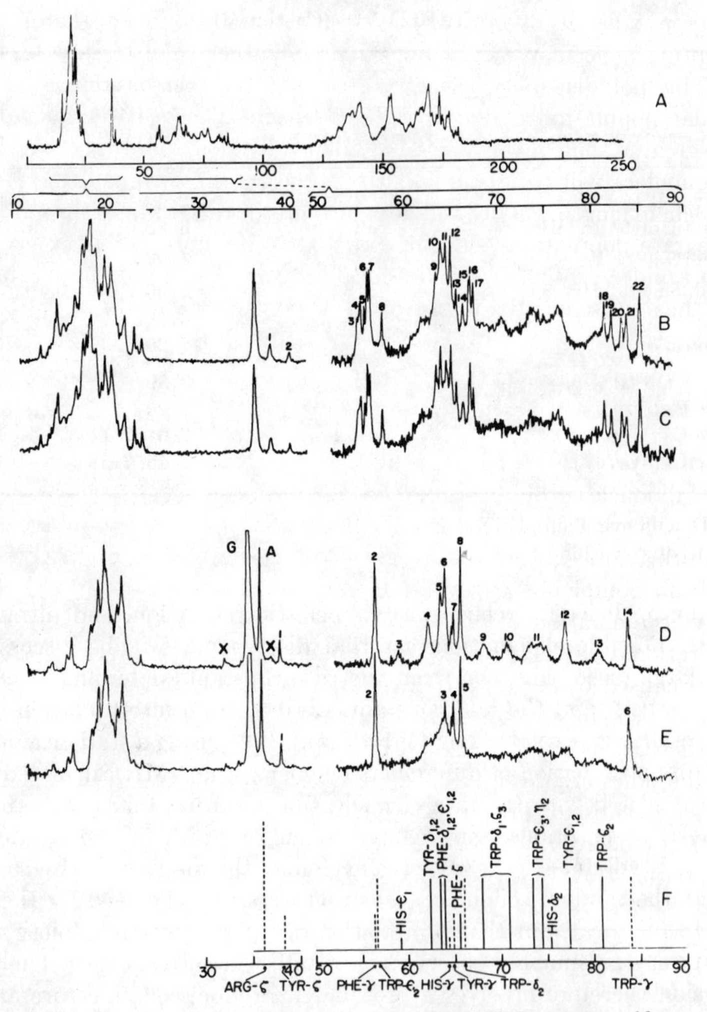

FIG. 7-4. Proton-decoupled natural abundance ^{13}C FT NMR spectra of hen egg white lysozyme ($C_{613}H_{950}N_{192}O_{186}S_{10}$) at 15.18 MHz in a 20 mm sample tube, with 4096 points in the time domain. Horizontal scale is in parts per million upfield from CS_2. (A) Fully proton-decoupled total spectrum of native lysozyme (about 20% w/v in 0.1 *M* NaCl, pH 4.1, 45°C), recorded with 250 ppm spectral width, 1.38 sec recycle time, and 32,768 accumulations (12 hr total time). (B) Fully proton-decoupled unsaturated carbon region of native lysozyme (about 25% w/v in 0.1 *M* NaCl, pH 4.0, 40°C), recorded with a 125 ppm spectral width, 1.09 sec recycle time, 131,072 accumulations (40 hr total time), 0.15 Hz digital broadening, and about 1 Hz resolution. Peaks 1–22 arise from the 28 nonprotonated aromatic carbons. (C) Same as B, but with noise-modulated, off-resonance proton decoupling. The decoupling frequency was centered at about 6 ppm upfield from tetramethylsilane. The random-noise modulation bandwidth was about 300

Another explanation ascribes the observation of double peaks to lack of polypeptide molecular weight homogeneity rather than to slow exchange (22). This polydispersity theory states that the two peaks arise from a molecular population of varying chain length. Those molecules of chain length greater than a certain value (dependent on temperature, solvent, and monomer) will be predominantly in the helical form and those with a shorter chain length will be predominantly in the random-coil form. Within the separate populations (leading to the two separate populations) there will be rapid ($\sim 10^7$ sec^{-1}) helix–coil interconversion. The polydispersity theory has subsequently found other supporters (36, 37).

7.1.2. Conformational Equilibrium and Denaturation of Proteins

The first proton magnetic resonance spectrum of a protein published, that of ribonuclease obtained at 40 MHz, contained only four broad peaks (38). It was apparent that a gross change in the protein would be required to produce a noticeable change in the NMR spectrum. The grossest change, aside from complete hydrolysis, is denaturation of the protein. Consequently, examination of the large changes induced in the spectrum on denaturation was the first use of protein NMR (1, 39–41). The large changes caused by denaturation reflect the loss of tertiary and, perhaps, secondary protein structure, giving a spectrum characteristic of a random-coil structure, as illustrated in Fig. 7-2 with 220 MHz spectra. Several detailed studies of proton NMR spectra using various methods of denaturation were subsequently reported for lysozyme (42–44), ribonuclease (7, 42), cytochrome c (42), α-chymotrypsin (45), and staphylococcal nuclease (46, 47).

Hz. (D) Fully proton-decoupled unsaturated carbon region of guanidine-denatured lysozyme (about 25% w/v in 0.1 *M* NaCl and 6.2 *M* guanidinium chloride, pH 4.3, 45°C), recorded with a 125 ppm spectral width, 1.09 sec recycle time, 65,536 accumulations (20 hr total time), 0.15 Hz digital broadening, and about 1 Hz resolution. The resonance of the 6.2 *M* guanidinium ion (denoted by G) is the only signal that is strong enough to yield detectable spinning side bands (denoted by X). Peak A is the resonance of the ζ carbons of the 11 arginine residues. Peaks 1–14 arise from the 87 aromatic carbons. (E) Same as D, but with noise-modulated off-resonance proton decoupling. Decoupling conditions were the same as for native lysozyme in C. Peaks 1–6 arise from the 28 nonprotonated aromatic carbons. (F) Simulated fully proton-decoupled ^{13}C spectrum of the unsaturated carbon region (except carbonyls) of lysozyme based on reported ^{13}C chemical shifts in peptides (see Table 7-2). Protonated and nonprotonated carbons are shown as solid and dashed lines, respectively. The chemical shifts of the histidine residue are those of the protonated form (13).

TABLE 7-2

TENTATIVE ASSIGNMENT OF CERTAIN ^{13}C RESONANCES[a]

AMINO ACID CARBON	CHEMICAL SHIFT (PPM)	
	PEPTIDE	AMINO ACID
Carbonyls	16.1–25.6	14.7–25.2
Arginine C^{ξ}		36.1
Tyrosine C^{ξ}	38.2	
Tryptophan $C^{\epsilon 2}$		56.4
Phenylalanine C^{γ}	56.4	
Histidine $C^{\epsilon 1}$	59.2	59.2
Tyrosine $C^{\delta 1}$, $C^{\delta 2}$	62.4	
Phenylalanine $C^{\epsilon 1}$, $C^{\epsilon 2}$	63.5	
Phenylalanine $C^{\delta 1}$, $C^{\delta 2}$	64.0	
Histidine C^{γ}	64.3	65.5
Tyrosine C^{γ}	64.9	
Phenylalanine C^{ξ}	65.6	
Tryptophan $C^{\delta 2}$		66.2
Tryptophan $C^{\delta 1}$, $C^{\delta 3}$		67.7
Tryptophan $C^{\delta 1}$, $C^{\delta 3}$		70.8
Tryptophan $C^{\epsilon 3}$, $C^{\eta 2}$		73.4
Tryptophan $C^{\epsilon 3}$, $C^{\eta 2}$		74.4
Histidine $C^{\delta 2}$	75.4	75.4
Tyrosine $C^{\epsilon 1}$, $C^{\epsilon 2}$	77.4	
Tryptophan $C^{\xi 2}$		80.9
Tryptophan C^{γ}		85.3
Threonine C^{β}	125.8	126.5
Serine C^{β}	131.6	132.3
Proline C^{α}	131.9	132.0
Valine C^{α}	134.1	132.1
Isoleucine C^{α}		132.7
Threonine C^{α}	133.7	132.1
Serine C^{α}	137.3	136.2
Tyrosine C^{α}	137.4	
Phenylalanine C^{α}	137.6	
Tryptophan C^{α}		137.6
Glutamine C^{α}		138.2
Methionine C^{α}		138.4
Arginine C^{α}		138.5
Glutamic acid C^{α}	138.9	137.9
Lysine C^{α}	138.9	138.3
Leucine C^{α}	140.0	138.9
Histidine C^{α}	140.2	139.2
Aspartic acid C^{α}		140.4
Cystine C^{α}		140.7
Asparagine C^{α}		140.9

TABLE 7-2—(Continued)

AMINO ACID CARBON	CHEMICAL SHIFT (PPM)	
	PEPTIDE	AMINO ACID
Alanine C^{α}		142.0
Glycine C^{α}	150.3	151.1
Arginine C^{δ}		152.1
Leucine C^{β}	153.1	152.8
Lysine C^{ϵ}	153.4	153.3
Phenylalanine C^{β}	156.0	
Aspartic acid C^{β}		156.0
Cystine C^{β}		156.2
Isoleucine C^{β}		156.5
Tyrosine C^{β}		156.8
Asparagine C^{β}		157.6
Glutamic acid C^{γ}	159.1	159.1
Glutamine C^{γ}		161.6
Methionine C^{β}		162.6
Valine C^{β}	162.6	163.4
Lysine C^{β}	162.7	162.7
Methionine C^{γ}		163.5
Proline C^{β}	163.6	163.9
Glutamic acid C^{β}	164.8	165.5
Arginine C^{β}		165.1
Glutamine C^{β}		166.2
Histidine C^{β}	166.6	166.2
Tryptophan C^{β}		166.3
Lysine C^{δ}	166.6	166.6
Isoleucine $C^{\gamma 1}$		167.9
Leucine C^{γ}	168.5	168.4
Proline C^{γ}	168.5	169.2
Arginine C^{γ}		168.7
Leucine $C^{\delta 1}$	170.6	170.6
Lysine C^{γ}	170.9	171.2
Leucine $C^{\delta 2}$	171.7	171.6
Threonine C^{γ}	174.1	173.1
Valine $C^{\gamma 1}$	175.0	174.7
Valine $C^{\gamma 2}$	176.0	176.0
Alanine C^{β}		176.3
Isoleucine $C^{\gamma 2}$		177.7
Methionine C^{δ}		178.4
Isoleucine C^{δ}		181.3

[a] All chemical shifts are expressed as parts per million upfield of CS_2. Carbon types are listed according to IUPAC-IUB biochemical nomenclature. From Allerhand *et al.* (13), Glushko *et al.* (14), and Horsley *et al.* (15).

Hollis *et al.* (45) examined the thermal denaturation of α-chymotrypsin (molecular weight $\sim$25,000) at pD 1.8 by carefully monitoring the 100 MHz peak intensity and linewidth of the methyl proton resonance region ($\sim$0.5–0.8 ppm from DSS). The effect of temperature on the spectrum is evident in Fig. 7-5. A distinct increase in peak intensity and decrease in linewidth with a transition temperature of $36 \pm 2°$ and a standard enthalpy change of 86 kcal/mole were noted for the enzyme at pD 1.8 in D_2O. The observations were interpreted in terms of a two-state process, i.e., native enzyme going to denatured enzyme.

Ribonuclease thermal denaturation and acid denaturation have been investigated by proton NMR at 220 MHz (42). Both denaturation processes result in the appearance of random-coil spectra. The differences between the spectra of thermally denatured ribonuclease and ribonuclease in CF_3COOD were attributed to the chemical shifts produced by protonation of the enzyme in the strongly acidic solution.

The extensive thermal and chemical denaturation studies of hen egg white (HEW) lysozyme via proton magnetic resonance observations have been reviewed in some detail by McDonald and Phillips (8). Thermal denaturation was viewed as a rapid two-state cooperative process. No

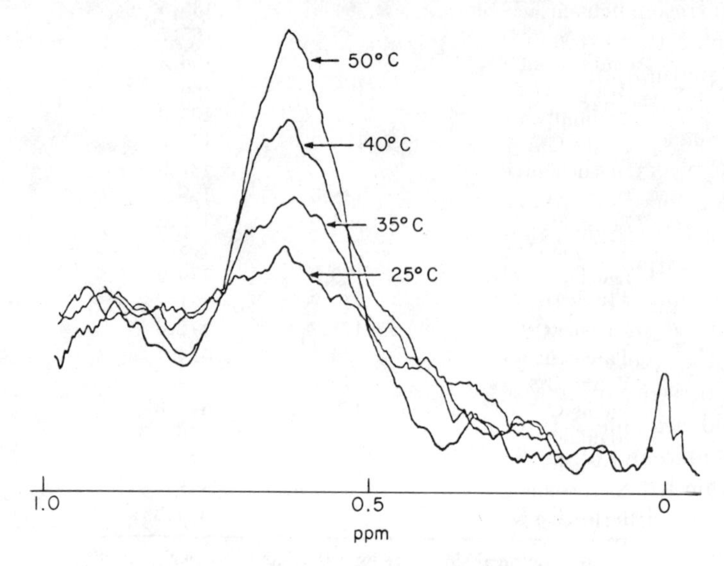

FIG. 7-5. Thermal denaturation of α-chymotrypsin (0.8% protein solution at pD 1.8 in D_2O). The spectra are the average of 25 scans. The chemical shift is relative to external sodium 2,2-dimethyl-2-silapentane-5-sulfonate (DSS) in concentrated HCl solution (45).

evidence for intermediates in the denaturation process was apparent in the spectra. The observed spectral linewidth therefore set limits on the existence of any possible intermediate forms; if an intermediate exists, it must have a lifetime $<5 \times 10^{-4}$ sec in order to satisfy the rapid exchange criterion.

McDonald *et al.* (43) also examined (in H_2O) the indole NH resonances of the tryptophan residues of HEW lysozyme, noting the effects of thermal denaturation. Five of the six resonances are resolved, with chemical shifts of 10–11 ppm from DSS (i.e., at a lower field than the histidine C2 proton resonances), in the native form. Raising the temperature to 73°C denatures the enzyme, giving a single indole NH peak with the intensity of all six protons. At temperatures just less than 73°C, both denatured and native enzyme coexist in solution, yielding spectra that are weighted superpositions of the native and denatured spectra with no apparent line broadening. The slow-exchange conditions apparent from the spectra and the length of time required to obtain a spectrum set limits of 1.4×10^{-3} sec^{-1} $< k_{ex} <$ 2×10^{2} sec^{-1} for the exchange rate between the native and denatured forms.

Bradbury and King (48) considered the question of whether or not denaturation processes for ribonuclease and lysozyme involve a single step or several steps. From the observation of changes in peak heights for six proton resonances (C2 histidine, peptide NH, phenylalanine phenyl, arginine NH, methionine methyl, and aliphatic methyl), it was concluded that denaturation of ribonuclease at pH 4.7 by addition of urea involves a single step. However, denaturation by addition of d_2-formic acid, hydrochloric acid, or potassium thiocyanate leads to two sharp methionine methyl peaks in the transition region. No methionine peak is observable in native ribonuclease. One methionine peak was assigned to the denatured form of the enzyme, and the additional peak, upfield of the usual methionine position, was attributed to a partially folded intermediate form of the enzyme. The conclusion, therefore, is that these denaturing agents promote a multistep process.

Another mode of denaturation was observed for hen egg white lysozyme in the presence of urea (48). It was noted that the heights of the histidine C2 and arginine NH proton resonances increased faster than those of methionine and aliphatic methyl with gradual addition of urea. This observation led to the suggestion that a multistep denaturation process was involved, with the surface moiety of lysozyme (containing arginine and histidine residues) unfolding before the hydrophobic core (containing methionine and aliphatic amino acid residues).

The reversible acid denaturation of staphylococcal nuclease was considered by Epstein *et al.* (47). The variation of the downfield (histidine

C2) region of the 220 MHz spectrum of staphylococcal nuclease with change in pH is shown in Fig. 7-6. At pH 5.04, the areas of all four histidine resonances are equal and the peaks are all resolved. With addition of acid, the peaks tend to coalesce into one large peak (intensity-four protons) at 8.44 ppm. The acid transition does not affect all four simultaneously, however. At pH 4.01, resonance H-4 has lost 40–50% of its area, whereas H-1 and H-3 still retain unit areas. The higher pH midpoint for H-4 implies that the enzyme moieties containing the individual histidines have different acid stabilities. The relatively sharp transitions, compared to simple protonation curves, were taken as evidence of cooperativity in the denaturation process.

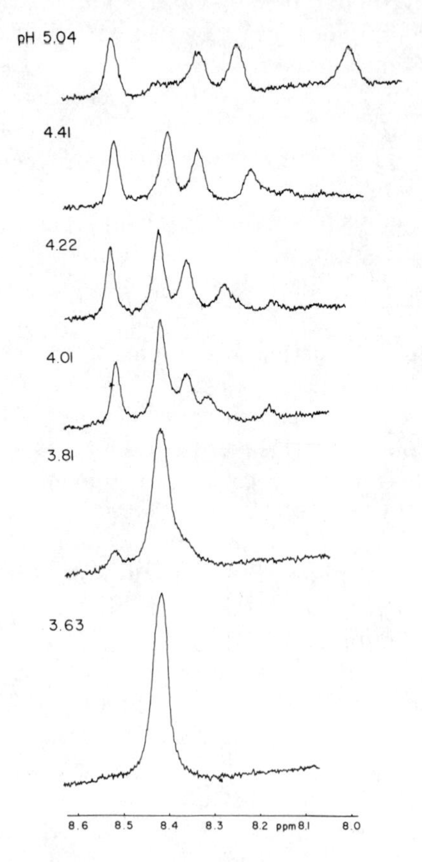

FIG. 7-6. 220 MHz proton magnetic resonance spectra of the four C2 histidine protons of staphylococcal nuclease as a function of added DCl. Each spectrum is the average of 100 scans. The origin of the small nontitratable resonance at pH 5.04 is not known. The small resonance at 8.15–8.20 ppm arising in the intermediate pH region is also of undetermined origin, although it may be related to the denaturation transition (47).

The spectra of Fig. 7-6 indicate that the conformational equilibrium between native and acid-denatured staphlococcal nuclease is slow. Using the chemical shift between the H-1 peak and the coalesced peak, it is possible to set an upper limit of 70 sec^{-1} for the exchange rate.

NMR is a very powerful tool for studying subtle conformational changes as well as gross conformational changes. Change in oxidation state, pH, and temperature and addition of substrates, inhibitors, or metal ions may promote small conformation changes that can be detected in the NMR spectra. Observations of such conformation changes will be noted at various places in following sections of this chapter.

7.1.3. Assignment of Resonances and Structural Interpretations

Assignment of resonances occurs on two levels. First, the resonance type (e.g., histidine C2 proton) must be determined. Second, it is desirable to identify a resonance with a particular residue in the protein (e.g., His-119). The first level is sufficient for certain studies, but specific assignments will yield specific structural information.

The following discussion is organized according to studies on individual proteins. Instead of a completely comprehensive description of the structural studies on each protein, however, the examples chosen for discussion were selected to show the type of structural (and sometimes kinetic) information that can be obtained from various kinds of NMR experiments with protein spectra.

A. *Ribonuclease*

As shown in the simulated spectra of Figs. 7-2 and 7-3, the only proton resonances that can generally be resolved in the spectrum of a protein in D_2O are the histidine C2 resonances. For this reason, the histidine resonances have received the most attention. It is also fortunate that histidine residues often play important roles in enzymic catalysis. In the case of bovine pancreatic ribonuclease A (molecular weight 13,700), it is known from x-ray diffraction studies that two (His-12 and His-119) of the four histidine residues in the enzyme are located in the active site cleft (49, 50) and, therefore, may be involved in the enzyme-catalyzed hydrolysis of ribonucleic acids as implied by earlier chemical evidence (51).

Three of the four histidine resonances of ribonuclease A were found to be resolved at 60 MHz (52), and all four were resolved at 100 MHz (53) as shown in Fig. 7-7. Subsequent studies have provided exact assignments for those four resonances (54). The capability of obtaining pH titration curves of the histidine proton chemical shifts has proved quite useful for identification purposes. The pH titration curves with calculated pK values

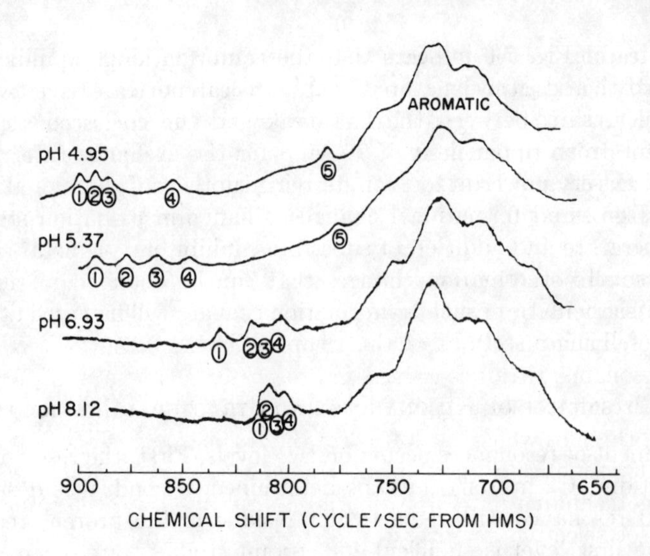

FIG. 7-7. 100 MHz proton NMR spectrum of the aromatic region of 0.012 *M* ribonuclease A in deuteroacetate buffer. Peaks 1–4 are imidazole C2 peaks of the four histidine residues and peak 5 is an imidazole C4 peak (53).

for bovine pancreatic ribonuclease A are shown in Fig. 7-8 (54). The broader line and anomalous position of peak 4 upfield from the others suggests that peak 4 is caused by His-48, which is known from x-ray diffraction work to be "buried" in the enzyme.

Chemical modification at the active site via reaction of the enzyme with iodoacetate caused both peaks 2 and 3 to shift when either His-12 or His-119 was carboxymethylated, whereas peaks 1 and 4 were unchanged (54). This is strong evidence for assigning peaks 2 and 3 to His-12 and His-119, which were in close proximity in the native enzyme. By elimination, peak 1 is assigned to His-105.

Distinction between His-12 and His-119 requires a subtle probe of the structural differences. It was found that isotopic substitution of deuterium for the C2 proton on His-12 provided the key. The *S*-peptide, which is formed by hydrolyzing the peptide bond between residues 20 and 21 with subtilisin and therefore contains His-12, was maintained at 40°C for 5 days at pH 7 in D_2O using a deuteroacetate buffer. The result of this treatment was that the C2 proton of His-12 was replaced by deuterium. The equivalent amount of *S*-protein (the other part of ribonuclease left after subtilisin treatment removes the *S*-peptide) was added to the deuterated *S*-peptide. The resulting *S'*-protein had only three histidine peaks, with pH titration curves that could be compared with those of the *S*-

protein, leading to the conclusion that peak 2 (cf. Fig. 7-7) is missing. Therefore, the assignment of peaks 2 and 3 are to His-12 and His-119, respectively, in the *S'*-protein. The assumption in the extension of this assignment from ribonuclease S to ribonuclease A is that the titration curves of His-12 and His-119 both decrease by 0.5 pH units (a *pK* of 6.7–6.2 for His-12 and 6.3–5.8 for His-119) going from the *S'*-protein to ribonuclease A.

A subsequent study by Ruterjans and Witzel (55) revealed deviations from the sigmoidal curves shown in Fig. 7-8 for the histidine C2 proton resonances. King and Bradbury (2) confirmed the deviations for the C-4 proton resonance titration curves and discussed the observed discrepancy from the results obtained by Meadows *et al.* (54). The C4 proton resonances, three of which are buried in the aromatic envelope with phenylalanine, tyrosine, and tryptophan resonances, were observed using the promising technique of NMR difference spectroscopy, which utilizes the principle of selective perturbation of only a few proton resonances in a spectrum via changes in solvent, pH, etc. (56). The difference spectrum is generated by subtracting the unperturbed spectrum from the perturbed spectrum. In principle, only those resonances which have been either

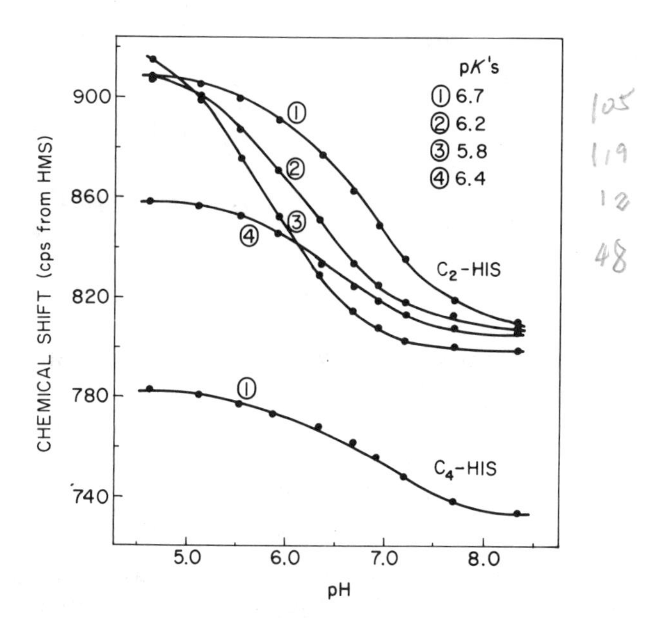

FIG. 7-8. Chemical shifts (100 MHz) of all four histidine C2 and one of the histidine C4 proton resonances as a function of pH for ribonuclease A at 32°C in D_2O using 0.2 *M* deuteroacetate buffer. The "pH" is the meter reading (54).

broadened (or narrowed) or shifted will show up in the difference spectrum. The aromatic region of ribonuclease A showing the difference spectrum resulting from spectral variations with pH change is presented in Fig. 7-9.

The influence of product inhibitors, the cytidine 2′-, 3′-, and 5′-monophosphates, on the histidine C2 resonances of ribonuclease A was investigated (57, 58). As expected, the binding of the inhibitors was found to perturb the His-12 and His-119 resonances, producing chemical shifts, line broadening, and shifts in the pH titration curves to higher pK values. For example, the complex formed with 3′-CMP revealed no change in the His-105 C2 peak but resulted in downfield shifts of 0.60 ppm at pH 5.5 and 0.82 ppm at pH 7.0, with an increased linewidth of 3–5 Hz for His-119 (57). Downfield shifts of 0.12 ppm at pH 5.5 and 0.69 ppm at pH 7.0 with no apparent linewidth increase resulted for the His-12 proton resonance in the inhibitor complex. It was also noted that the His-48 C2 resonance shifts slightly and its linewidth varies depending on 3′-CMP concentration, although His-48 is far from the active site. The noted

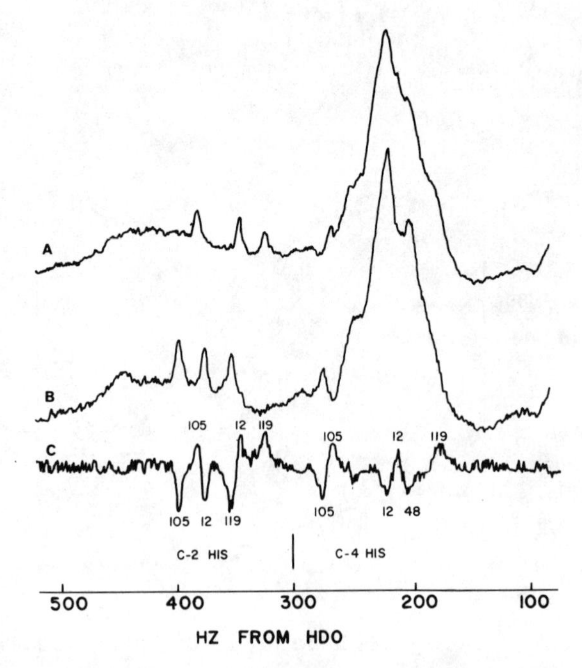

FIG. 7-9. Proton NMR spectra at 100 MHz and 34°C of 10% solutions in D_2O of ribonuclease A. pH values are meter readings uncorrected for deuterium isotope effects and were adjusted with 6 *M* DCl–HCl or NaOD. Each spectrum is the average of 30 scans downfield from HDO of the aromatic region obtained at a sweep rate of 5 Hz/sec. (A) pH 5.96; (B) pH 5.35; (C) difference spectrum = A − B. (2).

changes in NMR parameters for His-12 and His-119 have been interpreted in terms of the detailed mode of binding of the inhibitors (57, 58) and the results have been used in conjunction with independent x-ray diffraction data to propose a mechanism (59) that is an extension of an earlier proposal by Rabin and co-workers (60). The obvious assumption in using information from inhibitor complexes is that the active site geometry is not affected by the inhibitor; in certain cases it is conceivable that the inhibitor acts precisely by modifying the active site conformation.

The changes in the His-48 peak with inhibitor binding were taken as evidence for a conformational equilibrium,

$$EH' \rightleftarrows EH \rightleftarrows E + H^+ \tag{7-2}$$

This equilibrium is discussed in detail by Meadows *et al.* (58). In the case of 3′-CMP binding, it was estimated that the conformational interconversion rate is on the order of 60 sec^{-1} using the His-48 linewidth and the intermediate exchange equation.

The spin–lattice relaxation times of the histidine C2 protons of ribonuclease were measured using the $180°–\tau–90°$ pulse sequence with Fourier transform NMR (61). The surprising observation was that on addition of the inhibitors 3′-cytidylic acid or inorganic phosphate the T_1 values for His-119 and His-48 decreased, whereas those of His-12 and His-105 were unchanged. The ambiguity with respect to the previous discussion is obvious. Although the question was not resolved, it was suggested that the T_1 changes on addition of inhibitor were not the direct result of dipole–dipole interactions of the protons or phosphorus on the inhibitors. Instead, it was rationalized, the T_1 changes were either caused by a change in the side-chain motions of His-119 and His-48, or the inhibitor complexation causes a small shift in the positions of the C2 protons of those residues, forcing them to be nearer the protons on other residues and thus promoting relaxation.

The ^{13}C NMR spectrum of ribonuclease A has recently been obtained and is reproduced in Fig. 7-10 (14). The resonances were tentatively assigned on the basis of the NMR of model peptides and amino acids tabulated in Table 7-2. It is to be expected that some of the resonances that can be resolved in the ^{13}C spectrum will provide information about the environment of their amino acid residues just as the proton resonances of histidine have provided much information. Freedman *et al.* (62) have also assigned the ^{13}C resonances of the amino terminal 1-13, 1-15, and 1-20 peptides of ribonuclease A.

Initial proton NMR studies of the three hisitidines of ribonuclease T_1 have also been presented (63). The effect of pH, carboxymethylation, and addition of guanosine 3′-phosphate inhibitor were considered.

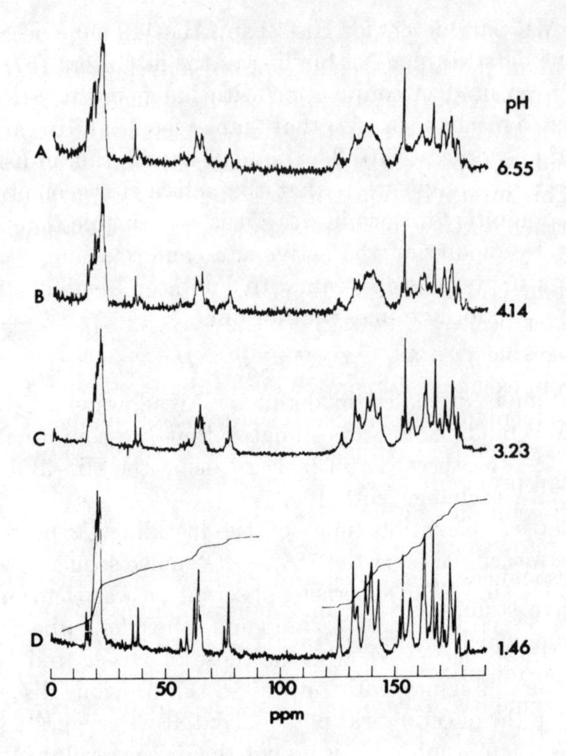

FIG. 7-10. Natural abundance ^{13}C NMR spectra of ribonuclease A. Resonance positions are given in parts per million upfield from CS_2. (A) pH 6.55, 23 mM, 16,384 scans, 1.36 sec recycle time. (B) pH 4.14, 19 mM, 16,384 scans, 1.36 sec recycle time. (C) pH 3.23, 20 mM, 53,100 scans, 0.68 sec recycle time. (D) pH 1.46, 19 mM, 31,284 scans, 1.36 sec recycle time. The probe temperature was 45°C. In (D) the integrated intensity in arbitrary units is shown in the superimposed curves (14).

B. Lysozyme

Both hen egg white and human lysozyme have been investigated by NMR, with the hen egg white enzyme (molecular weight 14,600) receiving most attention. Both isozymes have a single histidine, the C2 proton chemical shift of which has been monitored as a function of pH (53, 64). The relatively high pK value for the histidine on human lysozyme (7.1 at 32°C) implies that it is in a partially buried, negatively charged environment (64).

The function of lysozyme is to catalyze the hydrolysis of mucopolysaccharides. Mono-, di-, and tri-N-acetylglucosamines (NAG) inhibit the hydrolysis. It was observed that addition of the NAG inhibitors had no effect on the histidine resonance but that downfield shifts of certain peaks were apparent in the main aromatic proton envelope of both HEW (65) and human (64) lysozyme. It was therefore suggested that the single histidine is not at the active site (actually a cleft in the enzyme) and that aromatic residues, apparently tryptophans, are involved in binding the inhibitors. This interpretation is consistent with x-ray diffraction results and with further NMR experiments on the tryptophan indole NH proton resonances.

Most protein NMR investigations are carried out in D_2O to minimize the large water peak and to simplify the spectrum by eliminating peaks from exchangeable protons, e.g., peptide NH protons. However, the information from exchangeable protons is lost by employing D_2O as solvent. Using H_2O solvent, the resonances of slowly exchanging protons may be studied. It has been shown that studies of tryptophan indole NH proton resonances can provide valuable information about the active site of lysozyme in H_2O solution (43, 66, 67). All five of the indole NH proton resonances are resolved in the spectrum of human lysozyme and five of the six proton resonances are resolved in the spectrum of HEW lysozyme. The more extensive studies were made on tryptophan NH resonances of the HEW isozyme, which are shown in Fig. 7-11 to coalesce, forming a single peak with the intensity of six protons on thermal denaturation (66).

HEW lysozyme has six tryptophan residues at positions 28, 62, 63, 108, 111, and 123 (68). Tryp-62, Tryp-63, and Tryp-108 are in the cleft in the enzyme that accomodates the polysaccharide to be hydrolyzed (69). As illustrated in Fig. 7-11C, the resonances numbered I and II are caused by NH protons that exchange more slowly with D_2O than the others. These resonances are then assigned to Tryp-28 and Tryp-108, which are known (67) to be hydrogen bonded to the Tyr-23 and Leu-56 peptide carbonyl groups. The NH protons giving rise to peaks III and V exchange rapidly but the deuterium exchange rate is slowed when the inhibitor NAG is bound to the enzyme, as would be expected if those resonances were caused by the active site residues Tryp-62 and Tryp-63. The deuterium exchange rate for peak II was decreased even further in the presence of NAG, implying that peak II is caused by Tryp-108. The assignments of peaks III and V to Tryp-62 and Tryp-63 were strengthened by the observed chemical shifts induced in these peaks on binding of NAG and diNAG inhibitors. This is consistent with the formation of hydrogen bonds between these indole protons and oxygen atoms on the reducing residue of the β anomer of the inhibitor.

Chemical modification has also proved helpful in peak identification. Treatment of lysozyme with *N*-bromosuccinimide (NBS) selectively con-

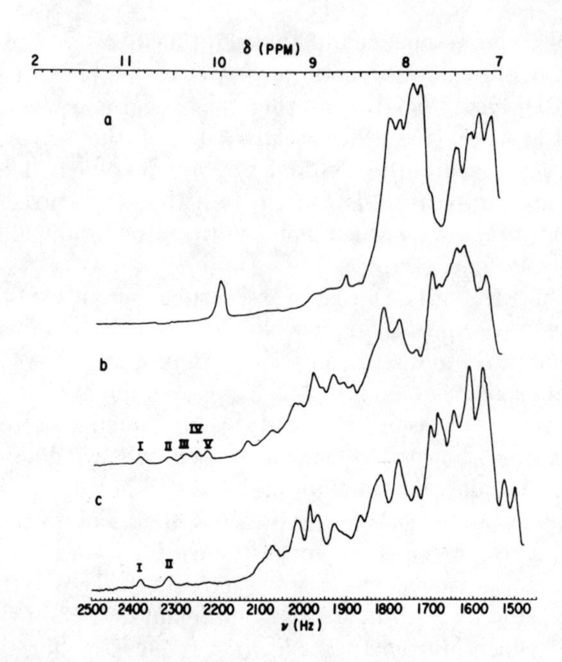

FIG. 7-11. 220 MHz proton NMR spectra of HEW lysozyme showing the tryptophan indole NH resonances (numbered): (a) Thermally denatured HEW lysozyme 15% (w/v) in H_2O at pH 3.3 and 72°C; (b) native lysozyme 15% (w/v) in H_2O at pH 3.3 and 35°C; (c) native lysozyme 15% (w/v) in D_2O at pD 3.3 and 35°C, 55 min after dissolution at 5°C. Chemical shifts were measured relative to DSS as internal standard. The pD (unbuffered) was the observed pH meter reading plus 0.40 (66).

verts Tryp-62 to its oxindole derivative. The NMR spectrum of NBS-treated lysozyme has peak III missing, suggesting that peak III arises from Tryp-62. It was also previously known that treatment of lysozyme with iodine oxidizes Tryp-108. The NMR spectrum of iodine-treated lysozyme revealed that peaks II and V had been perturbed, implying that they are caused by Tryp-108 and Tryp-63.

By piecing the evidence together, the assignments summarized in Table 7-3 can be made. Peak 4 was assigned to Tryp-123 by the process of elimination because the model of lysozyme constructed from x-ray diffraction experiments indicates that Tryp-28 and Tryp-111 are more "buried" than Tryp-123. The missing sixth peak, which is presumably lost amid the peptide NH peaks, upfield of the indole NH peaks, should therefore be caused by either Tryp-111 or Tryp-28.

At the other end of the proton magnetic resonance spectrum are the high-field methyl resonances. HEW lysozyme contains 40 methyl groups caused by eight leucine, six isoleucine, and six valine residues. According to Table 7-1, these methyl peaks will occur at 0.83–0.93 ppm (183–205 Hz at 220 MHz) in the random-coil protein. It has been observed, however, that the spectrum of native lysozyme has several peaks upfield of the main methyl envelope (8, 42, 43, 70), as shown in Fig. 7-2c.

Sternlicht and Wilson (70) calculated upfield shifts for many methyl resonances, using the x-ray crystallographic structure of lysozyme with consideration of ring current effects and electric field effects. McDonald and Phillips independently used ring current calculations (8, 42, 43) to assign high-field methyl (and some other) resonances. Where they overlap, the two studies are in good agreement. McDonald and Phillips also used the temperature dependence (42), effect of triNAG inhibitor binding (42), and effect of Co(II) binding (71) to aid their analysis. The addition of Co(II) to solutions of HEW lysozyme produced shifts in several of the upfield peaks; the shifts are probably pseudocontact in origin. These Co(II)-induced shifts permitted previously overlapping resonances to be resolved. The basis of the technique is the same as that governing the use of lanthanide shift reagents. It was shown that the probable binding site of Co(II) is provided by the carboxyl groups of Glu-35 and Asp-52. The assignments of the high-field peaks made on the basis of these studies are summarized in Table 7-4.

An attempt was made to use a nitroxide spin label, *N*-(1-oxyl-2,2,5,5-tetramethylpyrrolidinyl)maleimide, as a probe of the macromolecular structure of HEW lysozyme (72). Although the radical did covalently bind to the enzyme according to its EPR spectrum, the only apparent effect on the lysozyme spectrum was a general broadening. No selective effects were observed. A more successful spin-labeling effort was described by Wien *et al.* (73), who used a nitroxide spin-labeled saccharide bound to the active site of HEW lysozyme. It was noted that the presence of the free radical induced a slight broadening in the C2 proton resonance of His-15.

Based on initial binding studies of Gd(III) to HEW lysozyme and

TABLE 7-3

ASSIGNMENT OF THE TRYPTOPHAN INDOLE NH PEAKS OF HEW LYSOZYME[a]

PEAK	I	II	III	IV	V	VI
Residue	28(111)	108	62	123	63	111(28)

[a] Glickson *et al.* (43, 66, 67).

TABLE 7-4

HEW LYSOZYME RESONANCES OBSERVED TO HIGH-FIELD OF 1.5 PPM[a,b]

RESONANCE POSITION OR REGION [HZ (220 MHZ) FROM DSS]	ESTIMATED NUMBER OF COMPONENT PROTONS	ASSIGNMENT[c]	MAJOR SOURCE OF RING CURRENT SHIFT
−435	1	(Ile-98γ)	Tryp-63
−183	1		
−124	3	Leu-17 CH_3	Tryp-28 (Tyr-20 minor)
−64	1		
−38	3	Ile-98 δ-CH_3	Tryp-63
−8	3	Leu-17 CH_3	Tryp-28 (Tyr-20 minor)
0	3	Ile-98 γ-CH_3	Tryp-63, Tryp-108
15	3	(Leu-8 CH_3)	Phe-3
15	3	Met-105 CH_3	Tryp-28, Tryp-108 Tryp-111, Tyr-23
≈42	2		
57–104	10	2 or 3 methyl groups (Thr-51 CH_3, Ile-88 δ-CH_3) plus single protons	Thr-51, Tyr-53 Ile-88, Phe-3, His-15
104–163	23	Includes several methyl groups (Leu-8, Leu-56, Ile-55)	
163–194	16	Includes several methyl groups [Val-92, Leu-56, Ile-58, Ile-78 (2)]	
194–223	26		
224–300	86		
300–330	26		

[a] pD 5; 55°C.
[b] McDonald *et al.* (8, 43).
[c] Assignments in parentheses are considered less firmly established.

human leukemia lysozyme, it was recently proposed that lanthanide ions be used for determining the three-dimensional structure of an enzyme by observing the chemical shift and broadening effect of lanthanide ions on the proton resonances of the enzyme (74). The basis for this proposal is that some of the lanthanide ions [e.g., Gd(III)] produce line broadening of resonances (cf. Section 6.2.1), with the line broadening being a function of the metal–proton distance, whereas other lanthanide ions [e.g., Eu(III)] produce paramagnetic shifts of the enzyme resonances (cf. Section 3.3.3), with the shifts being a function of the metal–proton distance and the orientation of the metal–proton axis. Campbell *et al.* (74) used NMR difference spectroscopy to detect those resonances which has been broadened by Gd(III) binding.

The ^{13}C NMR spectrum of lysozyme has been obtained by Allerhand *et al.* (13) at 15 MHz and by Freedman *et al.* (62) at 25 MHz. The former workers combined the use of high concentrations, a 20 mm sample tube, and long-term Fourier transform averaging to obtain spectra with a high signal-to-noise ratio (cf. Fig. 7-4). The partial assignment of some of the nonprotonated aromatic carbons was made on the basis of a comparison with model peptides and amino acids (cf. Table 7-2). The partial assignments are given in Table 7-5. It is clear from the spectrum of the native enzyme in Fig. 7-4 and the assignments in Table 7-5 that all carbons of a given carbon type are not magnetically equivalent and, in certain cases, the resonances of single carbon types may be resolved. This resolution of magnetically nonequivalent resonances is a result of folding in the native enzyme. It also appeared to Freedman *et al.* (62) that the β-carbon resonances of the threonine residues (123–127 ppm from CS_2) were not all equivalent. It would seem from these preliminary results that ^{13}C NMR of small proteins should be a useful structural tool for monitoring the environment of different amino acid residues and should therefore prove complementary to proton NMR studies of these proteins.

C. Staphylococcal Nuclease

Being another small (molecular weight 16,500), available enzyme (which, incidentally, also catalyzes the cleavage of nucleic acids but with less specificity than ribonuclease), staphylococcal nuclease has received the attention of NMR spectroscopists–biochemists.

The spectrum of staphylococcal nuclease was simplified considerably by preparing deuterated analogs of the enzyme (46, 75). The only proton resonances remaining in the aromatic region were those of the single tryptophan, all those of the seven tyrosines and the C2 peaks of the four histidines. Markley *et al.* (75) observed five peaks in the histidine C2 region that titrated with pH change. Two of the peaks (labeled H2a and H2b in Fig. 7-12) were attributed to a single residue on the basis that (*1*) the area of the two peaks add to approximately one proton; (*2*) as the pH is increased from 6 to 7, the two peaks broaden and shift together, giving abnormal titration curves; and (*3*) the addition of calcium ion (which is required to activate the enzyme) at pH 7.8 increases the area of the H2b

TABLE 7-5

^{13}C CHEMICAL SHIFTS OF NONPROTONATED AROMATIC CARBONS OF HEW LYSOZYME[a,b]

NATIVE LYSOZYME[c]	DENATURED LYSOZYME[d]	SMALL PEPTIDE[e]	ASSIGNMENT[f]
37.55 (1)	38.10 (1)	38.2	Tyr C^{ζ}
39.56 (2)			
55.01 (3)[g]	56.53 (2)	56.4	Phe C^{γ}
55.25 (4)		56.7	Tryp $C^{\epsilon 2}$
55.74 (5)			
56.04 (6)			
56.29 (7)			
57.63 (8)			
63.37 (9)	64.04 (3)	64.3	His C^{γ}
63.92 (10)	64.96 (4)[h]	64.9	Tyr C^{γ}
64.41 (11)	65.75 (5)	66.0	Tryp $C^{\delta 2}$
64.96 (12)			
65.50 (13)			
66.24 (14)			
66.85 (15)			
67.03 (16)			
67.34 (17)			
81.37 (18)	83.81 (6)[i]	84.1	Tryp C^{γ}
82.11 (19)			
83.20 (20)			
83.75 (21)			
85.22 (22)			

[a] Allerhand *et al.* (13).

[b] Chemical shifts in ppm upfield from CS_2.

[c] Accurate to ±0.1 ppm. Numbers in parentheses are peak designations in Fig. 7-4B. Sample conditions are those of Fig. 7-4B,

[d] Accurate to ±0.1 ppm. Numbers in parentheses are peak designations in Fig. 7-4E. Sample conditions are those of Fig. 7-4D, E.

[e] See Table 7-2. The chemical shifts of the histidine residue are those of the protonated form.

[f] IUPAC-IUB nomenclature.

[g] Shoulder (see Fig. 7-4B, C).

[h] More intense resonance of a partly resolved doublet. The weaker component is at 64.8 ppm.

[i] This resonance shows some splitting.

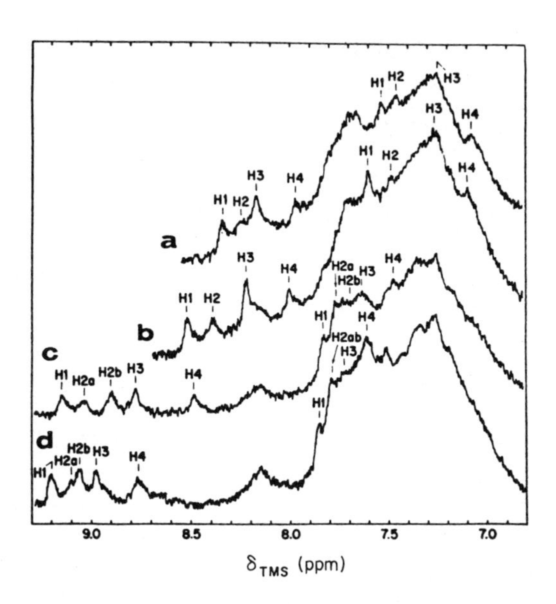

FIG. 7-12. Proton nuclear magnetic resonance aromatic spectral region of staphylococcal nuclease at four different pH values. Assignments: His C2 protons (low field) and C4 protons (high field), H1 through H4. (a) pH 7.42, 87 scans; (b) pH 7.00, 50 scans; (c) pH 5.33, 50 scans; and (d) pH 4.72, 50 scans (75).

peak at the expense of the H2a peak. These observations were interpreted as suggesting a highly localized (no other aromatic resonances were perturbed), slow conformational equilibrium affecting only the single H2 histidine. The equilibrium constant (calculated from the ratio of the areas of the H2a and H2b peaks) was calculated to be 0.5 at pH 4.7. A lower limit of 40 msec for the lifetime of the histidine in each conformation at pH 4.7 was calculated using the observed chemical shift between the H2a and H2b peaks and the slow-exchange equation. The equilibrium and kinetics were both pH dependent. Addition of the calcium ion activator or 3′,5-thymidine diphosphate (pdTp) inhibitor shifts the equilibrium in favor of one conformational form, as indicated by an increase in the area of the H2b peak.

The addition of pdTp also produced other changes in the aromatic spectral region (76). Titration of nuclease with pdTp caused shifts in three of the tyrosine resonances. One of the binding curves was anomalous, leading Markley and Jardetzky (76) to infer that a second conformational equilibrium localized at or near one tyrosine exists for nuclease only in the presence of the inhibitor; calcium has no effect on this equilibrium.

Niebor *et al.* (77) have recently studied the interactions of the rare earth ions Gd(III) and Dy(III) with nuclease. The rationale behind the use of these ions was to replace calcium ion, which activates the enzyme but has few observable physical properties, with the paramagnetic lanthanide ions, which have several interesting spectroscopic properties. Linewidth measurements of the four histidine C2 proton resonances of nuclease in the presence of Gd(III) or Dy(III) yielded ratios of metal-to-proton distances in reasonable accord with those determined by x-ray crystallography. A detailed structural interpretation with nuclease containing lanthanide ions must be considered cautiously, however, because the lanthanide ions inactivate rather than activate nuclease.

D. Carbonic Anhydrase

Several isozymes exist for carbonic anhydrase (molecular weight $\sim$30,000), an enzyme that catalyzes the reversible hydration of carbon dioxide. Cohen *et al.* (78) have examined the pH titration behavior of the histidine C2 proton resonances of the B and C isozymes of human carbonic anhydrase and the B isozyme of bovine carbonic anhydrase. At pH 5.93, it was possible to completely resolve all eight histidine C2 resonances of human carbonic anhydrase C. Addition of saturating amounts of the inhibitor acetazolamide to the carbonic anhydrases produced many changes in the spectrum, implying the existence of inhibitor-induced conformational changes.

E. α_s-Casein B and Phosvitin

Phosphorus-31 NMR spectroscopy has been used by Ho *et al.* (79) to investigate the chemical nature of the phosphorus in phosvitin (molecular weight $\gtrsim$45,000) from hen egg yolk and in bovine α_s-casein B (molecular weight 27,300). In both proteins, the phosphate groups are apparently quite mobile, with linewidths sufficiently narrow that the phosphorus–proton splitting of 5–8 Hz is resolved. The pH behavior of the ^{31}P chemical shifts and the phosphorus–proton coupling constants of the proteins were compared with those parameters for model compounds (including *O*-phosphoserine). On the basis of the comparison, it was proposed that the phosphate groups on both phosvitin and α_s-casein B are present as serine monoesters. In the case of phosvitin, this suggestion was hypothesized *a priori* because of the nearly 1:1 ratio of serine to phosphate (121 serine residues and 119 phosphate groups).

F. Human Serum Lipoproteins

Hamilton *et al.* (80) have obtained the natural abundance ^{13}C spectra of very low density (0.95–1.019), low density (1.019–1.063), and high density

(1.063–1.21) lipoprotein fractions. Aside from the broad protein carbonyl peak, the other peaks were identified by comparison with ^{13}C spectra of cholesteryl acetate, lecithin, and triglycerides. The lack of prominent protein peaks was ascribed to two effects. First, there is a much greater diversity of carbon types in the protein than in the lipid moiety. Second, most protein carbons have short transverse relaxation times and, consequently, broad lines. In contrast, many lipid carbons have narrow lines as a result of short correlation times or lack of a directly bonded proton. Differences in spectra from the very low density, low density, and high density lipoprotein fractions were attributable to differences in lipid composition (80).

G. *Flavodoxin*

Flavodoxins are small flavoproteins (MW 15,000–22,000) that replace ferredoxins as low-potential electron carriers in a number of organisms. The purified proteins all contain one flavin mononucleotide (FMN) per molecule, with no other prosthetic groups or bound metal ions. The presence of the FMN ring system leads to three possible oxidation states for flavodoxin. The fully oxidized and fully reduced states are diamagnetic, whereas the semiquinone form is paramagnetic. The dominant effect of the unpaired electron in the flavodoxin semiquinone is manifested in nuclear relaxation, as monitored by linewidths, rather than contact shifts because its electron spin relaxation time is very long; a value of the order of 10^{-8} sec can be estimated from the electron paramagnetic resonance linewidth. The flavin free radical selectively broadens the resonances of protons on amino acid residues located near the FMN ring, permitting, in principle, the identification of amino acid residues in the vicinity of the active site. The semiquinone form of flavodoxin therefore provides a natural site-specific "spin label" of the active site environment without the complications arising from introducing an artificial spin label.

James *et al.* (81) examined the proton NMR spectra of flavodoxins derived from *Peptostreptococcus elsdenii* and *Clostridium* MP for each of the three oxidation states. It was found that some of the resonances in the semiquinone spectra of the two flavodoxin species were broadened beyond detection relative to the resonances in the oxidized and reduced forms (cf. Figs. 7-13 and 7-14). Aside from the broadening of some resonances in the semiquinone spectra because of the unpaired electron, there are no evident differences among the oxidized, reduced, and semiquinone spectra for either species. The implication is that there is no apparent conformational change with variation in oxidation state.

Certain peaks (marked by the dashed vertical lines in Figs. 7-13 and

7-14) that disappear in the semiquinone spectra are shifted upfield from their usual resonance positions. The upfield shift is a consequence of ring current effects from the FMN ring system, and the disappearance is a consequence of proximity to the unpaired electron of the FMN in the semiquinone state. By comparison with x-ray crystallographic models, tentative assignments of the disappearing peaks may be made. The peaks at 5.95 ppm and 6.39 ppm in the *Clostridium MP* flavodoxin spectra probably

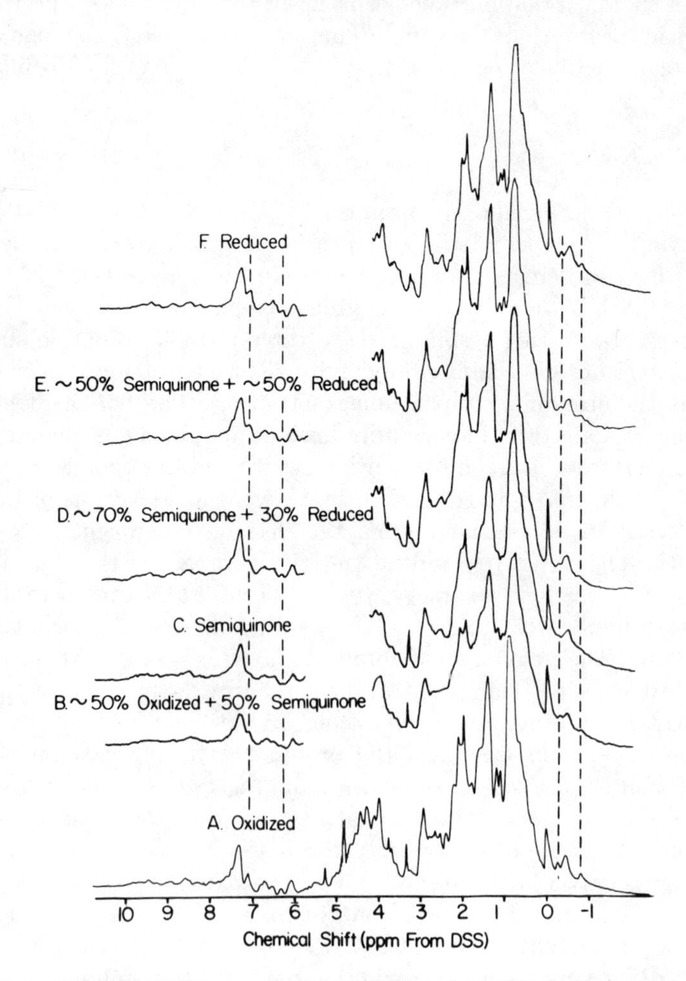

FIG. 7-13. 220 MHz NMR spectra of 4.8 m*M* flavodoxin (*Peptostreptococcus elsdenii*) in pD 7.5 potassium phosphate buffer under various conditions of oxidation or reduction. Each spectrum is the Fourier transform of the sum of 3000 transients with a 0.2 sec acquisition time. The vertical scale is not the same for all spectra (81).

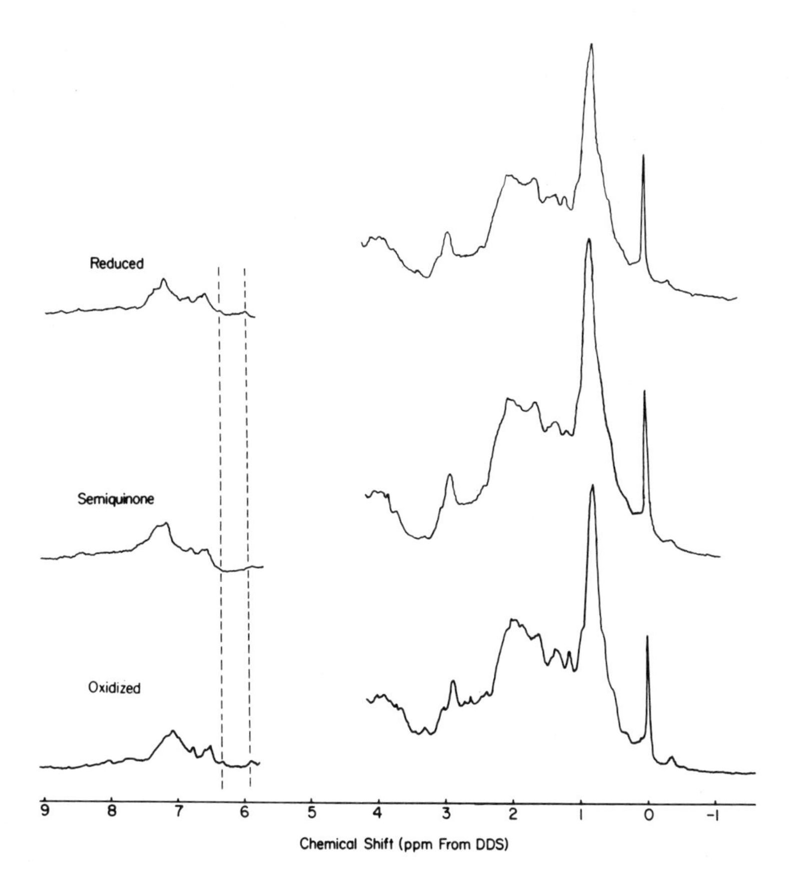

FIG. 7-14. 220 MHz proton NMR spectra of 4.5 m*M* flavodoxin (*Clostridium MP*) in pD 7.5 potassium phosphate buffer in three different oxidation states. Each spectrum is an average of ten CW scans (81).

arise from Tryp-90, which is stacked with the FMN ring system. It was also suggested that the aromatic peaks broadened beyond detection in the *P. elsdenii* semiquinone spectra may be caused by Tryp-91 (6.32 ppm and 7.10 ppm). In both species, the broadening of the remaining Tryp-90 or Tryp-92 resonances is not detected because of overlap with other aromatic peaks.

Peaks in the *P. elsdenii* spectra upfield of DSS (cf. Fig. 7-13) are located at -0.27 ppm, -0.43 ppm, and -0.81 ppm and have intensity ratios of 3:11:3, respectively, according to intensity measurements made with a continuous wave spectrum. The two peaks at -0.27 ppm and -0.81 ppm are absent in the semiquinone spectrum. From the x-ray crystallographic model, it was suggested that these two resonances are caused by the Ala-56 methyl and Met-57 methyl.

The perturbed resonances lose intensity rather than broaden as the semiquinone form is approached from either the oxidized or reduced state. This indicates that there is slow exchange between oxidized and semiquinone forms and between reduced and semiquinone forms. Using the slow-exchange equation and the linewidth of the perturbed resonances, an upper limit of 50 sec^{-1} may be placed on the electron exchange rate.

H. Heme Proteins

The unpaired electron in the flavodoxin semiquinone has a long electron spin relaxation time and, consequently, promotes broadening in the resonances of nuclei proximate to the FMN ring. In contrast, the unpaired electron on the iron in heme proteins generally has a short electron spin relaxation time and, consequently, produces large contact or pseudocontact shifts in the resonances of some protein nuclei (cf. Section 3.3).

The large contact shifts of some proton resonances observed with heme proteins provide valuable information about the structure, subunit interactions, and ligand complexation of the heme proteins (6). The contact shift itself may give information about the electronic structure of the heme group (82–85). The resonances of protons near the paramagnetic iron are shifted via hyperfine interactions to regions of the spectrum away from the envelope of broad overlapping peaks usually observed in protein NMR spectra. The hyperfine shifted resonances can easily be resolved and used as probes of changes in the heme environment in hemoproteins (86–88).

The iron–porphyrin (heme) structure, which characterizes heme proteins and provides the active site, is shown in Fig. 7-15; the figure caption describes some of the structural differences. Important features relating to function are the state of the iron, the nature of the sixth axial ligand, the nature of the ring side groups, and polypeptide–heme interactions. These structural features may be manifest in the NMR spectra of the heme proteins.

As discussed in Section 3.3, the contact shifts depend on the oxidation state and spin state of the iron atom (cf. Eq. 3-5). The spin state, in turn, depends on the nature of attached ligands. Some ligands produce a low-spin state, whereas others produce a high-spin state. The low-spin ferrous state is diamagnetic [Fe(II), $S = 0$], whereas the high-spin ferrous state [Fe(II), $S = 2$], low-spin ferric state [Fe(III), $S = \frac{1}{2}$] and high-spin ferric state [Fe(III), $S = \frac{5}{2}$] are paramagnetic. In addition to changing the value of S, a change in spin state leads to a different value of τ_s, the

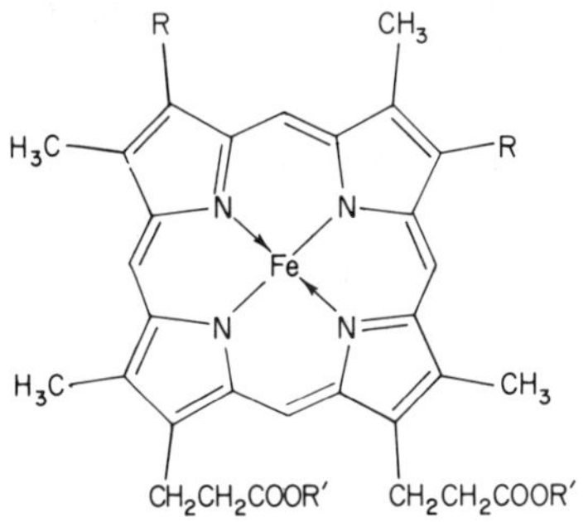

FIG. 7-15. Structure of iron–porphyrin complexes. Protoheme IX (Proto): R = —CH=CH$_2$; Deuteroheme IX (Deut): R = —H; Mesoheme IX (Meso): R = —CH$_2$—CH$_3$; cytochromes: R = —CH(CH$_3$) S—polypeptide chain. R′ = —H in the heme groups of hemoglobin, myoglobin, and cytochrome c. A fifth axial ligand is often a histidine residue from a polypeptide chain, e.g., in native hemoglobin, myoglobin, and cytochrome c. The sixth ligand, if any, varies depending on the function and state of the heme protein.

electron spin relaxation time (cf. Table 6-1). Both oxidized and reduced cytochrome c are in the low-spin state. Native hemoglobin (Hb) and myoglobin (Mb) are in the high-spin ferrous state but, on complexation with O_2, convert to the diamagnetic low-spin ferrous state.

The effects of other structural features on the NMR spectra have also been examined. It has been apparent that the electron spin densities (causing contact shifts) vary considerably from one heme protein to another, and the variation is also considerable at different locations in the same heme. A proton NMR study of spectra for several myoglobins from different species, where some of the amino acids are different because of evolutionary mutations, and for chemically modified myoglobins was made (89). Substitution or modification of amino acids outside the van der Waals radius from the porphyrin ring had no apparent affect on the spin density distribution. In contrast, the nature of the amino acid bound in the axial position of the heme iron is quite important (82). The hyperfine shifted resonances of cyanoferrimyoglobin, cyanoferrideuteromyoglobin, cyanoferrimesomyoglobin, and cyanoferricytochrome c, which all have a cyanide ion and a histidine residue in axial positions, are nearly the same in spite of the variations in ring substituents (see Fig. 7-15) and differences in the polypeptide chains of the proteins. The NMR spectra of these molecules is distinctly different from those of cyanoferriprotoporphyrin IX, cyanoferrideuteroporphyrin IX, and cyanoferrimesoporphyrin IX, which each have two cyanide ions as the axial ligands, and from ferricytochrome c, which has a histidine and a methionine in the axial positions.

Cytochrome c. As indicated in Fig. 7-15, the side chains R of cytochrome c (molecular weight 12,400) have a thioether linkage holding the heme to the single polypeptide chain. The sulfur atoms for the thioethers come from Cys-14 and Cys-17 of the polypeptide. One of the axial ligands is His-18. Although there are more than 30 species variations for cytochrome c (90), the Cys-14, Cys-17, and His-18 residues are always present. The cytochromes c of all species variants function as mitochondrial electron carriers in the respiratory pathway.

The first contact shifts in proteins were observed in cytochrome c and myoglobin by Kowalsky (40), who ascribed the shifts (up to 35 ppm) to unpaired electron spin density around the porphyrin ring resulting from $d\pi$–$p\pi$ bonding with the low-spin paramagnetic iron. Both positive and negative (upfield and downfield) contact shifts were observed owing to negative and positive spin densities on carbon atoms of the porphyrin ring as a result of the nonalternant π-electron system of the five-membered pyrrole ring constituents of prophyrin.

Further studies on cytochrome c have been concerned largely with assigning a resonance to the sixth axial ligand and identifying that ligand. The evidence accumulated to date indicates that the sixth ligand is methionine-80 coordinated to iron through the thioether sulfur (91–94).

The spectrum of oxidized cytochrome c, ferricytochrome c, is shown in Fig. 7-16. (It should be noted that a positive shift now denotes a peak upfield from DSS, contrary to previous discussion in this chapter). Most of the positive and negative contact-shifted resonances can be accounted for by porphyrin ring constituents (91). However, the peak at +23.2 ppm with a five-proton intensity cannot be attributed to any heme groups basis of symmetry considerations for the electronic wave functions of heme and because no large negative spin density is expected on the porphyrin ring carbon atoms (cf. Eq. 3-9). The implication is that the peak is caused by the sixth ligand, which, in consideration of all evidence, is most probably Met-80, the five-proton intensity being caused by an accidental degeneracy of the Met-80 methyl and γ-methylene proton resonances. The lineshape implies that the resonance is in fact composed of two overlapping lines. The addition of cyanide to ferricytochrome c results in marked changes in the spectrum (cf. Fig. 7-16) as cyanide replaces the sixth ligand, converting the iron to a high-spin state. There are no resonances of cyanoferricytochrome c above 2 ppm with an intensity greater than one proton. This further substantiates the Met-80 assignment. It was also noted that the reduction of cyanoferricytochrome c with ascorbic acid, giving diamagnetic ferrocytochrome c, caused the disappearance of all contact shifted resonances and the appearance of four new resonances between 1.9 and 3.7

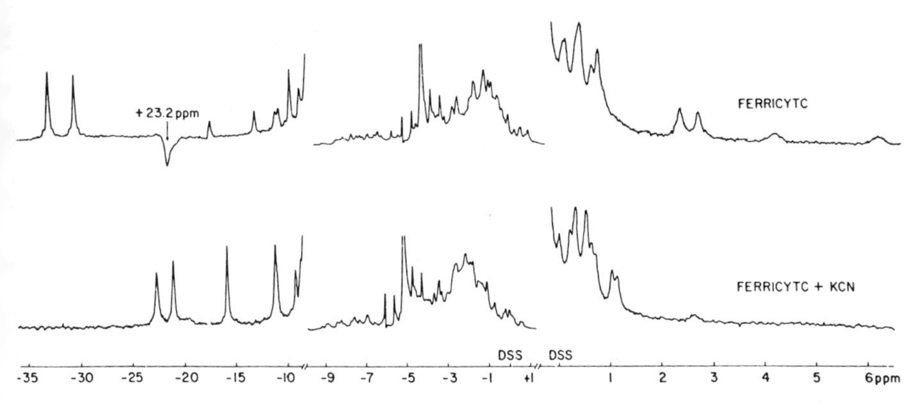

FIG. 7-16. Proton NMR spectrum at 220 MHz (pD 7.0, 35°C) of ferricytochrome c and cyanoferricytochrome c. No DSS was added to these samples. The sharp lines between −4 and −6 ppm correspond to the HDO resonance and its first and second spinning side bands. The vertical and horizontal scales are different for the three parts of the spectrum. The high-field line at +23.2 ppm is observed as an inverted resonance of the center band of the spectrum. (The HR-220 spectrometer operates with a 10 kHz field modulation. Usually one observes the first upfield side band. If large hyperfine shifts occur, parts of the center band and the different side bands of the spectrum overlap.) (91.)

ppm, including a peak with a three-proton intensity at about 3.2 ppm (91).

McDonald *et al.* (92) examined the spectra of nine mammalian-type ferrocytochromes c and found little species variation, obtaining ferrocytochrome c spectra nearly identical to that observed by Wüthrich (91). On the basis that other aromatic systems in a protein do not generate ring current fields sufficient to shift resonances more than 2 ppm upfield, it was concluded that the three resonances in the 2.0–3.6 ppm region upfield of DSS must be caused by protons that are very near the porphyrin and that are normal to the rings; namely, the resonances must be caused by the axial ligands. The unperturbed resonances of the fifth ligand (histidine) are too far downfield (7–8 ppm below DSS) to be shifted upfield of DSS. The three resonances in the 2.0–3.6 ppm region were therefore attributed to the sixth ligand and could be accounted for by presuming that Met-80 is that ligand. The sharp three-proton peak was ascribed to the methyl group and the two single-proton peaks were ascribed to the γ-methylene protons, which may be found in different magnetic environments as a result of sterically hindered rotation. The conclusion is that Met-80 is the sixth ligand in both the oxidized and the reduced states.

An elegant set of double resonance experiments was performed by Redfield and Gupta (93, 94) on horse cytochrome c. With these experiments, it was possible to map the position of a resonance caused by a particular proton in the oxidized state onto its position in the diamagnetic reduced state and to estimate the electron exchange rate between oxidized and reduced forms as approximately 10^4 $mole^{-1}sec^{-1}$. The double resonance experiment was carried out with a sample composed roughly of half oxidized and half reduced cytochrome c. The spectrum of the mixture is that of a superposition of the oxidized and reduced state spectra. When the sample is irradiated with a strong field at the frequency of one of the contact shifted resonances of the oxidized species, the energy levels for that nuclear spin will tend to become equally populated and the resonance is saturated. The rate of interconversion of oxidized and reduced forms is sufficiently rapid that the nuclear energy level population is still in a state of non-equilibrium when oxidized cytochrome c is converted to the reduced form. Therefore, when the NMR spectrum is monitored immediately following saturation of a resonance in the oxidized state, a decrease in intensity is observed for a resonance in the reduced state spectrum. The saturation transfer (cross-saturation) then allows the resonance for a particular chemical group in the oxidized species to be correlated with the resonance of the same group in the reduced species. A schematic diagram showing the correspondance between oxidized and reduced state resonances is given in Fig. 7-17. The assignment, for example, of the Met-80 methyl to the +23.2 ppm peak in ferricytochrome c made by Wüthrich (91) and the assignment of the same methyl to the +3.2 ppm peak in ferrocytochrome c by McDonald *et al.* (92) were corroborated by the saturation transfer experiments. A summary of the methyl resonance correlations is given in Table 7-6.

A saturation transfer experiment using a sample composed of half ferricytochrome c and half azidoferricytochrome c was used to confirm that the methyl peaks at −7.2 and −10.3 ppm (i.e., downfield of DSS) were indeed heme-ring methyl resonances because they moved further downfield to −17.3 and −14.8 ppm (cf. Fig. 7-17) in the azide complex (93). Resonances at such a low field can be assigned only to ring methyls (91).

Largely on the basis of the NMR studies showing evidence of Met-80 coordination to the heme iron, Dickerson and co-workers (95) revised their earlier interpretation of x-ray diffraction data and proposed a structure for cytochrome c with Met-80 as the sixth ligand.

Wüthrich *et al.* (96) examined the NMR spectra of cytochromes c modified by formylation of tryptophan, alkylation of methionine, or polymerization by ethanol treatment. It was found that the existence of methionine coordination could be correlated with the respiratory activity of

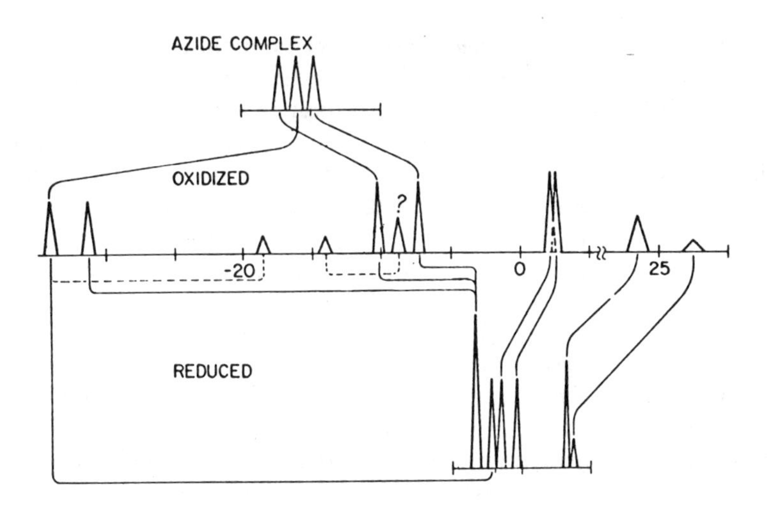

FIG. 7-17. Schematic representation of all the resonances of horse cytochrome c for which cross-saturation (saturation transfer) has been observed. The resonances at −18.5, −14, and +27.4 in the oxidized and +3.8 in the reduced state have intensities characteristic of a single spin; the intensity of the resonance at −8.5 ppm is not known and all the others are methyl resonances. Lines are drawn between resonances that are found experimentally to be connected by cross-saturation. Solid lines indicate cross-saturation occuring via exchange; broken lines indicate that occuring via interspin cross-relaxation within the same molecule (93).

the modified cytochromes c. The electron carrier properties of cytochrome c are maintained only if methionine is coordinated to the heme iron in both the oxidized and the reduced form.

The proton NMR spectra of four cytochromes c derived from photosynthetic organisms have also been obtained (97). All four have a peak corresponding to the methionine methyl resonance, 3.0–3.4 ppm upfield, and the two bacterial cytochromes c_2 have additional resonances in the 1.9–2.4 ppm range, which have been tentatively assigned to leucine and valine residues.

Myoglobin. The function of myoglobin (molecular weight 17,800) is to bind and store oxygen until it is needed for biological oxidation. The mode of oxygen binding and the changes in the protein on binding have been of considerable interest.

The spectrum of cyanometmyoglobin is illustrated in Fig. 3-10. Wüthrich *et al.* (83, 84, 89) have dealt with the problem of peak assignments. The inverse temperature dependence, predicted by Eq. 3-5, of the four ring methyls (see Fig. 7-15) and peak intensities were used to assign the low-field hyperfine shifted resonances of the heme group in sperm whale cyanometmyoglobin (83) and human cyanomethemoglobin (84). Table 7-7 compares the observed contact shifts, hyperfine coupling constants (calculated from Eq. 3-5), and the spin densities on the ring carbons adjacent to the four methyl groups (calculated from Eq. 3-9) for cyanomethemoglobin, cyanomyoglobin, and cyanoprotoporphyrin IX iron(III). Although contact and pseudocontact shifts cannot be distinguished on the basis of temperature dependence, it was shown theoretically that pseudocontact contributions to the shifts are small (85).

Although the large downfield shifts of the paramagnetic cyanide complexes cannot be caused by ring current effects (because of the observed temperature dependence), the resonances of diamagnetic oxymyoglobin that are not in the main envelope of the protein spectrum are shifted by ring currents. Shulman *et al.* (98) compared the spectra of oxymyoglobin and deoxymyoglobin (Fig. 7-18) noting ring current shifted peaks (from

TABLE 7-6

SUMMARY OF METHYL RESONANCE CROSS-SATURATION (SATURATION TRANSFER) EXPERIMENTS[a]

IDENTIFICATION	OXIDIZED STATE POSITION (PPM)	REDUCED STATE POSITION (PPM)
a	23.4	3.3
b_1	−34.0	−2.1
b_2	−31.3	−3.8
b_3	−10.3	−3.5
b_4	−7.2	−3.4
c_1	2.2	−1.4
c_2	2.6	−0.3

[a] Positions are parts per million from DSS (positive values denote upfield shifts) reference in both states and were obtained at room temperature (27°C). The oxidized state resonance was always strongly irradiated; cross-saturation occurred at the frequency indicated under the reduced state. Except for the first resonance tabulated, all the diamagnetic resonance positions inferred here occur in places in the spectrum where they are swamped by other overlapping resonances, if studied conventionally. Their observation was facilitated by automatically taking a difference spectrum with and without the strong irradiation. Identification: a, methionine methyl; b, porphyrin ring methyl; c, porphyrin side-chain methyl. From Gupta and Redfield (94).

TABLE 7-7

OBSERVED NMR CONTACT SHIFTS AND HYPERFINE COUPLING CONSTANTS OF THE FOUR RING METHYLS, AND SPIN DENSITIES ON THE RING CARBONS ADJACENT TO THE METHYL GROUPS IN CYANOPROTOPORPHYRIN IX IRON (III) AND THE HEME GROUPS OF SPERM WHALE MYOGLOBIN AND HUMAN CYANOMETHEMOGLOBIN[a]

CYANOMETHEMOGLOBIN			CYANOMETMYOGLOBIN			CYANOPROTOPORPHYRIN IX IRON(III)[e]		
δ_c[b]	A (10^5 Hz)[c]	ρ_C[d]	δ_c[b]	A (10^5 Hz)[c]	ρ_C[d]	δ_c[b]	A (10^5 Hz)[c]	ρ_C[d]
−18.6	6.8	0.90	−23.2	8.5	1.12	−12.95	4.7	0.63
−17.55	6.4	0.85	−14.4	5.3	0.70	−12.45	4.5	0.60
−12.1	4.4	0.58	−8.95	3.3	0.44	−8.8	3.2	0.43
−11.9	4.3	0.57	−8.5	3.1	0.41	−6.95	2.5	0.33

[a] Wüthrich *et al.* (84).

[b] δ_c (ppm) is the difference in the positions of corresponding resonances in the paramagnetic compounds and diamagnetic protoporphyrin IX dimethyl ester iron(II) at 25°C. Negative values denote downfield shifts.

[c] Derived from δ_c through Eq. 3-5, assuming the shift is caused entirely by the Fermi contact interaction.

[d] Derived from A through Eq. 3-9 with $Q = 7.5 \times 10^7$ Hz. ρ_c is given as the percentage of one unpaired electron.

[e] Obtained in a mixed solvent of 80% d_5-pyridine and 20% deuterium oxide.

the lack of temperature dependence) that did not correspond in the two spectra. It was inferred on this basis that oxygen binding produces some conformational changes affecting the proximity and orientation of some protons relative to aromatic groups.

The proton spectra of the cyanide complexes of myoglobins derived from porpoise, horse, and harbor seal have been examined by Wüthrich *et al.* (89). The contact shifted resonances in the three spectra were quite similar. The species variation in amino acid substitutions (35 amino acid variations) for the three species apparently did not greatly alter the polypeptide chain conformation, especially in the all-important heme vicinity.

The ^{13}C spectrum of carboxymyoglobin has recently been obtained (11). It was noted that the chemical shift (relative to CS_2) for the ^{13}CO resonance of carbon monoxide was −12 ppm, in better agreement with that of $Fe(CO)^5$ carbonyl (−19 ppm) than that of CO in water (+12 ppm).

Hemoglobin. Hemoglobin (molecular weight 64,500), which functions as an oxygen and carbon dioxide carrier, is more complicated than cytochrome c and myoglobin, being composed of four subunits with a heme

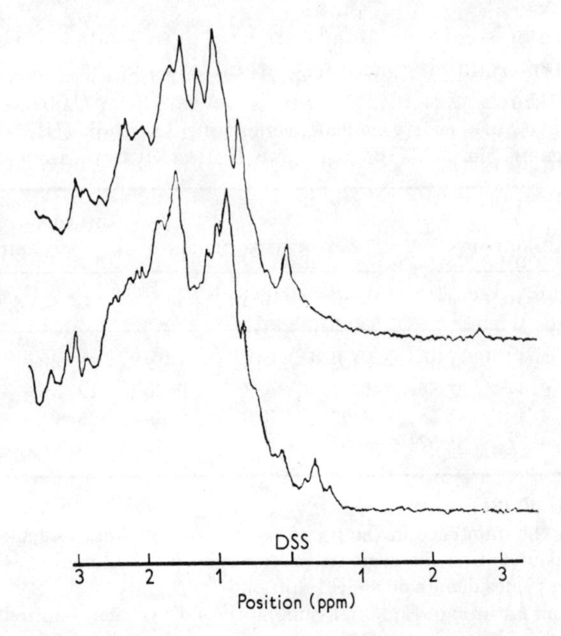

FIG. 7-18. A comparison of the upfield proton NMR region of oxy-(top) and deoxymyoglobin (bottom) at 25°C. Note in the figure that the aliphatic proton regions are grossly similar but nonetheless differ in many details (98).

group in each subunit. The two α subunits each contain 141 amino acid residues, although the subunits may not be identical. The two β subunits are not necessarily identical, but each contains 146 residues. The three-dimensional structure is arranged such that the unlike chains interact considerably via hydrophobic bonding but the like chains do not interact to any great extent. This may be represented schematically as $\substack{\alpha\beta \\ \beta\alpha}$.

It has been known for some time that binding of oxygen to hemoglobin results in a conformational change and that the oxygen binding exhibits cooperativity; i.e., the oxygen affinity increases as each oxygen is bound. The free energy of the subunit interactions leading to the observed cooperativity is approximately 3 kcal/mole, which amounts to about 10% of the total free energy change for complete oxygenation of the hemoglobin tetramer (99).

NMR has been brought to bear on the question of subunit interactions (cooperativity) as well as on the determination of the quaternary state of hemoglobin and the structural changes induced on ligand binding. Information concerning conformational changes at hemes induced by ligand binding on a neighboring heme is of special interest.

The spectra of isolated α- and β-hemoglobin subunits have been obtained as a basis for examining subunit interactions by comparison with the spectrum of the tetramer (88, 100) (cf. Fig. 7-19). It is apparent from Fig. 7-19 that the spectrum of the isolated α subunit differs from that of the β subunit, and the spectrum of the cyanomethemoglobin tetramer is not simply a superposition of the two subunit spectra (100). The fact that the spectrum of the tetramer cannot be represented by superposing the α- and β-monomer spectra was taken as evidence for subunit interactions in the tetramer. The spectral differences were also indicative of conformational changes within each subunit as the tetramer is formed. The subunit interactions obviously play an important role in the spin density distribu-

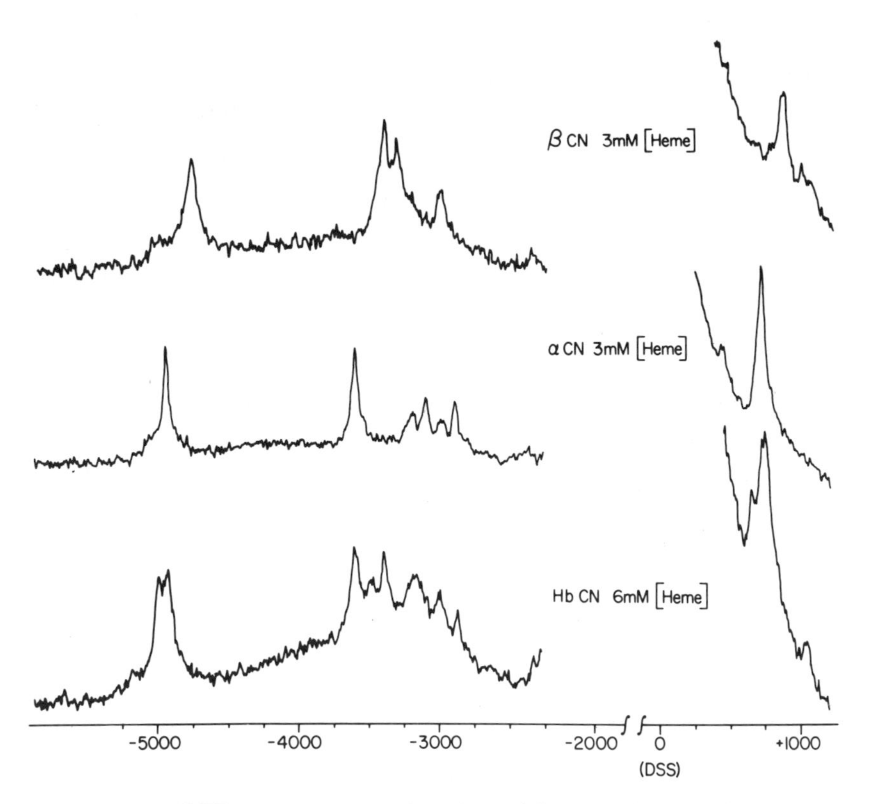

FIG. 7-19. 220 MHz proton spectra of methemoglobin cyanide and the metcyano complexes of the α and β subunits. The spectra are time averages of 50,000 transients (HbCN), 30,000 transients (β-CN), and 20,000 transients (α-CN) at 0.1 sec acquisition time. Resonances are referred to disodium 2,2-dimethyl-2-silapentanesulfonic acid (DSS) at zero on the sale (100).

tion on the heme, although the hemes are not located on the subunit interface.

In contrast to the situation with myoglobin, it has been observed that the contact shifted resonances of hemoglobin are species dependent. Yamane *et al.* (101) found the contact shifted resonances of several mammalian cyanoferrihemoglobins to vary, although the amino acid substitutions were not located near the hemes. It has also been observed that when the hemes are extricated from the β subunits, the contact shifts occurring in the α subunits are modified (102). Davis *et al.* (103) have also noted that amino acid variations in mutant hemoglobins affect the hyperfine shifted resonances by examining spectra of human deoxyhemoglobins A, F, Chesapeake, and Zurich. In the case of hemoglobin Chesapeake, it was observed that the heme groups in the different subunits were magnetically nonequivalent. In contrast to these observed differences, the spectra of hemoglobin Capetown and hemoglobin A were found to be the same (104). Apparently, amino acid variations either may or may not affect the heme moiety.

The hyperfine–shifted resonances in the region -10 to -25 ppm (i.e., downfield of DSS) have also been used to provide evidence for the nonequivalence of the α and β subunits in hemoglobins. Davis *et al.* (87) found separate porphyrin ring methyl resonances for the α and either β or

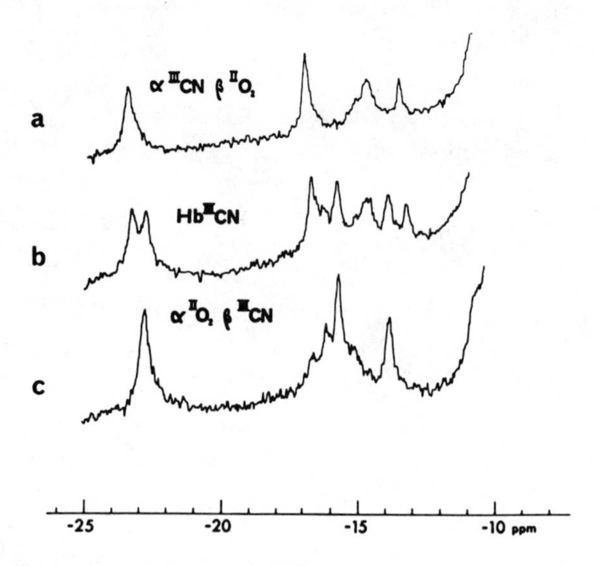

FIG. 7-20. Comparison of the proton NMR spectra of mixed-state hemoglobins and cyanoferrihemoglobin. Spectra obtained for solutions in pH 6.6 phosphate–D_2O buffer at 20°C (105).

γ chains in azide (N_3^-) complexes of human adult ($\alpha_2\beta_2$), human fetal ($\alpha_2\gamma_2$), and horse ($\alpha'_2\beta_2'$) methemoglobins. The proteins were titrated with azide with the observation that the α subunit has less affinity for azide than the β or γ subunits.

Further definitive evidence for subunit nonequivalence was provided by the study of valence hybrid hemoglobins. Ogawa and Shulman (105) observed the downfield resonances in cyanide complexes of mixed-state hemoglobins as shown in Fig. 7-20 and noted, for the two mixed-state hemoglobins $\alpha^{III}CN\beta^{II}O_2$ and $\alpha^{II}O_2\beta^{III}CN$, that the hyperfine shifted resonances from the paramagnetic ferric subunits were $\sim$1 ppm apart in the α and β subunits. This shows that the hemes in the α and β subunits do not have the same electronic distribution. That this chemical shift difference is not the result of some artifact was shown by superposition of the two mixed-state spectra, resulting in a spectrum indistinguishable from that of oxidized cyanoferrihemoglobin. Lindstrom and Ho (106) have recently shown via changes in the contact shifted resonances that oxygen is preferentially bound to the α subunit in the presence of an excess (15:1) of diphosphoglycerate, which is an allosteric effector of hemoglobin.

I. Non-Heme Iron Proteins

Some of the proteins that function as electron carriers in biological oxidation processes are small iron–sulfur proteins containing no hemes. The iron ions, however, still retain the ability to promote paramagnetic shifts.

Rubredoxin. The rubredoxins are small (molecular weight 6400) proteins containing a single non-heme iron with no acid-labile sulfur atoms. Phillips *et al.* (107) obtained the proton NMR spectra (cf. Fig. 7-21) of reduced and oxidized *C. pasteurianum* rubredoxin, which contains high-spin iron(II) in the reduced state and high-spin iron(III) in the oxidized state. Using the temperature dependence criterion, no resonances in the spectrum of the oxidized species could be unambiguously identified as being paramagnetic shifted. In contrast, paramagnetic shifts could be discerned in the reduced state spectrum, although unlike ferredoxin, which exhibits resonances extending to -29 ppm (see below), reduced rubredoxin exhibits observable resonances in the relatively narrow range of -10.5–4.5 ppm relative to the DSS reference (cf. Fig. 7-21). The observed upfield shifts were ascribed to a pseudocontact effect. The four sulfur atoms (from cysteines) coordinated to the iron(II) do not form a perfect tetrahedron. Therefore, the ligand field environment around the iron is sufficiently asymmetric to cause anisotropy in the electronic g tensor. Such anisotropy has in fact been observed in the electron spin resonance spectrum of reduced rubredoxin, and such anisotropy in the g factor can yield

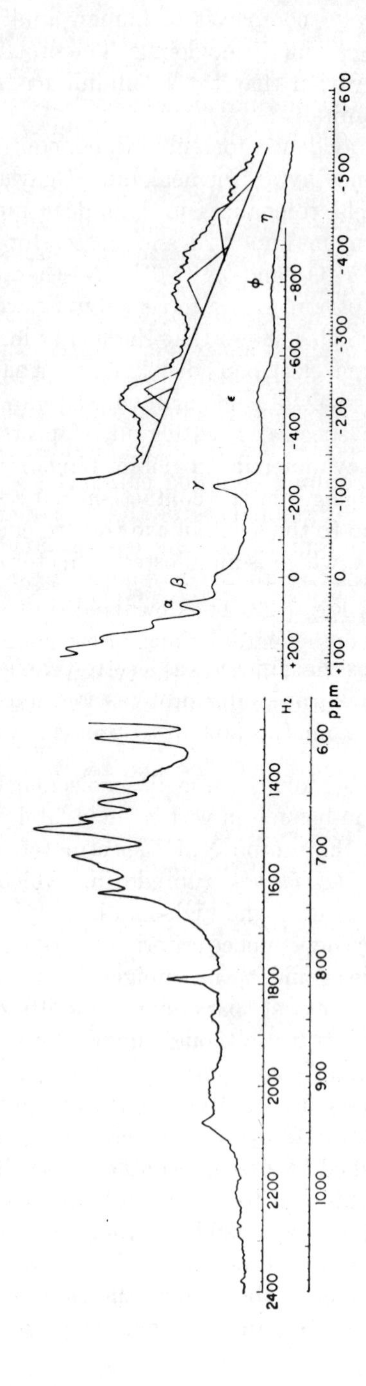

FIG. 7-21. 220 MHz proton NMR spectra of the low-field and high-field regions of reduced *C. pasteurianum* rubredoxin (5.45 m*M*) obtained at 23°C, pD 7.8 in D_2O, averaging 300 spectra for the low-field region and 36 spectra for the high-field region using a CAT. The fourfold enlarged portion of the high-field spectrum is underlain with triangles of width and amplitude appropriate for data analysis. Chemical shifts are relative to DSS (107).

pseudocontact shifts. The observation that the linewidths increased as the shift upfield increased (cf. Fig. 7-21) is consistent with a pseudocontact origin, inasmuch as the line broadening has an r^{-6} dependence (cf. Eq. 2-60) and the pseudocontact shift has an r^{-3} dependence (cf. Eq. 3-10).

Ferredoxin. Another electron carrier is ferredoxin (molecular weight ~6000), which contains eight non-heme iron atoms coordinated to the sulfur atoms of the constituent cysteines and to acid-labile sulfur atoms. The temperature dependence of the proton NMR spectrum has indicated that the many resonances extending to −17 ppm and −29 ppm (downfield of DSS) in the oxidized and reduced ferredoxins, respectively, result from paramagnetic shifts (108, 109). The large shifts have been attributed to β-CH_2 protons of cysteine residues, which coordinate the iron ions.

Carbon-13 NMR spectra of ferredoxin obtained from *Clostridium acidiurici* grown on a medium containing enriched [1-^{13}C]glycine have been studied (110). The ^{13}C resonances of the 2′,6′-ring carbons of the two tyrosine residues were found to be shifted downfield in both oxidized and reduced ferredoxin relative to the resonance frequency of free *N*-acetyltyrosine, as shown in Fig. 7-22. The downfield shifts are caused by contact interactions, pseudocontact interactions, or a combination of the two, as a result of the proximity (<4–5 Å) of the tyrosine residues to the iron ion in ferredoxin. It was shown that the observed ^{13}C peaks were caused by two tyrosine residues in magnetically equivalent environments by experiments utilizing selective proton decoupling and by ^1H NMR experiments with ferredoxin incorporating [3′,5′-$^2H_2$]tyrosine. The results of the investigation suggested that the two tyrosine residues were close to the two iron–sulfur clusters in both the reduced and the oxidized species. It was also noted that, unlike protons, the ^{13}C contact shifted resonances show little temperature dependence. In fact, no temperature dependence over the range 5°–35°C was detected.

7.1.4. Isotope-Labeled Proteins

As has been seen, one of the problems in examining the NMR spectra of biopolymers is the large number of overlapping resonances. It has been mentioned in the preceding discussion of staphylococcal nuclease that the use of partially deuterated protein greatly simplifies the aromatic proton resonance region, enabling studies of the remaining resonances to be carried out. The general idea of this selective labeling technique is that spectra are very much simplified if only one or a few resonances can be observed. Isotope labeling greatly extends the realm of possible NMR investigations to proteins that could not be previously studied because of their compli-

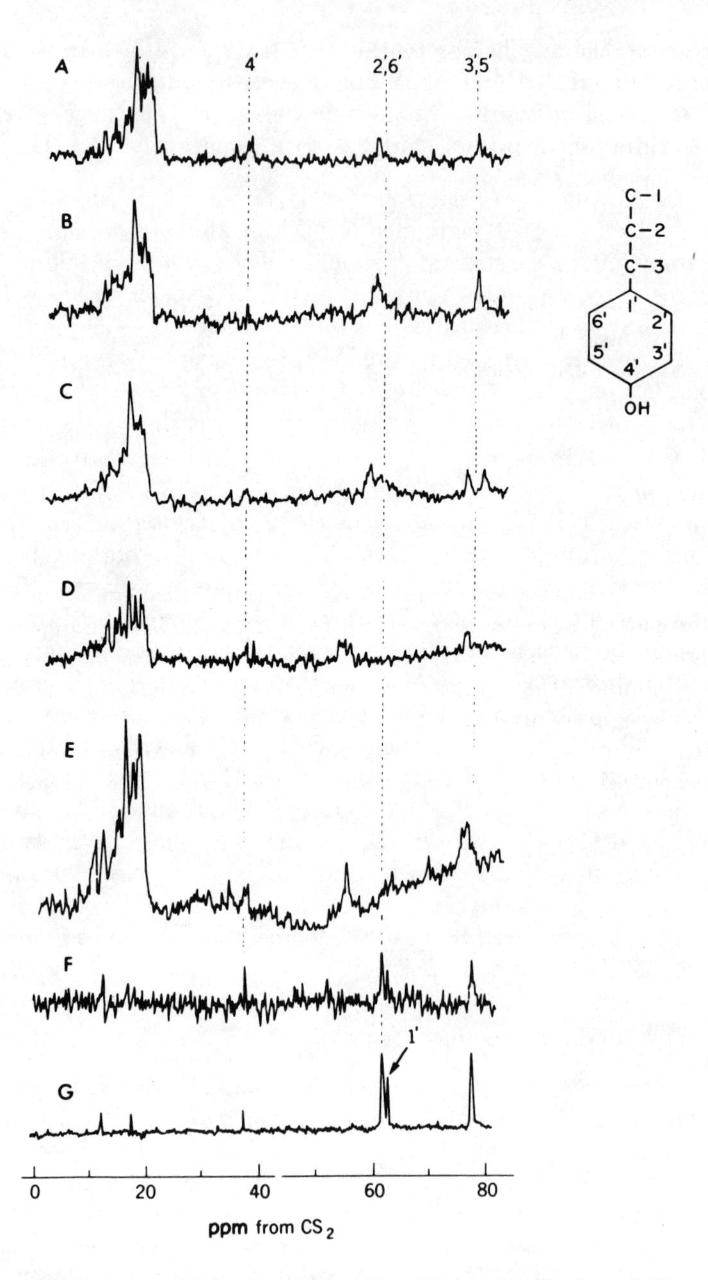

FIG. 7-22. Carbon-13 NMR (8–10°C) of *C. acidi-urici* ferredoxin and *N*-acetyltyrosine. (Sample; Preparation): (A) Oxidized ferredoxin (15 mg/ml), all protons decoupled. (B) Oxidized ferredoxin (30 mg/ml), 3′,5′-tyrosyl protons decouplied. (C) Oxidized

cated spectra. Selective isotope labeling for NMR experiments has been accomplished for ^{2}H, ^{19}F, and ^{13}C and will presumably be carried out with ^{15}N and ^{17}O in the future.

The identification of the His-12 C2 proton resonance in the spectrum of bovine pancreatic ribonuclease was predicted on its gradual disappearance over a period of a few days in the spectrum of ribonuclease because of deuterium exchange with heavy water (54). A more generalized procedure, however, is to grow an organism in a medium with heavy water incorporating a selective few protonated amino acids. The protein isolated from the organism will then be mostly deuterated, having only a few protonated amino acid residues. This general procedure is certainly feasible in the case of algae, bacterial molds, yeast, and other simple living organisms. It will no doubt also be utilized for certain more complicated organisms.

Crespi *et al.* (111–113) have utilized the general deuterium labeling procedure to prepare partially deuterated proteins. The spectrum of [^{2}H]phycocyanin incorporating protonated L-leucine, L-methionine, L-phenylalanine, L-alanine, and L-valine has been obtained following isolation of the partially protonated protein from a culture of the blue-green alga *Phormidium luridum* grown in heavy water with the above-named protonated amino acids (112). Protonated [^{1}H]flavin mononucleotide (^{1}H-FMN) was exchanged with the ^{2}H-FMN prosthetic group of a fully deuterated [^{2}H]flavoprotein isolated from the alga *Synechococcus lividus*, and the chemical shifts of the bound FMN proton resonances were observed (113).

The preparation of deuterated analogs of staphylococcal nuclease (46, 114, 155) has proved especially valuable, as already discussed. The

---

ferredoxin (30 mg/ml), 3′,5′-tyrosyl protons incompletely decoupled. (D) Methyl viologen-reduced ferredoxin (30 mg/ml), all protons decoupled. (E) Methyl viologen-reduced ferredoxin (30 mg/ml), 3′,5′-tyrosyl protons decoupled. (F) *N*-acetyltyrosine (25 m*M*) plus 100 m*M* Zn-reduced methyl viologen, all protons decoupled. (G) *N*-acetyltyrosine (250 m*M*), all protons decoupled. Double resonance experiments with the following decoupling conditions: (A and D) proton noise decoupling frequency = 100.061300 MHz; (C) coherent decoupling using ^{1}H decoupling field strength of about 550 Hz. With this decoupling field strength and the carbon resonance splittings observed in (C), the 2′,6′- and 3′,5′-tyrosine ring proton resonances were calculated to occur within a few tenths of a ppm of that observed for *N*-acetyltyrosine (F and G). This is confirmed in (B), where no detectable splittings occur. In reduced ferredoxin (E), the 2′,6′- and 3′,5′-ring proton resonances are within 1 ppm of that of the oxidized ferredoxin (B). The observed shifted resonances represent an averaged position between that of the oxidized and that of the fully reduced ferredoxins, which indicates a rapid electron exchange between oxidized and reduced forms of the protein (110).

spectrum was simplified considerably by using the deuterated enzyme, with only tyrosine (2′,6′ ring), histidine (C2), tryptophan, methionine, aspartic acid (β protons), asparagine (β protons), glutamic acid (γ protons), and glutamine (γ protons) as the protonated residues.

The preparation of deuterated proteins has been the subject of two recent reviews (116, 117). Katz and Crespi (116) have also described the use of deuterated organisms and NMR in studying the biosynthesis of clavine alkaloids and cholorophylls a and b.

Labeling a protein with ^{19}F provides a nuclear probe of high sensitivity. Huestis and Raftery (118–120) have labeled hemoglobin with ^{19}F by trifluoroacetonylating the β93 cysteine. Changes in the ^{19}F chemical shift caused by ligand binding depend on which subunit is complexed. The binding of *n*-butyl isocyanide (118), carbon monoxide (119), and oxygen (120) have been studied in this manner. Results of these studies indicate that initial oxygen molecules preferentially bind to the α subunits. The appearance of two additional ^{19}F peaks in the spectrum of partially oxygenated Hb^{TFA} reveals the existence of intermediate species (120).

The general enrichment of all constituent carbon atoms in a protein should be possible by isolating the protein from an organism grown in a medium enriched in $^{13}CO_2$. This would partially alleviate the sensitivity problem with ^{13}C NMR. The selective ^{13}C labeling of one or a few residues in a protein, however, would aid both sensitivity and peak resolution. This selective labeling has been accomplished. Saunders and Offord (121) removed the aminoterminal phenylalanine of the B chain of porcine insulin, replaced it with [1-^{13}C]glycine or [2-^{13}C]glycine, and observed the ^{13}C spectrum of the resulting selectively enriched protein. Nigen *et al.* (122) have also employed chemical modification, preparing carboxymethylated myoglobin and ribonuclease A using enriched [2-^{13}C]bromoacetic acid.

A more natural ^{13}C label was employed by Browne *et al.* (123), who examined the tryptophan synthetase α subunit (molecular weight 29,000) prepared from *E. coli* grown on a medium containing [2-^{13}C]-L-histidine. The ^{13}C spectrum of the enzyme subunit is shown in Fig. 7-23. The peak at -95 ppm (relative to the glycine α carbon) arises from the four enriched histidine C2 resonances. The other peaks are caused by natural abundance ^{13}C nuclei, resonating at about -135 ppm for the carbonyls, -116.5 ppm for the ϵ-arginine and tyrosine C-4 carbons, and -89 ppm for several aromatic resonances.

The tryptophan synthetase α subunit was also labeled with deuterium at the histidine C2 position in an attempt to reduce the histidine ^{13}C linewidth caused by ^{13}C–^{1}H dipolar coupling. It was observed, however, that the linewidth was not reduced, because the quadrupole relaxation of

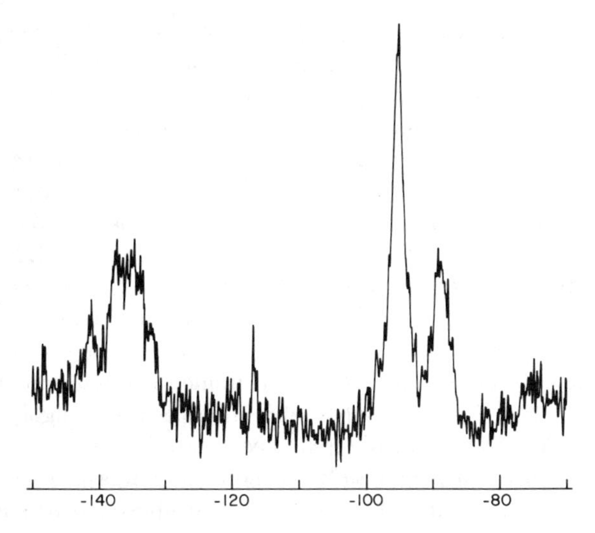

FIG. 7-23. ^{13}C NMR spectrum (102,000 accumulations) of [2-^{13}C]histidine-labled tryptophan synthetase α subunit (2.9×10^{-3} M) in D_2O at 10°C determined at 24 kGauss under conditions of proton noise decoupling. Scale: ppm relative to glycine α carbon. The 90° rf pulses were spaced at 0.51 sec intervals. Signals are artificially broadened 2.5 Hz by pretransform exponential filtering (123).

the deuterium, which is coordinated to ^{13}C, causes the ^{13}C linewidth to be broadened by scalar relaxation of the second kind. This broadening should be eliminated by using a strong rf field for decoupling deuterium.

7.2. Nucleic Acids and Polynucleotides

Proton nuclear magnetic resonance is uniquely qualified for the detailed studies of base stacking in nucleosides, nucleotides, and oligonucleotides, as discussed in Section 3.1.2, A. NMR studies have also been valuable in determining the conformation in these small molecules, indicating that the bases prefer the anti conformation with respect to the sugar–base torsion axis (124–126).

In the nucleotide macromolecules (nucleic acids), however, NMR has proved less useful than with proteins. This is simply a result of the respective monomeric compositions. Proteins are composed largely of 20 amino acids, but nucleic acids are composed for the most part of four different nucleotides (which in fact have many similar characteristics).

The featureless proton NMR spectra of very high molecular weight DNA and RNA are modified by line narrowing and chemical shifts when the molecules are denatured (127–130). McDonald *et al.* (127) observed that the resonance intensities of the resulting single-stranded nucleic acids could be used to quantitatively investigate hybridization, "melting," and partial ordering of the bases in the single strand. Schweizer (129) found relatively narrow lines resulting from the denaturation of transfer RNA (tRNA) in aqueous solutions by addition of dimethyl sulfoxide. Two well-resolved (7 Hz separation) peaks were observed in the 220 MHz spectrum of single-stranded DNA obtained by heating an aqueous DNA solution above the "melting" temperature (130). The two peaks originate from thymine methyl protons in different magnetic environments. The non-equivalent environments result from having either a purine or a pyrimidine nucleotide as the 5′ neighbor of thymidine. The relative areas of the two peaks may therefore be used to obtain the nearest neighbor base frequency ratios in DNA from different species.

As might be expected, the spectra of homopolymers give fairly narrow proton NMR lines that may be examined in studies of base stacking and partial ordering (127, 131). Mantsch and Smith (132) have also studied the ^{13}C Fourier transform NMR spectrum of polyuridylic acid (molecular weight 130,000), comparing it with the spectra of uridine, 2′-uridine monophosphate, 3′-uridine monophosphate, and 5′-uridine monophosphate. The narrow lines of the poly U ^{13}C spectrum permitted the observation of $^{31}POC^{13}C$ coupling constants, which in conjunction with the chemical shifts implied that poly U has certain conformational preferences.

Glasel *et al.* (133) used a novel approach to investigate homopolymers and copolymers. They measured the deuteron spin–lattice relaxation time (via progressive saturation) of poly-7-methylguanylic acid-d_3 and mixtures of PMG-d_3 with polycytidylic acid or polyuridylic acid. The deuterium T_1 measurements were related to the motional freedom in the polymers. An implication of these studies is that selective deuteration in nucleic acids may prove useful as it has in protein studies.

In natural nucleic acids, transfer RNA would be expected to be the most favorable case for study because of its relatively low molecular weight (~30,000). Smith *et al.* (134) studied the 100 MHz proton NMR spectra of unfractionated yeast tRNA as a function of NaCl concentration. Two broad bands consisting of (I) the resonances of the ribose 1′, uracil 5, and cytidine 5 protons at about 6 ppm, and (II) the resonances of the adenine 2 and 8, guanine 8, uracil 6, and cytidine 6 protons at about 8 ppm from DSS were found to depend on salt concentration. The peak broadening observed at higher NaCl concentrations was attributed to salt-induced aggregation of the tRNA molecules.

In consideration of the unresolved proton NMR peaks in unfractionated tRNA, two further steps may be taken in quest of a resolved peak from tRNA. The first step is to use an amino acid-specific tRNA. The second is to focus on the resonances of the "rare" bases found on tRNA which, in fact, are not really very rare, generally comprising 10–15% of the bases. This approach has been used with moderate success (135–137). Smith *et al.* (135) examined the 220 MHz spectrum of alanine tRNA above the "melting" temperature and detected the dihydro resonance of dihydrouridine and the methyl resonances of some other "rare" bases in the 2–3 ppm region (downfield from DSS).

Even with only 10–15% of the tRNA composed of "rare" nucleotides, the upfield resonances still feature broad overlapping lines. Koehler and Schmidt (137) therefore chose to investigate the tyrosine-specific tRNA from *E. coli* because it contains only four residues that should give resonances in the high-field region: ribothymidine (rT) at position 63, 2′-*O*-methylguanosine (2′-*O*-MeG) at position 17, a modified guanosine (G*) of unknown structure at position 35, and N^6-(Δ^2-isopentenyl)-2-methylthioadenosine (ms2iPA) at position 38. The resonances from the residues at positions 35 and 38 may prove to be useful monitors of tRNA functions and codon–anticodon binding processes because position 35 is the "wobble" residue on the anticodon loop. The proton NMR upfield region of $tRNA^{Tyr}_{coli}$ is shown in Fig. 7-24. The upfield region of the spectrum of the suppressor $tRNA^{Tyr}_{sut}$ (see Fig. 7-24D) was used to aid in identification of the upfield resonances because it differs from the wild type $tRNA^{Tyr}_{coli}$ in that cytosine replaces G* and ms2iPA lacks the thiomethyl group. Consideration of peak intensities, chemical shift effects, and differences between the two tRNA's led to the tentative assignment of peak 4 to the methylthio resonance of ms2iPA, peak 3 to the two isopentenyl methyl groups of ms2iPA, and peak 1 (possibly) to protons on G*.

Komoroski and Allerhand (138) have demonstrated the feasibility of examining ^{13}C NMR spectra of nucleic acids by obtaining the spectrum of unfractionated yeast tRNA and poly A in the presence of Mg^{2+}, as shown in Fig. 7-25. The lines could be easily assigned by comparison with the ^{13}C spectra of the mononucleotides. Only the ribose 4′ carbon resonance was shifted to any extent ($\sim$1.5 ppm upfield) relative to its position in the mononucleotide spectrum. The spectra also reflected the "melting" and aggregation of tRNA. Lines were broader at lower temperatures (27°–60°C range) or higher concentrations of tRNA reflecting the aggregation of tRNA. At temperatures above 60°C, the linewidth decreased markedly as a result of tRNA "melting." The ^{13}C spin–lattice relaxation times of the 4′ carbons of tRNA were measured as a function of temperature. The ^{13}C T_1 values were essentially independent of temperature

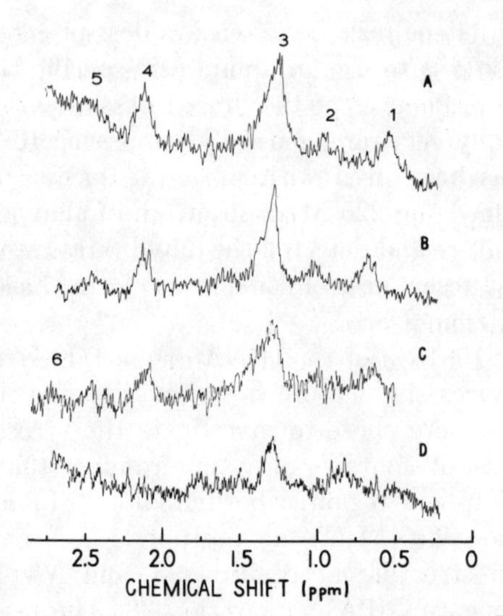

FIG. 7-24. 220 MHz spectra of the high-field region of $tRNA^{Tyr}_{coli}$. (*a*) $tRNA^{Tyr}_{coli}$; 25 mg/ml in D_2O, pH 4.5, 0.015 *M* $MgCl_2$, 0.015 *M* d_4-acetate, 28°C, 900 scans. (B) Same as (A) except 43°C, 900 scans. (C) $tRNA^{Tyr}_{coli}$; 21 mg/ml in D_2O, pH 7.0, 0.02 *M* $MgCl_2$, 0.2 *M* NaCl, 0.01 *M* Na_2HPO_4, 28°C, 900 scans. (D) $tRNA^{Tyr}_{su^+_{III}}$; ~20 mg/ml in D_2O, pH 7.0, 0.02 *M* $MgCl_2$, 0.2 *M* NaCl, 0.01 *M* Na_2HPO_4, 28°C, 953 scans (137).

below 60°C but increased rapidly above 60°C as a result of the increased segmental motion of the polyner manifesting the "melting" of tRNA.

Guéron (139) has obtained the ^{31}P NMR spectra of unfractionated tRNA, glutamic acid-specific tRNA from *E. coli*, and phenylalanine-specific tRNA from yeast. Surprisingly, the spectra did not change appreciably on heating to 70°C. A change in pD from 6.3 to 8.5 produced a large shift in part of the ^{31}P resonances of $tRNA^{Phe}_{yeast}$, although neither phosphates nor bases titrate in this range.

One of the more promising applications of NMR has come as a result of running the proton NMR spectra of specific tRNA's in H_2O rather than the usual D_2O. This "trick" enables the resonances of exchangeable protons to be observed. As discussed in Section 3.1.2, the resonances of protons involved in hydrogen bonds are shifted downfield. This phenomenon has permitted the resonances for protons in the Watson-Crick hydrogen bonds, which are responsible for maintenance of the cloverleaf secondary structure (cf. Fig. 7-26) of tRNA, to be observed downfield of the other resonances in the spectra of aqueous solutions of $tRNA^{Phe}_{yeast}$, $tRNA^{Phe}_{coli}$, $tRNA^{fMet}_{coli}$, and yeast $tRNA^{Lev}_{3}$ (140–142), Assignments of resonances to particular proton types has been made on the basis of comparisons with model systems, pH dependence, and temperature dependence. The 220 MHz proton NMR spectrum of yeast phenylalanine tRNA in aqueous solution is shown in Fig. 7-27 (141). A partial assignment of the resonances in the region 6–7 ppm may be ascribed to free amino protons. The resonances in the broad hump at 7–9 ppm are caused by non-hydrogen bonded adenine H8 and pyrimidine H6 protons and to hydrogen bonded amino protons. The small peaks at 9.5–11 ppm cannot be unequivocably assigned but may be caused by hydrogen bonds involved in the tertiary structure of the $tRNA^{Phe}_{yeast}$. The resonances in the region of 11.5–15 ppm are caused by protons in the hydrogen bonds of the A–U and G–C Watson-Crick

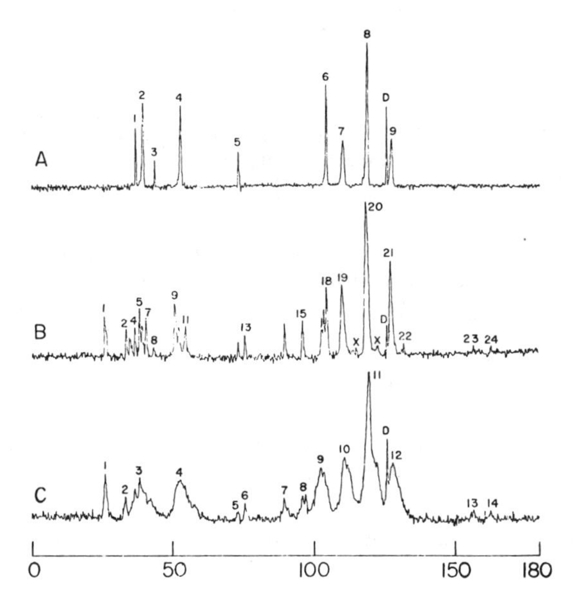

FIG. 7-25. Proton-decoupled natural abundance ^{13}C NMR spectra of aqueous nucleic acids. Spectra were recorded at 15.18 MHz, with 2048 points in the time domain and 250 ppm sweep widths. Resolved peaks are numbered consecutively. Dioxane (peak D) was added as an internal standard for chemical shift measurements. Horizontal scale is in ppm upfield from CS_2. (A) Polyadenylic acid, potassium salt, 87.5 mg/ml, pH 6.92, 59°C, and 16,384 accumulations, with a recycle time of 0.68 sec. (B) Unfractionated tRNA from Baker's yeast (undialyzed), 200 mg/ml, 80°C, 8192 accumulations, and a recycle time of 2.7 sec. 1030 points in the "tail" of the time domain spectrum were set to zero. Peaks marked X are from decomposition products. Peak 22 is an impurity and disappears after dialysis. In order to minimize decomposition, the spectra were recorded in four batches of 2048 scans, on fresh samples, and added digitally. (C) Unfractionated tRNA from Baker's yeast (dialyzed), 200 mg/ml, 52°C, 131,072 accumulations, and a recycle time of 0.34 sec. 1033 points in the "tail" of the time domain spectrum were set to zero (138).

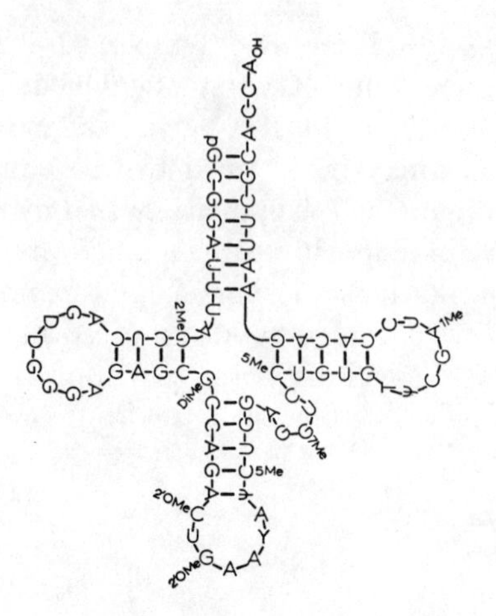

FIG. 7-26. The cloverleaf model for yeast phenylalanine tRNA showing the secondary structure resulting from hydrogen bonding between the complementary base pairs adenine–uracil and guanine–cytosine (141).

base pairs, with the A–U proton resonances being at lower fields. The total number of base pairs observed with NMR was 20 ± 2, which agrees with that predicted from the cloverleaf model, 20 ± 1 (cf. Fig. 7-26). The number of A–U and G–C base pairs calculated from the NMR spectrum was 7–8 and 11–12, respectively, which again agrees well with the 8 and 12 respective pairs expected from the cloverleaf model. We note that the recent modification of the tRNA structure on the basis of new x-ray diffraction data leaves intact the cloverleaf secondary structure shown in Fig. 7-26 (143).

The resonances originating from the protons involved in the hydrogen

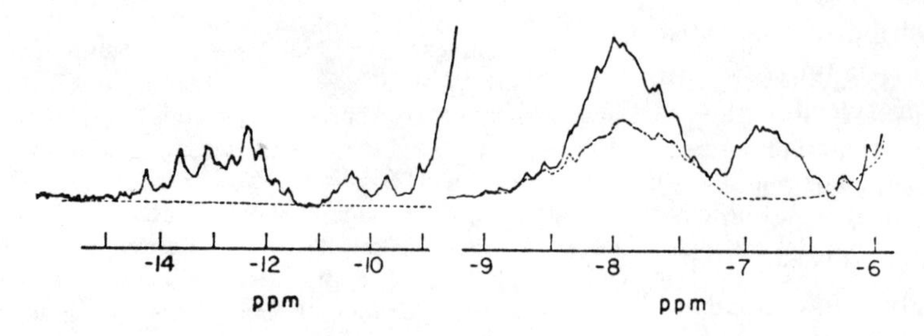

FIG. 7-27. Proton NMR spectrum of an aqueous solution of $tRNA^{Phe}_{yeast}$ at 220 MHz. The solid line is the spectrum obtained with H_2O solvent; the dashed line, D_2O. Conditions: Temperature, 35°C; tRNA concentration, ~50 mg/ml; NaCl, 0.1 M; Mg^{2+}, 58 mM; pH 5.0. Chemical shift (ppm) from DSS (141).

bonds responsible for the secondary structure of the yeast 5 S RNA have also been studied (144). The 5 S RNA is a relatively small macromolecule (molecular weight ~45,000) which is necessary for maintaining the functional integrity of the ribosome. Wong *et al.* (144) estimated from the intensity of the downfield resonances that the 5 S RNA molecule in the presence of Mg^{2+} has only 28 ± 3 base pairs, of which 85% are guanine–cytosine base pairs.

7.3. Polysaccharides

Very few attempts have been made to study the carbohydrate polymers, polysaccharides. The reason polysaccharides have been generally ignored is not entirely clear because many are either water soluble or are capable of readily forming small particle dispersions. The general drawback would appear to be high molecular weights, with relatively few nonexchangeable protons in the case of most polysaccharides. Nevertheless, some structural and conformational studies would be expected to be fruitful.

Pasika and Cragg (145, 146) have examined the proton NMR spectra of dextrans, which are composed of glucose residues bound in α-1,6-glycosidic linkages. Depending on the source of dextran, the polymer may be branched to a greater or lesser extent. It was observed that the proton spectrum of a branched dextran contains a peak not detected with linear dextrans. The additional peak arises from non-1,6 linkages, which form the branch points in branched dextrans (145). The intensity ratio of C1 resonances at 1,6 linkages to non-1,6 linkages was used to give a quantitive measure of the degree of branching in good agreement with periodate oxidation results. Substitution of sulfate groups on dextran was found to change the proton NMR spectrum, with the appearance of a new peak (146). It was inferred from the spectrum that sulfate substitution occurs at the C2 position of the anhydroglucose monomeric unit.

Friebolin *et al.* (147) examined the proton NMR spectra of 2,3,6-tri-*O*-acetylcellulose and 2,3,6-tri-*O*-acetylamylose in $CDCl_3$. The spectra gave relatively sharp lines (permitting observation of spin–spin splitting) in which the resonances of each of the glucose protons is clearly resolved (cf. Fig. 7-28). The coupling constant $J_{12} = 8$ Hz observed in the case of tri-*O*-acetylcellulose indicates the monomeric units are connected by a β-glycosidic linkage. The observed coupling constant $J_{12} \approx 3$ Hz for tri-*O*-acetylamylose indicates that the glucose units are α-glycosidically linked (cf. Section 4.3.2). Further studies have been concerned with the configuration and conformation of the monomeric units in cellulose derivatives (148), mannan derivations (149), amylose derivatives, and dextran

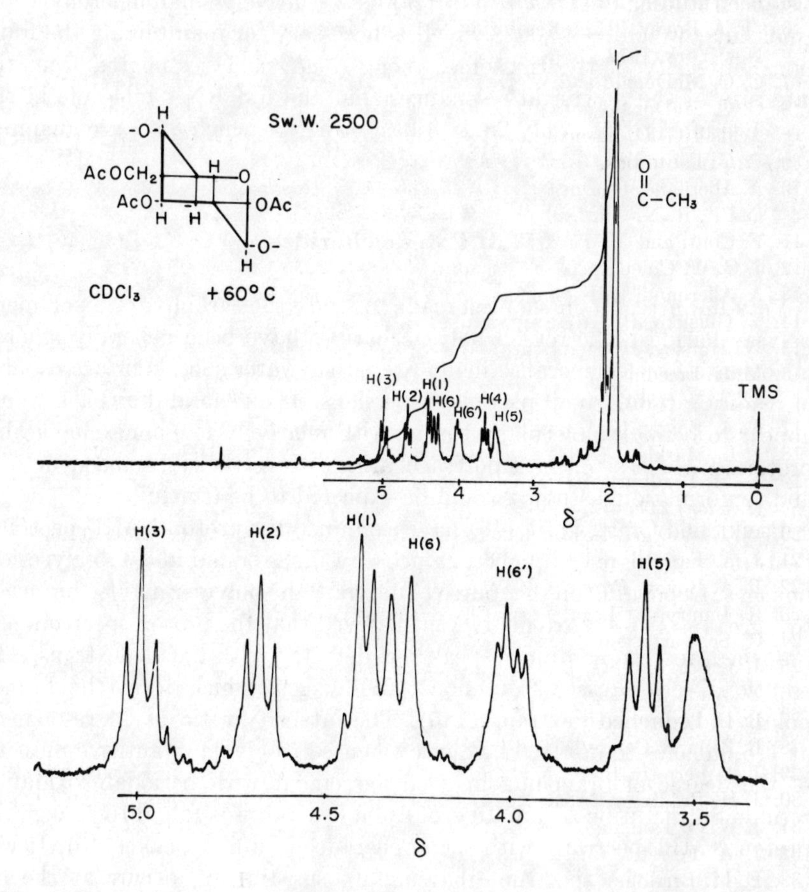

FIG. 7-28. 220 MHz proton NMR spectrum of 2,3,6-tri-*O*-acetylcellulose having a degree of polymerization ~200 (molecular weight ~60,000) in $CDCl_3$ at 60°C (147)

derivatives (150), as well as the type of glycosidic bonds in those polysaccharides.

References

1. A. Kowalsky, *J. Biol. Chem.* **237**, 1807 (1962).
2. N. L. R. King and J. H. Bradbury, *Nature* (*London*) **229**, 404 (1971).
3. R. K. Gupta and A. G. Redfield, *Biochem. Biophys. Res. Commun.* **41**, 273 (1970).
4. J. H. Bradbury, B. E. Chapman, and N. L. R. King, *Int. J. Protein Res.* **3**, 351 (1971).

5. G. C. K. Roberts and O. Jardetzky, *Advan. Protein Chem.* **24**, 447 (1970).
6. K. Wüthrich, *Struct. Bonding* (*Berlin*) **8**, 53 (1970).
6a. F. A. Bovey, "High Resolution NMR of Macromolecules," p. 337. Academic Press, New York, 1972.
7. C. C. McDonald and W. D. Phillips, *J. Amer. Chem. Soc.* **91**, 1513 (1969).
8. C. C. McDonald and W. D. Phillips, *in* "Fine Structure of Proteins and Nucleic Acids" (G. D. Fasman and S. N. Timasheff, eds.), p. 1. Dekker, New York, 1970.
9. P. C. Lauterbur, *Appl. Spectrosc.* **24**, 450 (1970).
10. A. Allerhand, D. Doddrell, V. Glushko, D. W. Cochran, E. Wenkert, P. J. Lawson, and F. R. N. Gurd, *J. Amer. Chem. Soc.* **93**, 544 (1971).
11. F. Conti and M. Paci, *FEBS* (*Fed. Eur. Biochem. Soc.*) *Lett.* **17**, 149 (1971).
12. J. C. W. Chien and J. F. Brandts, *Nature* (*London*) **230**, 209 (1971).
13. A. Allerhand, R. F. Childers, and E. Oldfield, *Biochemistry* **12**, 1335 (1973).
14. V. Glushko, P. J. Lawson, and F. R. N. Gurd, *J. Biol. Chem.* **247**, 3176 (1972).
15. W. Horsley, H. Sternlicht, and J. S. Cohen, *J. Amer. Chem. Soc.* **92**, 680 (1970).
16. M. H. Freedman, J. S. Cohen, and I. M. Chaiken, *Biochem. Biophys. Res. Commun.*, **42**, 1148 (1971).
17. F. A. Bovey, G. V. D. Tiers, and G. Filipovich, *J. Polym. Sci.* **38**, 73 (1959).
18. J. L. Markley, D. H. Meadows, and O. Jardetzky, *J. Mol. Biol.* **27**, 25 (1967).
19. E. M. Bradbury, C. Crane-Robinson, H. Goldman, and H. W. E. Rattle, *Nature* (*London*) **217**, 812 (1968).
20. W. E. Stewart, L. Mandelkern, and R. E. Glick, *Biochemistry* **6**, 143 (1967).
21. J. A. Ferretti and L. Paolillo, *Biopolymers* **7**, 155 (1969).
22. F. A. Bovey, *Pure Appl. Chem.* **16**, 417 (1968).
23. R. Lumry, R. Legare, and W. G. Miller, *Biopolymers* **2**, 484 (1964).
24. G. Schwartz and J. Seelig, *Biopolymers* **6**, 1263 (1968).
25. J. J. Burke, G. G. Hammes, and T. B. Lewis, *J. Chem. Phys.* **42**, 3520 (1969).
26. Y. Wada, H. Sasabe, and M. Tomono, *Biopolymers* **5**, 887 (1967).
27. R. C. Parker, L. J. Slutsky, and K. R. Applegate, *J. Phys. Chem.* **72**, 3477 (1968).
28. R. Zana, S. Candau, and R. Cerf, *J. Chem. Phys.* **60**, 869 (1963).
29. J. A. Ferretti, B. W. Ninham, and V. A. Parsegian, *Macromolecules* **3**, 30 (1970).
30. J. H. Bradbury, M. D. Fenn, and A. G. Moritz, *Aust. J. Chem.* **22**, 2443 (1969).
31. J. W. O. Tam and I. M. Koltz, *J. Amer. Chem. Soc.* **93**, 1313 (1971).
32. J. A. Ferretti, *Chem. Commun.* p. 1030 (1967).
33. E. M. Bradbury, C. Crane-Robinson, H. Goldman, H. W. E. Rattle, and R. M. Stephens, *J. Mol. Biol.* **29**, 507 (1967).
34. F. Conti and A. M. Liquori, *J. Mol. Biol.* **33**, 953 (1968).
35. J. H. Bradbury and M. D. Fenn, *Aust. J. Chem.* **22**, 357 (1969).
36. E. M. Bradbury, C. Crane-Robinson, and H. W. E. Rattle, *Polymer* **11**, 277 (1970).
37. R. Ullman, *Biopolymers* **9**, 471 (1970).
38. M. Saunders, A. Wishnia, and J. G. Kirkwood, *J. Amer. Chem. Soc.* **79**, 3289 (1957).
39. M. Saunders and A. Wishnia, *Ann. N. Y. Acad. Sci.* **70**, 870 (1958).
40. A. Kowalsky, *Biochemistry* **4**, 2382 (1965).
41. M. Mandel, *Proc. Nat. Acad. Sci. U.S.* **52**, 736 (1964).
42. C. C. McDonald and W. D. Phillips, *J. Amer. Chem. Soc.* **89**, 6332 (1972).
43. C. C. McDonald, W. D. Phillips, and J. D. Glickson, *J. Amer. Chem. Soc.* **93**, 235 (1971).
44. C. C. McDonald and W. D. Phillips, *Biochem. Biophys. Res. Commun.* **35**, 43 (1969).

45. D. P. Hollis, G. McDonald and R. L. Biltonen, *Proc. Nat. Acad. Sci. U.S.* **58,** 758 (1967).
46. I. Putter, J. L. Markley, and O. Jardetzky, *Proc. Nat. Acad. Sci. U.S.* **65,** 395 (1970).
47. H. F. Epstein, A. N. Schechter, and J. S. Cohen, *Proc. Nat. Acad. Sci. U.S.* **68,** 2042 (1971).
48. J. H. Bradbury and N. L. R. King, *Nature (London)* **223,** 1154 (1969).
49. G. Kartha, J. Bello, and D. Harker, *Nature (London)* **213,** 862 (1967).
50. H. W. Wyckoff, D. Tsernoglou, A. W. Hanson, J. R. Knox, B. Lee, and F. M. Richards, *J. Biol. Chem.* **245,** 305 (1970).
51. A. M. Crestfield, W. H. Stein, and S. Moore, *J. Biol. Chem.* **238,** 2421 (1963).
52. J. H. Bradbury and H. A. Scheraga, *J. Amer. Chem. Soc.* **88,** 4240 (1966).
53. D. H. Meadows, J. L. Markley, J. S. Cohen, and O. Jardetzky, *Proc. Nat. Acad. Sci. U.S.* **58,** 1307 (1967).
54. D. H. Meadows, O. Jardetzky, R. M. Epand, H. H. Ruterjans, and H. A. Scheraga, *Proc. Nat. Acad. Sci. U.S.* **60,** 766 (1968).
55. H. H. Ruterjans and H. Witzel, *Eur. J. Biochem.* **9,** 118 (1969).
56. B. Bak, E. J. Pedersen, and F. Sundby, *J. Biol. Chem.* **242,** 2637 (1967).
57. D. H. Meadows and O. Jardetzky, *Proc. Nat. Acad. Sci. U.S.* **61,** 406 (1968).
58. D. H. Meadows. G. C. K. Roberts, and O. Jardetzky, *J. Mol. Biol.* **44,** 491 (1969).
59. G. C. K. Roberts, E. A. Dennis, D. H. Meadows, J. S. Cohen, and O. Jardetzky, *Proc. Nat. Acad. Sci. U.S.* **62,** 1151 (1969).
60. D. Findlay, D. G. Herries, A. P. Mathias, B. R. Rabin, and C. A. Ross, *Biochem. J.* **85,** 152 (1962).
61. F. W. Benz, G. C. K. Roberts, J. Feeney, and R. R. Ison, *Biochim. Biophys. Acta* **278,** 233 (1972).
62. M. H. Freedman, J. R. Lyerla, Jr., I. M. Chaiken, and J. S. Cohen, *Eur. J. Biochem.* **32,** 215 (1973).
63. H. Ruterjans and O. Pongs, *Eur. J. Biochem.* **18,** 313 (1971).
64. J. S. Cohen, *Nature (London)* **223,** 43 (1969).
65. J. S. Cohen and O. Jardetzky, *Proc. Nat. Acad. Sci. U.S.* **60,** 92 (1968).
66. J. D. Glickson, C. C. McDonald, and W. D. Phillips, *Biochem. Biophys. Res. Commun.* **35,** 492 (1969).
67. J. D. Glickson, W. D. Phillips, and J. A. Rupley, *J. Amer. Chem. Soc.* **93,** 4031 (1971).
68. R. E. Canfield and A. K. Liu, *J. Biol. Chem.* **240,** 1997 (1965).
69. C. C. F. Blake, G. A. Mair, A. C. T. North, D. C. Phillips, and V. R. Sarma, *Proc. Roy. Soc. Ser. B* **167,** 365 (1967).
70. H. Sternlicht and D. Wilson, *Biochemistry* **6,** 2881 (1967).
71. C. C. McDonald and W. D. Phillips, *Biochem. Biophys. Res. Commun.* **35,** 43 (1969).
72. H. Sternlicht and E. Wheeler, *in* "Magnetic Resonance in Biological Systems" (A. Ehrenberg, B. G. Malmström, and T. Vänngard, eds.), p. 325. Pergamon, Oxford, 1967.
73. R. W. Wien, J. D. Morrisett, and H. M. McConnell, *Biochemistry* **11,** 3707 (1972).
74. I. D. Campbell, C. M. Dobson, R. J. P. Williams, and A. V. Xavier, *Proc. Rare Earth Res. Conf., 10th, 1973* Vol. II, p. 791 (1973).
75. J. L. Markley, M. N. Williams, and O. Jardetzky, *Proc. Nat. Acad. Sci. U.S.* **65,** 645 (1970).

76. J. L. Markley and O. Jardetzky, *J. Mol. Biol.* **50,** 223 (1970).
77. E. Niebor, D. East, J. S. Cohen, B. Furie, and A. N. Schechter, *Proc. Rare Earth Res. Conf., 10th, 1973* Vol. II, p. 763 (1973).
78. J. S. Cohen, C. T. Yim, M. Kandel, A. G. Gornall, S. I. Kandel, and M. H. Freedman, *Biochemistry* **11,** 327 (1972).
79. C. Ho, J. A. Magnuson, J. B. Wilson, N. S. Magnuson, and R. J. Kurland, *Biochemistry* **8,** 2074 (1969).
80. J. A. Hamilton, C. Talkowski, E. Williams, E. M. Avila, A. Allerhand, E. H. Cordes, and G. Camejo, *Science* **180,** 193 (1973).
81. T. L. James, M. L. Ludwig, and M. Cohn, *Proc. Nat. Acad. Sci. U.S.* **70,** 3292 (1973).
82. K. Wüthrich and R. G. Shulman, *Phys. Today* **23,** 43 (1970).
83. K. Wüthrich, R. G. Shulman, and J. Peisach, *Proc. Nat. Acad. Sci. U.S.* **60,** 373 (1968).
84. K. Wüthrich, R. G. Shulman, and T. Yamane, *Proc. Nat. Acad. Sci. U.S.* **61,** 1199 (1968).
85. R. G. Shulman, S. H. Glarum, and M. Karplus, *J. Mol. Biol.* **57,** 93 (1971).
86. D. G. Davis, N. L. Mock, V. R. Laman, and C. Ho, *J. Mol. Biol.* **40,** 311 (1969).
87. D. G. Davis, S. Charache, and C. Ho, *Proc. Nat. Acad. Sci. U.S.* **63,** 1403 (1969).
88. R. G. Shulman, S. Ogawa, K. Wüthrich, T. Yamane, J. Peisach, and W. E. Blumberg, *Science* **165,** 251 (1969).
89. K. Wüthrich, R. G. Shulman, T. Yamane, B. J. Wyluda, T. E. Hugli, and F. R. N. Gurd, *J. Biol. Chem.* **245,** 1947 (1970).
90. M. O. Dayhoff, "Atlas of Protein Sequence and Structure," Vol. 4. Nat. Biomed. Res. Found., Silver Spring, Maryland, 1969.
91. K. Wüthrich, *Proc. Nat. Acad. Sci. U.S.* **63,** 1071 (1969).
92. C. C. McDonald, W. D. Phillips, and S. N. Vinogradov, *Biochem. Biophys. Res. Commun.* **36,** 442 (1969).
93. A. G. Redfield and R. K. Gupta, *Cold Spring Harbor Symp. Quant. Biol.* **36,** 405 (1971).
94. R. K. Gupta and A. G. Redfield, *Science* **169,** 1204 (1970).
95. T. Takano, R. Swanson, O. B. Kallai, and R. E. Dickerson, *Cold Spring Harbor Symp. Quant. Biol.* **36,** 397 (1971).
96. K. Wüthrich, I. Aviram, and A. Schejter, *Biochim. Biophys. Acta* **253,** 98 (1971).
97. G. E. Krejcarek, L. Turner, and K. Dus, *Biochem. Biophys. Res. Commun* **42,** 983 (1971).
98. R. G. Shulman, K. Wüthrich, T. Yamane, D. J. Patel, and W. E. Blumberg, *J. Mol. Biol.* **53,** 143 (1970).
99. J. Wyman, *Advan. Protein Chem.* **19,** 223 (1964).
100. R. Hershberg and T. Asakura, *J. Mol. Biol.* **70,** 735 (1972).
101. T. Yamane, K. Wüthrich, R. G. Shulman, and S. Ogawa, *J. Mol. Biol.* **49,** 197 (1970).
102. K. H. Winterhalter and D. A. Deranleau, *Biochemistry* **6,** 3136 (1967).
103. D. G. Davis, N. H. Mock, T. R. Lindstrom, S. Charache, and C. Ho, *Biochem. Biophys. Res. Commun.* **40,** 343 (1970).
104. S. Ogawa, R. G. Shulman, P. A. M. Kynoch, and H. Lehman, *Nature (London)* **225,** 1042 (1970).
105. S. Ogawa and R. G. Shulman, *Biochem. Biophys. Res. Commun.* **42,** 9 (1971).
106. T. R. Lindstrom and C. Ho, *Proc. Nat. Acad. Sci. U.S.* **69,** 1707 (1972).

107. W. D. Phillips, M. Poe, J. F. Weiher, C. C. McDonald, and W. J. Lovenberg, *Nature* (*London*) **227,** 574 (1970).
108. M. Poe, W. D. Phillips, C. C. McDonald, and W. J. Lovenberg, *Proc. Nat. Acad. Sci. U.S.* **65,** 797 (1970).
109. M. Poe, W. D. Phillips, C. C. McDonald, and W. H. Orme-Johnson, *Biochem. Biophys. Res. Commun.* **42,** 705 (1971).
110. E. L. Packer, H. Sternlicht, and J. C. Rabinowitz, *Proc. Nat. Acad. Sci. U.S.* **69,** 3278 (1972).
111. H. L. Crespi, R. M. Rosenberg, and J. J. Katz, *Science* **161,** 795 (1968).
112. H. L. Crespi and J. J. Katz, *Nature* (*London*) **224,** 560 (1969).
113. H. L. Crespi, and J. R. Norris, and J. J. Katz, *Nature* (*London*), *New Biol.* **236,** 178 (1972).
114. J. L. Markley, I. Putter, and O. Jardetzky, *Science* **161,** 1249 (1968).
115. I. Putter, A. Barreto, J. L. Markley, and O. Jardetzky, *Proc. Nat. Acad. Sci. U.S.* **64,** 1396 (1969).
116. J. J. Katz and H. L. Crespi, *Recent Advan. Phytochem.* **2,** 1 (1969).
117. H. L. Crespi and J. J. Katz, "Methods in Enzymology" (C. H. W. Hirs and S. N. Timasheff, eds.), Vol. 26, part C, p. 627. Academic Press, New York, 1973.
118. W. H. Huestis and M. A. Raftery, *Biochem. Biophys. Res. Commun.* **48,** 678 (1972).
119. W. H. Huestis and M. A. Raftery, *Biochemistry* **11,** 1648 (1972).
120. W. H. Huestis and M. A. Raftery, *Biochem. Biophys. Res. Commun.* **49,** 1358 (1972).
121. D. J. Saunders and R. E. Offord, *FEBS* (*Fed. Eur. Biochem. Soc.*) *Lett.* **26,** 286 (1972).
122. A. M. Nigen, P. Keim, R. C. Marshall, J. S. Morrow, and F. R. N. Gurd, *J. Biol. Chem.* **247,** 4100 (1972).
123. D. T. Browne, G. D. Kenyon, E. L. Packer, H. Sternlicht, and D. M. Wilson, *J. Amer. Chem. Soc.* **95,** 1316 (1973).
124. I. C. P. Smith, B. J. Blackburn, and T. Yamane, *Can. J. Chem.* **47,** 513 (1969).
125. P. O. P. Ts'o, N. S. Kondo, M. P. Schweitzer, and D. P. Hollis, *Biochemistry* **8,** 997 (1969).
126. J. H. Prestegard and S. I. Chan, *J. Amer. Chem. Soc.* **91,** 2843 (1969).
127. C. C. McDonald, W. D. Phillips, and S. Penman, *Science* **144,** 1234 (1964).
128. J. P. McTague, V. Ross, and J. H. Gibbs, *Biopolymers* **2,** 163 (1964).
129. M. P. Schweizer, *Biochem. Biophys. Res. Commun.* **36,** 871 (1969).
130. C. C. McDonald, W. D. Phillips, and J. Lazar, *J. Amer. Chem. Soc.* **89,** 4166 (1967).
131. Z. M. Bekker and Y. N. Molin, *Biofizika* **14,** 740 (1969).
132. H. H. Mantsch and I. C. P. Smith, *Biochem. Biophys. Res. Commun.* **46,** 808 (1972).
133. J. A. Glasel, S. Hendler, and P. R. Srinivasan, *Proc. Nat. Acad. Sci. U.S.* **60,** 1038 (1968).
134. I. C. P. Smith, T. Yamane, and R. G. Shulman, *Science* **159,** 1360 (1968).
135. I. C. P. Smith, T. Yamane, and R. G. Shulman, *Can. J. Biochem.* **47,** 480 (1969).
136. J. E. Crawford, S. I. Chan, and M. P. Schweizer, *Biochem. Biophys. Res. Commun.* **44,** 1 (1971).
137. K. M. Koehler and P. G. Schmidt, *Biochem. Biophys. Res. Commun.* **50,** 370 (1973).
138. R. A. Komoroski and A. Allerhand, *Proc. Nat. Acad. Sci. U.S.* **69,** 1804 (1972).
139. M. Guéron, *FEBS* (*Fed. Eur. Biochem Soc.*) *Lett.* **19,** 264 (1971).

140. D. R. Kearns, D. J. Patel, R. G. Shulman, and T. Yamane, *J. Mol. Biol.* **61,** 265 (1971).
141. Y. P. Wong, D. R. Kearns, B. R. Reid, and R. G. Shulman, *J. Mol. Biol.* **72,** 725 (1972).
142. Y. P. Wong, D. R. Kearns, R. G. Shulman, T. Yamane, S. Chang, J. G. Chirikjian, and J. R. Fresco, *J. Mol. Biol.* **74,** 403 (1973).
143. S. H. Kim, G. Quigley, F. L. Suddath, A. McPherson, D. Sneden, J. J. Kim, J. Weinzierl, and A. Rich, *J. Mol. Biol.* **75,** 421 (1973).
144. Y. P. Wong, D. R. Kearns, B. R. Reid, and R. G. Shulman, *J. Mol. Biol.* **72,** 725 (1972).
145. W. M. Pasika and L. H. Cragg, *Can. J. Chem.* **41,** 293 (1963).
146. W. M. Pasika and L. H. Cragg, *Can. J. Chem.* **41,** 777 (1963).
147. H. Friebolin, G. Keilich, and E. Siefert, *Angew. Chem.* **81,** 791 (1969).
148. H. Friebolin, G. Keilich, and E. Siefert, *Org. Magn. Resonance* **2,** 457 (1970).
149. N. Frank, H. Friebolin, G. Keilich, J. P. Merle, and E. Siefert, *Org. Magn. Resonance* **4,** 725 (1972).
150. G. Keilich, E. Siefert, and H. Friebolin, *Org. Magn. Resonance* **3,** 31 (1971).

CHAPTER 8

MOLECULAR DYNAMICS IN BIOLOGICAL AND BIOCHEMICAL SYSTEMS

Nuclear magnetic resonance should be ideally suited for investigating molecular dynamics in membranes, biological tissue, and biochemical model systems. The motional freedom of molecules in such systems may, in principle, be amenable to NMR linewidth, relaxation time, and self-diffusion coefficient measurements. One of the chief features of using NMR for such studies is that the system need not be perturbed in any manner. Another feature is that the motion of various nuclei in the system may be monitored, providing information about the motions of different molecules and, in fact, the motions of different parts of the same molecule. The various rotational and translational motions may, in principle, also be differentiated, although in practice it may be difficult to distinguish between them.

This chapter will be concerned with the dynamics of lipids in membranes, the dynamic state of cellular water, and the state of ionic solutes (namely sodium ion) in biological tissue. These topics have been quite controversial.

8.1. Membranes and Membrane Models

A living cell is surrounded by a semipermeable membrane that serves to regulate the movement of nutrients, respiratory gases, and wastes into and out of the cell. In addition, the specialized organelles within a cell are also enclosed by membranes. These membranes are capable of controlling the flux of solutes in such a way that a substance can be transported across a membrane from a dilute solution to a more concentrated one. This "active transport" process is characteristic of a viable biological membrane. The substances that are transported by any particular membrane depend on the function of the organelle encased by the membrane. The specialized functions of different membranes reflect the complexity of membranes.

Presumably, an understanding of cell membrane structure will aid in understanding the functions of cell membranes. NMR is a valuable technique for investigating cell membrane structure. It is particularly amenable to elucidating the dynamic state of lipids, lipid–lipid interactions, and lipid–protein interactions.

Membranes are composed largely of lipids and proteins with some polysaccharides present. Although the lipid composition of different membranes may vary, the lipids generally provide the gross structural features whereas the constituent proteins provide the functional diversity of membranes. Some of the common lipid molecules found in membranes are illustrated in Table 8-1. Most of the membrane lipids are amphipathic, i.e., one end of the molecule is hydrophilic, or water soluble, and the other end of the molecule is hydrophobic, or water insoluble. The amphipathic character of these lipids is manifest in the membrane structure exhibiting a phospholipid bilayer (cf. Fig. 8-1). A recent review of membrane ultrastructure (1) has concluded, based largely on x-ray diffraction evidence (2), that the bilayer is a general, although not an ubiquitous, structural unit. The membrane proteins are embedded in or on the bilayer in various ways. The presence of the lipid bilayer as a structural feature of most membranes has provided the justification for use of phospholipid bilayers as membrane models.

This section will be concerned with the various applications of NMR to investigating molecular dynamics (molecular mobility and local ordering) in membranes and model membrane systems. Chapman (3, 4) has presented reviews of the subject, reflecting largely the extensive work by himself and co-workers. Linewidth, spin–lattice (T_1) relaxation time, and spin–spin (T_2) relaxation time measurements on ^{1}H, ^{2}H, ^{31}P, and ^{13}C nuclei have proved especially useful for such studies. These NMR parameters have been examined as a function of several variables. Temperature

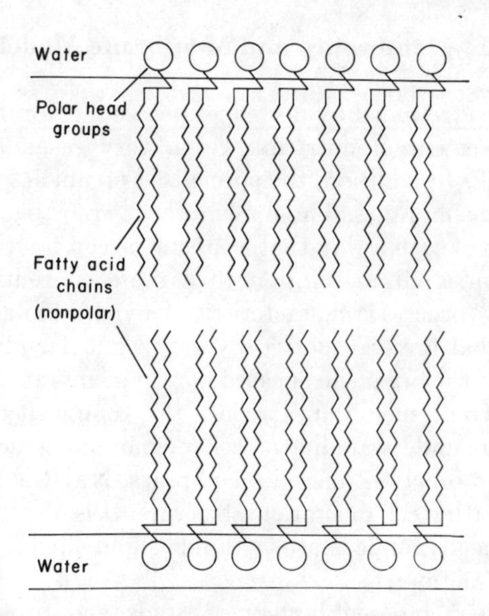

FIG. 8-1. Schematic cross-section of a phospholipid bilayer with the polar head groups protruding into the water and the hydrophobic long-chain hydrocarbon moieties comprising the interior of the bilayer.

variation, presence (and amount) of water, and ion content in the aqueous phase give rise to phase transitions or more subtle changes in lipid structure. The sample composition is another major factor in determining dynamic structure. The nature of the phospholipid head group plays a role, as does the hydrocarbon moiety (long or short chain, saturated or unsaturated chain). The amount and nature of other membrane constituents (e.g., cholesterol or protein) and additives (e.g., antibiotics) is also very important.

Another factor to be examined is a result of experimental technique. It has been observed that sonication, especially in the case of model membrane systems, yields spectra with better resolution. The high-resolution spectra are considerably more convenient for study, but then the total effect of sonication on the lipid system must be considered carefully.

With the exception of ^{2}H, which has a large electric quadrupole relaxation contribution, the nuclear relaxation mechanism in lipid systems is predominantly dipolar in origin. As discussed in Section 2.4, dipolar relaxation may be separated into intramolecular and intermolecular contributions. Both sources show an r^{-6} dependence for the distance between the relaxing nucleus and a nucleus causing relaxation. The observed relaxation is a sum

TABLE 8-1

STRUCTURE OF SOME LIPIDS FREQUENTLY OCCURRING IN CELL MEMBRANES[a]

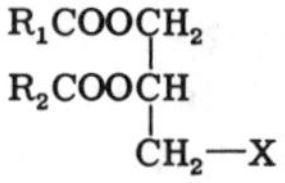

R_1COOCH_2
R_2COOCH
$CH_2—X$

Phosphatidylglycerols where X is

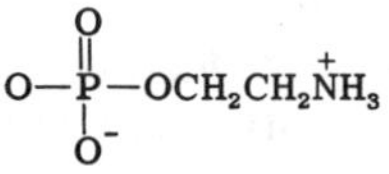

$—O—P(=O)(O^-)—OCH_2CH_2\overset{+}{N}H_3$

Phosphatidylethanolamine

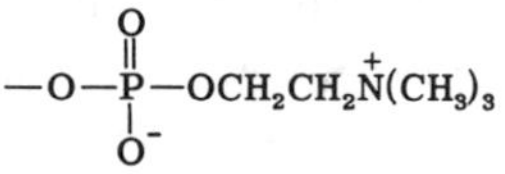

$—O—P(=O)(O^-)—OCH_2CH_2\overset{+}{N}(CH_3)_3$

Phosphatidylcholine (lecithin)

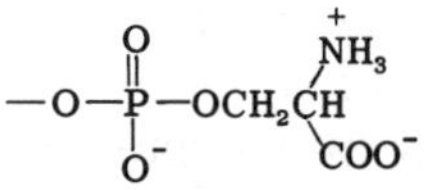

$—O—P(=O)(O^-)—OCH_2CH(\overset{+}{N}H_3)COO^-$

Phosphatidylserine

$—O—P(=O)(OH)—O—$ (inositol ring with HO, HO, HO, HO, OH)

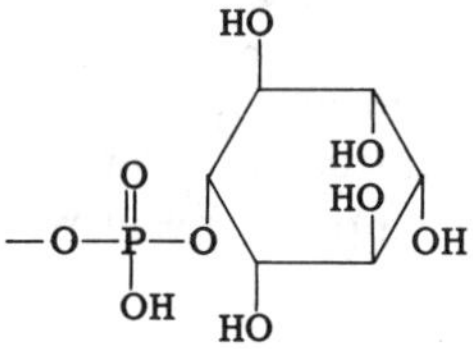

Phosphatidylinositol

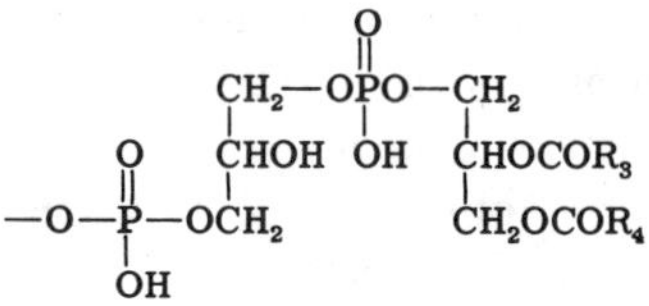

$—O—P(=O)(OH)—OCH_2—CHOH—CH_2—OP(=O)(OH)O—CH_2—CHOCOR_3—CH_2OCOR_4$

Diphosphatidylglycerol

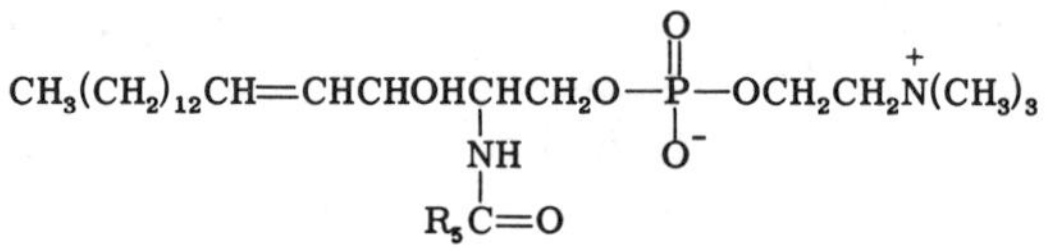

$CH_3(CH_2)_{12}CH{=}CHCHOHCH(NHC(=O)R_5)CH_2O—P(=O)(O^-)—OCH_2CH_2\overset{+}{N}(CH_3)_3$

Sphingomyelin

HO (steroid ring structure)

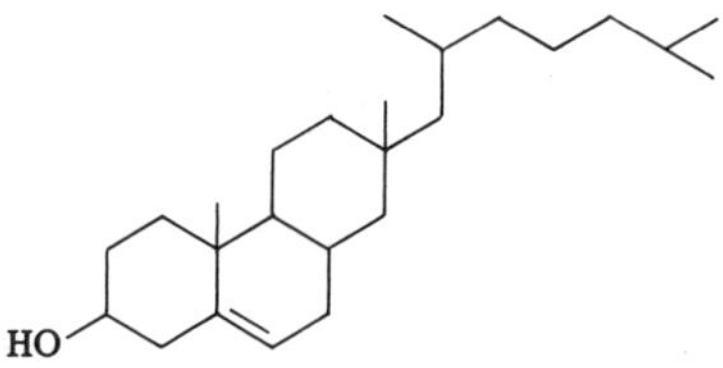

Cholesterol

---

[a] R_1, R_2, R_3, R_4, and R_5 are typically straight-chain alkyl groups with chain lengths C_{14} to C_{20} with most being C_{16} or C_{18}. The alkyl groups may be saturated or may contain one or more double bonds. The double bonds (usually cis) are rarely found in that portion of the chain between the carboxyl group and the ninth carbon atom. If a single double bond is present, it will generally occur in the 9, 10 position.

of contributions from all relaxation-promoting nuclei on the same molecule (intramolecular) and on other molecules (intermolecular). The intermolecular contributions are related to translational molecular motion via the translational correlation time τ_{tr} (cf. Eq. 2-56), and the intramolecular contributions are related to the rotational motion of the molecule or parts of the molecule by the rotational correlation time τ_r corresponding to that motion (cf. Eqs. 2-52 through 2-55).

For any specific relaxation time measurement, there may be several rotational and translational motions (with corresponding correlation times) serving to modulate the dipole–dipole interactions and thus effecting nuclear relaxation in the lipid system. One particular correlation time may be able to account for the relaxation in certain cases but generally several motions must be considered. There are several motions that might be considered for a system as complex as a phospholipid bilayer. Some of these motions are illustrated in Fig. 8-2. Any one of the motions illustrated will obviously be more important for some nuclei than for others.

It might be expected that a phospholipid with a single gauche configuration as shown in B1 of Fig. 8-2 would become less probable as the position of the gauche bond proceeds toward the head group. This decreasing probability is a result of the steric hindrance imposed by nearby molecules on the askew chain. The steric hindrance will increase as we progress toward the head group and more of the chain is displaced by a single gauche configuration. However, because the energy of activation is low (~0.5 kcal/mole), there is rapid interconversion of the rotamers. Therefore, it is quite probable that more than one gauche configuration exists in a chain at any instant. A second gauche configuration may serve to minimize the displacement of the first. A β-coupled configuration, as shown in B3 of Fig. 8-2, will yield a minimal displacement.

Another motion that may be important can be described as a rotational oscillation in the same mode as the trans–gauche interconversion but one in which the interconversion is not completed.

An additional motion must be considered for sonicated phospholipid vesicles; that is, rotation of the vesicle itself. The sonicated vesicles are about 250 Å in diameter. According to the Debye-Stokes theory, a 250 Å sphere will have a rotational correlation time on the order of 10^{-6} sec.

In membranes and most model membrane systems, the motion of the phospholipid molecules must generally be regarded as anisotropic. This is most easily visualized for nuclei on methyl groups. The isotropic motion of the phospholipid molecules may be very slow, with a correlation time of several seconds, where the term "isotropic" denotes a motion in which the chain axis of the molecule is able to spatially reorient in all possible direc-

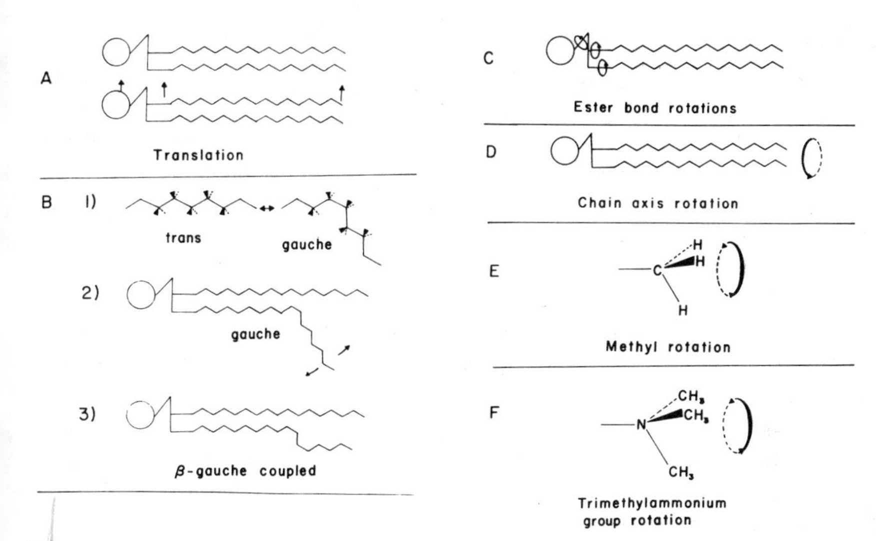

FIG. 8-2. Some of the possible motions in phospholipids modulating dipole–dipoel interactions and thus contributing to nuclear relaxation. The long-chain hydrocarbon moieties are represented as fully saturated; they could be unsaturated as well. (A) Translational motion as one phospholipid molecule moves relative to the other molecules. This movement affects intermolecular interactions. (B) Rapid rotation about methylene carbon–carbon bonds resulting in (1) rapid interconversion of trans and gauche configurations. (2) The trans–gauche interconversion affects intramolecular and intermolecular interactions for nuclei between the gauche bond and the methyl terminus. (3) More than one gauche configuration may be present at one time. The β-coupled configuration is one such possibility that manages to preserve the straight-chain configuration with only a small perturbation. (C) Rapid rotation about bonds in carboxylate and phosphate esters. (D) Rotation of the whole phospholipid molecule about its long axis. (E) Methyl group rotation for methyls at the hydrocarbon chain terminus and in choline groups (F) Trimethylammonium group rotation in cholines of lecithin and sphingomyelin. The rotational motions promote intramolecular relaxation.

tions. In contrast to this slow motion, the methyl group rotation (cf. Fig. 8-2E) is rapid and may have a correlation time as small as 10^{-11} sec. The methyl group rotation, however, is anisotropic in space because reorientation about the C_3 axis of rotation does not permit the internuclear vectors to be reoriented in all possible directions in space except on a time scale much longer than that of the rotational motion. It is sufficient for the purposes of this discussion to consider anisotropic motion in which the methyl group rotational correlation time τ_c for motion about the C_3 axis is much less than the correlation time τ_{c_1} for the motion of that axis.

According to Abragam (5), the nuclear magnetic resonance linewidth $W_{1/2}$ for dipolar relaxed nuclei in an anisotropically rotating system may be described in simplified form by

$$W_{1/2}^2 = W_{1/2}''^2 + W_{1/2}'^2(2/\pi)\tan^{-1}(\alpha W_{1/2}\tau_c) \qquad (8\text{-}1)$$

when τ_{c_1} is very long. The rigid lattice linewidth in the absence of any motion is simply the sum $W_{1/2}' + W_{1/2}''$, where $W_{1/2}''$ is the isotropic dipolar contribution and $W_{1/2}'$ is the anisotropic contribution, which may be reduced by rapid anisotropic reorientation. The numerical factor α is approximately equal to one. According to Eq. 8-1, as the methyl group on the phospholipid molecule rotates faster, the linewidth will decrease to $W_{1/2}''$, the residual isotropic dipolar linewidth. The linewidth can be reduced further if isotropic motions are sufficiently rapid to decrease the isotropic contribution from its rigid lattice value. A more detailed consideration yields information about lineshapes (5).

Woessner (6) derived equations for T_1 and T_2 for protons undergoing anisotropic motion in which the spin–spin vector (e.g., between two protons in a methyl group) reorients about an axis with a correlation time τ_c, and the axis may randomly reorient with a correlation time τ_{c_1}. In the case of methyl protons in phospholipids where $\tau_{c_1} \gg \tau_c$, it is possible to obtain the following expressions from Woessner's equation for the T_1 and T_2 contributions to methyl proton relaxation caused by the other protons in the methyl group:

$$\frac{1}{T_1} = \frac{45}{8}\frac{\gamma^4\hbar^2}{r^6}\left[\frac{1}{3}\frac{\tau_{c_1}}{1+\omega_o^2\tau_{c_1}^2} + \frac{4}{3}\frac{\tau_{c_1}}{1+4\omega_o^2\tau_{c_1}^2} + \frac{1}{4}\frac{\tau_c}{1+\omega_o^2\tau_c^2/16} + \frac{\tau_c}{1+\omega_o^2\tau_c^2/4}\right] \qquad (8\text{-}2)$$

$$\frac{1}{T_2} = \frac{9}{8}\frac{\gamma^4\hbar^2}{r^6}\left[\frac{1}{10}\tau_{c_1} + \frac{1}{6}\frac{\tau_{c_1}}{1+\omega_o^2\tau_{c_1}^2} + \frac{1}{15}\frac{\tau_{c_1}}{1+4\omega_o^2\tau_{c_1}^2} + \frac{1}{8}\frac{\tau_c}{1+\omega_o^2\tau_c^2/16} + \frac{1}{20}\frac{\tau_c}{1+\omega_o^2\tau_c^2/4}\right] \qquad (8\text{-}3)$$

where γ is the gyromagnetic ratio, $\hbar$ is Planck's constant divided by 2π, r is the distance between the interacting nuclei, and ω_o is the angular precession frequency of the nucleus.

Equations 8-2 and 8-3 reveal that a plot of $1/T_1$ versus correlation time would have two maxima, corresponding to $\omega_o\tau_{c_1} \sim 1$ and $\omega_o\tau_c \sim 1$. If τ_{c_1} is large ($\tau_{c_1} \gg \tau_c$), i.e., $\gg \omega_o^{-1}$, T_1 relaxation will be governed by the relatively rapid motion corresponding to τ_c, and T_2 relaxation will be governed by the relatively slow motion corresponding to τ_{c_1}. That is the main point to be derived from Eq. 8-2 and 8-3. Linewidth and T_2 measurements may be related to one motion, whereas T_1 is related to another motion. If the point is reached where T_2 relaxation is governed solely by τ_{c1}, the anisotropic motion plays little role in the T_2 relaxation and the isotropic terms in Eqs. 8-1 and 8-3 dominate.

Cobb and Johnson (7) have applied a density matrix treatment to the calculation of lineshapes in solids or slowly moving molecules having a rotating three-spin group (e.g., methyl). Lineshapes as a function of τ_c, the correlation time for methyl group rotation, were calculated. Some of the features of the calculated lineshapes closely resemble some phospholipid methyl resonances. The approach is limited, however, by the assumption of no intermolecular interactions.

The spin–lattice relaxation of ^{13}C and D nuclei in anisotropically rotating $^{13}CH_3$, $^{13}CD_3$, and CD_3 has been considered by Noggle (8). Expressions similar in form to Eq. 8-2 were obtained. It would appear that the best way of obtaining information about anisotropic methyl rotation is to examine the ^{13}C and D NMR of $^{13}CH_3$ and CD_3 groups, respectively.

In spite of the many complexities of motions even in model membrane systems, several attempts have been made to interpret the observed NMR parameters (linewidths and relaxation times) in terms of specific rotational and translational motions. The interpretations of the observed parameters have generated much controversy.

8.1.1. Membrane Models (Lipids)

We will be concerned here only with lipid systems that can be considered as membrane models. NMR studies on some other lipids have been reviewed by Chapman and Salsbury (3).

A. *Anhydrous Phospholipids*

Broad-line proton NMR spectra of anhydrous phosphatidylethanolamines (9) and phosphatidylcholines (10) have been examined over a wide range of temperature from liquid nitrogen temperatures to about 150°C. At low temperatures, the solid phospholipids exhibit very broad lines of width 15–16 Gauss (~60–65 kHz) so the resonances from different chemical groups are not distinguished. As illustrated in Fig. 8-3 for phosphatidylethanolamine, as the temperature is raised, the lines gradually narrow until, at a particular transition temperature, the linewidth rapidly decreases to about 400 Hz. The sharp linewidth decrease marks the transition from a crystalline to a mesomorphic phase. The transition temperature is a function of the water content (none for the present discussion) and

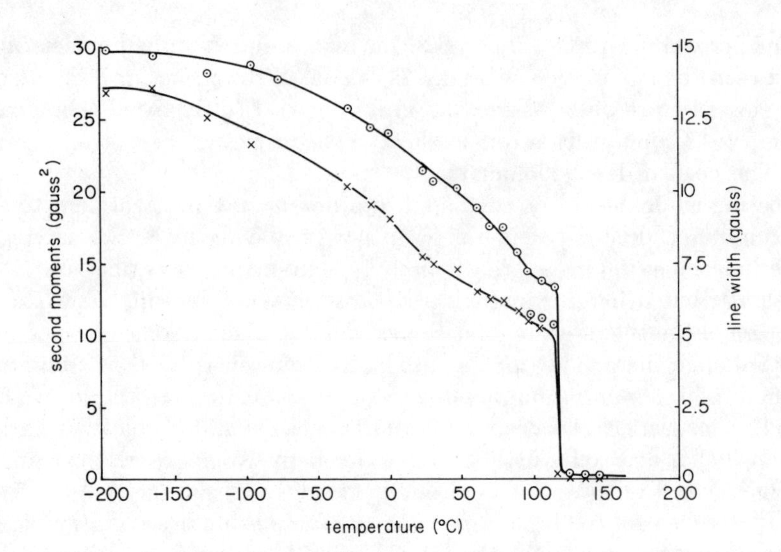

FIG. 8-3. Linewidth (○) and second moment (×) data as functions of temperature for 1,2-dimyristoyl-DL-phosphatidylethanolamine. The second moment provides a description; of the lineshape of a broad line (cf. reference 5, p. 106ff.) (9).

the content of unsaturated fatty acid chains. Increasing amounts of unsaturation (or of water) result in a lower transition temperature.

The sharp decrease in linewidth is indicative of greater molecular motion in the fatty acid chains of the phospholipid in the mesomorphic phase. The increased motion tends to average the dipole–dipole interactions toward zero. Both rotational and translational motions are presumably hastened.

The temperature dependence of the proton NMR linewidth of anhydrous 1,2-distearoyl-L-phosphatidylcholine has also been investigated (10). A reduction in linewidth at about −150°C, not observed with phosphatidylethanolamine, was attributed to reorientation of the $N(CH_3)_3^+$ protons. The crystalline-to-liquid crystalline phase transition was similar to that observed with phosphatidylethanolamine, but the lineshape in the liquid crystalline phase was more complicated, possessing a broad component in addition to the intense relatively narrow peak.

NMR spectra of phospholipids in nonaqueous solvents have also been examined and interpreted in terms of structure and lipid–lipid interactions (11, 12).

B. *Unsonicated Aqueous Lipid Dispersions*

Proton NMR. The addition of water (D_2O) to anhydrous lecithin markedly reduces the proton NMR linewidth (13) in both the gel and the liquid crystalline phases and lowers the temperature required for transition between the two phases (cf. Fig. 8-4). The narrower line (2.7–0.4 Gauss) in the gel phase was assigned to the $N(CH_3)_3^+$ protons and the broader line (9–4 Gauss) was assigned to the rest of the protons.

The peak in the liquid crystalline phase of lecithin with linewidth decreasing from about 2.4 Gauss to 1.5 Gauss on addition of 1 mole of D_2O per mole of lecithin was assigned to protons of the glycerol and choline methylenes in the polar head group. An intense line of width $\sim$0.1 Gauss was attributed to the rapidly reorienting protons of the fatty acid chains and choline methyl groups in the liquid crystalline phase.

Although there is little narrowing effect on addition of D_2O in excess of 1 mole of D_2O per mole of lecithin in the liquid crystalline phase, the linewidths for the gel phase continue to narrow until a molar ratio of greater than five is achieved (cf. Fig. 8-4). The narrowing is accompanied by a decrease in the transition temperature as D_2O is added.

It was also noticed by Veksli *et al.* (13) that the narrower line in the gel-phase spectrum narrowed considerably just below the transition temperature. This was interpreted to mean that the choline methyl group gains greater rotational mobility prior to reaching the transition temperature. This observation could explain the small peak obtained in the differential scanning calorimetry curve at a temperature just preceding the main endothermic transition.

Biological activity depends on the fatty acid chain composition of phospholipids in membranes and, perhaps consequently, on the fluidity of the phospholipid bilayer. Kohler *et al.* (14) have considered the fatty acid chain composition in comparing proton NMR spectra of liquid crystalline yeast and egg yolk lecithin dispersions. It was observed that the linewidths and relaxation rates of unsonicated yeast lecithin are smaller than those of egg yolk lecithin (cf. Fig. 8-5). This observation supports the hypothesis that the interior hydrocarbon region of yeast lecithin is more fluid than egg yolk lecithin by virtue of its higher percentage of unsaturated and shorter hydrocarbon chains.

The proton NMR spectrum of lecithin isolated from yeast grown on D_2O exhibits narrower linewidths than does its fully protonated counterpart, as shown in Fig. 8-5, illustrating the importance of dipolar interactions in causing T_2 relaxation. The appearance of several well-resolved peaks in the spectrum of the deuterated yeast lecithin suggests the potential for utilizing partially deuterated organisms. If deuterated species are used, verification that the deuteration has not altered the crystal structure of the bilayer should be obtained.

The effects of heavy water concentration and fatty acid chain composi-

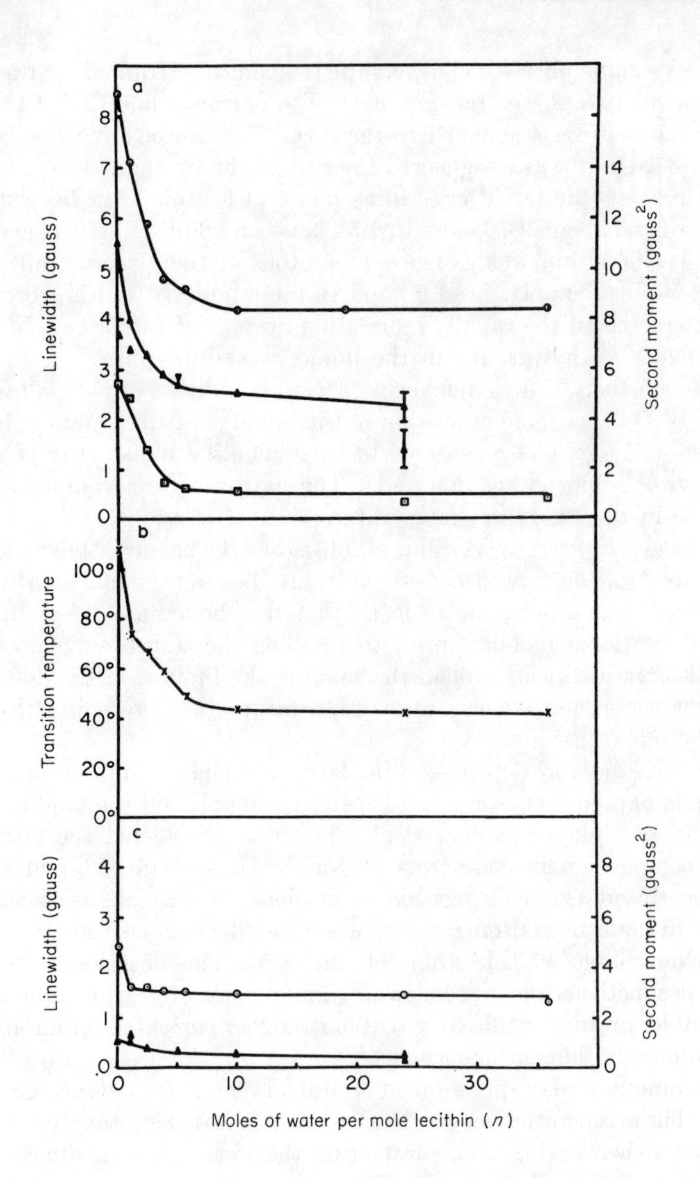

FIG. 8-4. Variation of linewidth and second moment for dipalmitoyl lecithin as a function of water content (moles D_2O per mole of 1,2-dipalmitoyl lecithin). (a) Linewidth of broad (⊙) and intermediate (⊡) components and the second moment (△) in the gel phase at 24°C. (b) The transition temperature T_c (×). (c) The linewidth (⊙) and second moment (△) of the broad component in the liquid crystalline phase ($T > T_1$). The corresponding parameters for the perdeuterated analog are shown by ▲. The symbol ▽ indicates the use of H_2O (13).

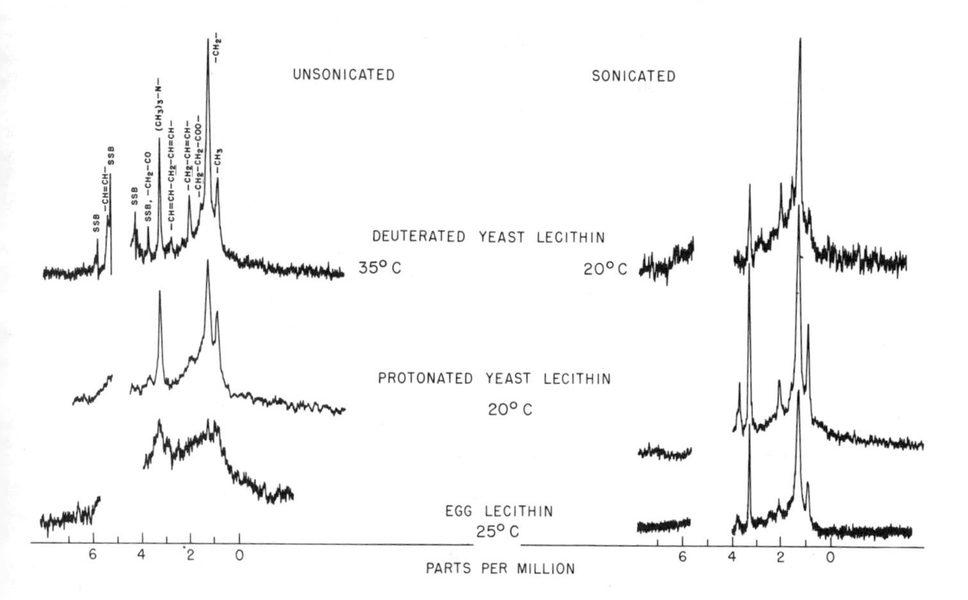

FIG. 8-5. 220 MHz proton NMR spectra of yeast and egg lecithin dispersions in D_2O. The spectrum of the unsonicated deuterated species was recorded using a larger sample than was used for the others. The changes in relative peak intensities between the protonated and deuterated yeast lecithin samples reflect variations in the amount of deuterium incorporated in the different positions of the deuterated species. All shifts are relative to TMS. Residual HOD peaks are not shown. SSB is a spinning side band of HOD (14).

tion on the measured proton spin–lattice (T_1) relaxation time measurements in aqueous lecithin dispersions were examined by Daycock *et al.* (15) over a large temperature range. Using the 180°–τ–90° and 90°–τ–90° two pulse methods at 40 MHz, the T_1 decay of the composite signal for all detectable protons revealed no evidence of multiple relaxation times. A minimum in the plot of T_1 versus temperature was observed in all cases. The minimum occurred below the gel–liquid crystalline transition temperature. The temperature of the T_1 minimum increased, and the activation energy for the T_1 process below the transition temperature increased as the D_2O concentration increased. These observations are summarized in Table 8-2. As shown in Fig. 8-6 and Table 8-3, the temperature minimum in T_1 also depends on the fatty acid chain length.

From the T_1 minimum below the transition, a correlation time $\tau_c \sim \omega_0^{-1} \sim 10^{-8}$ sec was estimated for the motion governing the T_1 relaxation process at the temperature of the minimum (15). Although it is not en-

TABLE 8-2

EFFECT OF D_2O CONCENTRATION ON OBSERVED TEMPERATURE DEPENDENCE OF T_1 FOR DIPALMITOYL LECITHIN[a]

MOLES OF D_2O PER MOLE OF LECITHIN	TEMPERATURE OF T_1 MINIMUM (°C)	ACTIVATION ENERGY (KCAL/MOLE) BELOW TRANSITION	ACTIVATION ENERGY (KCAL/MOLE) ABOVE TRANSITION
0	—	1.4	4.5
1	−46	1.0	4.0
3.7	−26	2.5	4.3
24	−5	3.2	4.8

[a] Daycock *et al.* (15).

tirely unequivocal, the motion was attributed to either rotation of the choline methyl groups (cf. Fig. 8-2E) or the entire trimethylammonium group (cf. Fig. 8-2F). Although not fully understood, the dependence of the activation energy on water content and chain length was thought to be associated with changes in the conformation of the polar head group.

The single value of T_1 obtained for all protons was taken as an indication of a spin diffusion process enabling the whole molecule to relax by a single relaxation mechanism (15). Spin diffusion (16) is not uncommon in

TABLE 8-3

EFFECT OF FATTY ACID CHAIN LENGTH ON THE TEMPERATURE DEPENDENCE OF T_1 FOR LECITHIN DISPERSIONS CONTAINING 40% D_2O BY WEIGHT[a]

FATTY ACID CHAINS IN LECITHIN	TEMPERATURE OF T_1 MINIMUM (°C)	ACTIVATION ENERGY (KCAL/MOLE) BELOW TRANSITION	ACTIVATION ENERGY (KCAL/MOLE) ABOVE TRANSITION
Dilauryl (C_{12})	−17		4.6
Dimyristoyl (C_{14})	−13		5.2 (4.8)[b]
Dipalmitoyl (C_{16})	−5	3.2	4.8
Distearoyl (C_{18})	+5	3.8	5.0
Dibihenoyl (C_{22})	+13	5.0	4.6

[a] Daycock *et al.* (15).
[b] Obtained for a perdeuterated sample.

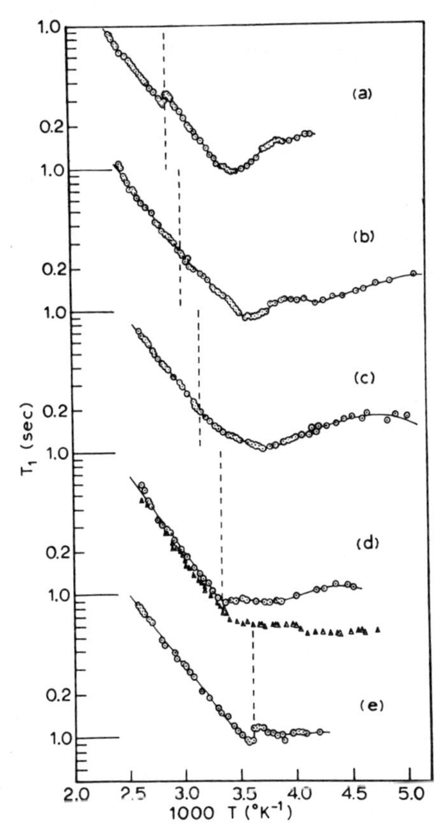

FIG. 8-6. Log T_1 versus reciprocal temperature for lecithins containing 40% (by weight D_2O. The vertical dashed line locates the chain melting point. (a) Dibehenoyl lecithin (C_{22}). (b) Distearoyl lecithin (C_{18}). (c) Dipalmitoyl lecithin (C_{16}). (d) Dimyristoyl lecithin (C_{14}). The results for a perdeuterated sample are shown as ($\triangle$). (e) Dilauryl lecithin (C_{12}) (15).

solids and operates as follows. If one part of a spin system is strongly coupled to the lattice (thus relaxing very efficiently) and the remainder of the spin system is weakly coupled to the lattice (thus relaxing inefficiently in the absence of a spin diffusion process), the strongly coupled part may act as an energy sink for the remainder of the system. Spin exchange processes conduct energy from various parts of the spin system to the efficiently relaxing part, i.e., the energy sink, and thus equalize T_1 over the entire system. Spin exchange requires overlap of the resonance lines. A rapid equilibration of energy within the spin system (achieved when $T_2 \ll T_1$) is also necessary for spin diffusion to be viable. In the case of lecithin, the proton group with the relatively efficient relaxation mechanism (presumably the choline methyl protons) may act as an energy sink for the remaining protons in the molecule.

As discussed in Section 5.7, $T_{1\rho}$, the spin–lattice relaxation time in the rotating frame, may be used to prove relatively slow motions. $T_{1\rho}$ reaches a minimum value when $\tau_c\omega_1 \sim 1$, where $\omega_1 = \gamma H_1$ and H_1 is the amplitude of the rf pulse. ω_1 will generally be a few orders of magnitude smaller than the resonance frequency ω_0, and $T_{1\rho}$ will therefore be more sensitive than T_1 to motions with a large correlation time.

Salsbury *et al.* (17) have taken advantage of the properties of $T_{1\rho}$, measuring $T_{1\rho}$ for the monohydrate of dipalmitoyl lecithin over the temperature range $-210°$–$170°C$. At various temperatures, minima in $T_{1\rho}$ were observed, corresponding to a correlation time of about 10^{-5} sec (ω_1^{-1}). The motions and temperatures for the minima were ascribed to rotation of the terminal group of the alkyl chain ($-190°C$), $N(CH_3)_3^+$ methyl group rotation ($-100°C$), and low-amplitude oscillatory motion of the fatty acid chain methylenes ($-23°C$). These minima do not correspond to phase transitions, which must therefore be associated with higher frequency motions.

The explanation of several observed proton NMR spectral characteristics of unsonicated liquid crystalline (above the transition temperature) lipid–D_2O systems has consumed much effort. Some of these characteristics are as follows. Most of the observed lines are broad but methyl proton resonances are relatively narrow, having linewidths of about 100–600 Hz. The lineshape is neither Lorentzian nor Gaussian but does exhibit a broad base sometimes referred to as "super-Lorentzian wings." The value of T_2 estimated from the linewidth is on the order of 1 msec, but the value of T_2 obtained from a Carr-Purcell pulsed NMR experiment is about two orders of magnitude larger than this. T_1 measurements yield values of several hundred milliseconds, the value increasing with an increase in temperature indicative of the short correlation time limit. Some reports have also claimed that linewidths are frequency dependent. It has been observed that the free induction decay following a 90° rf pulse is nonexponential. Because the observed spectral features are related, the following discussion considers the observations and accompanying explanations in roughly chronological order of development.

Lecithin has received the most attention. To relate the following discussion to something concrete, the proton NMR spectrum of lecithin is shown in Fig. 8-7A. The delayed Fourier transform (DFT) spectrum and the methyl proton assignments are shown in Fig. 8-7B. The DFT spectrum tends to eliminate broad peaks caused by protons that relax quickly (18). By switching out the receiver for 0.50 msec following each pulse, the con-

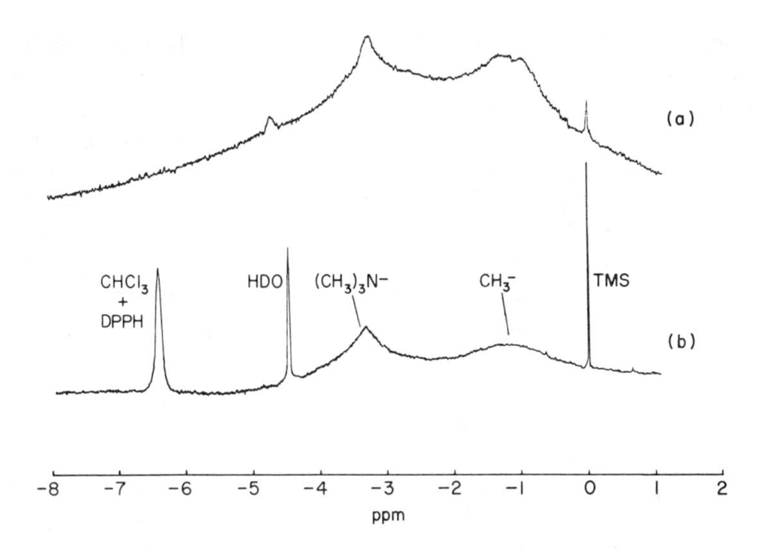

FIG. 8-7. Proton NMR spectra of unsonicated aqueous lecithin dispersions. (a) 220 MHz CW spectrum of egg lecithin at 29°C. (b) 100 MHz delayed (0.50 msec) Fourier transform spectrum of dipalmitoyl lecithin at 60°C. Both samples are in the liquid crystalline phase at the stated temperatures. The peak at −6.42 ppm (downfield) is an intensity standard of chloroform doped with the free radical 2,2-diphenyl-1-picryl-hydrazyl. External standard TMS is a 10% solution in carbon tetrachloride (18).

tribution to the Fourier transform spectrum from any proton with $T_2 \lesssim$ 0.50 msec will be eliminated. The methylene protons will be the major contributor to the broad peak in Fig. 8-7A, which has been eliminated in Fig. 8-7B.

Penkett *et al.* (19) originally reported the linewidths of egg lecithin in D_2O to be field dependent. It was suggested that the broad lines and field dependence may be a result of a chemical shift anisotropy effect.

Contrary to that suggestion, evidence that the linewidths in aqueous lipid systems are largely a result of dipolar broadening was provided by DeVries and Berendsen (20), who studied liquid crystalline potassium oleate–D_2O systems in which the bilayers were oriented between two glass slides. It was found that the proton NMR linewidth was a function of the angle between the normal to the glass slide and the magnetic field direction, the minimum linewidth occurring at an angle of 55°. The dipolar broadening in a rigid lattice (21) contains a factor $(3 \cos^2 \theta - 1)$ and therefore approaches zero as $(3 \cos^2 \theta - 1)$ approaches zero, i.e., when $\theta = \cos^{-1} \sqrt{1/3} = 54.7°$ (sometimes referrred to as the "magic angle"). The minimum linewidth obtained by DeVries and Berendsen at $\theta = 55°$ corresponds to the dipolar coupling between protons on the same CH_2 group, because the primary component of that coupling is perpendicular to the glass slide.

Finer *et al.* (22) also used the technique of DeVries and Berendsen (20), in which a sample is macroscopically oriented between two glass slides. It was observed (cf. Fig. 8-8) that the dipolar broadening was minimized by orienting the normal to the plane of the egg lecithin bilayers at an angle of 54°44′ to the external magnetic field. The linewidth was reduced to 350 Hz, compared to a value of 600 Hz for a randomly oriented sample. The broad "wings" were also greatly reduced (cf. Fig. 8-8).

Kaufman *et al.* (23) compared the T_2 relaxation times of erythrocyte ghost lipids obtained from the free induction decay (FID) with those from the Meiboom-Gill (MG) modification of the Carr-Purcell (CP) method (cf. Sections 2.1.3 and 5.7.2). The values of T_2 obtained were 1.4 msec from the FID method, in agreement with the linewidth ($W_{1/2} = 1/\pi T_2$) in the CW spectrum, and 75 msec from the CPMG method. These observations were interpreted as meaning the small FID and consequent large linewidths result from local magnetic field inhomogeneities in the sample. Because T_2 obtained from the CPMG method should not, in general, depend on the

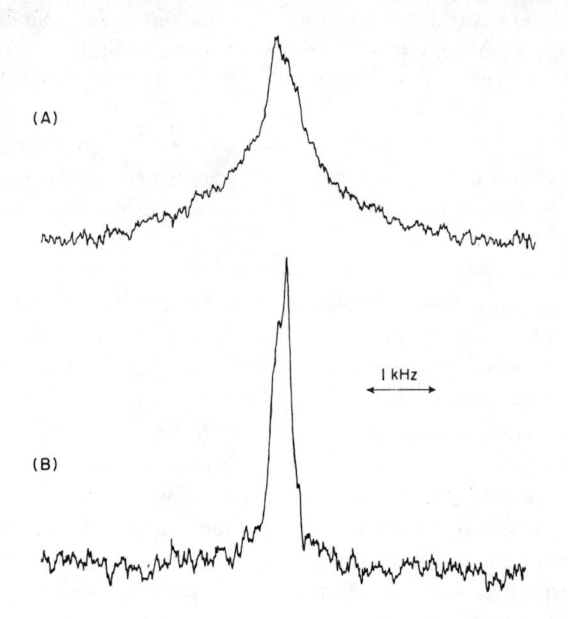

FIG. 8-8. 60 MHz proton NMR spectra of (A) 60 wt % egg lecithin–D_2O sample randomly oriented and (B) 60 wt % egg lecithin–D_2O sample with the normal to the plane of the bilayers oriented at 54°44′ relative to the external magnetic field (22).

magnetic field inhomogeneities (cf. Section 5.7.2), the motions in the sample should be characterized by T_2 from CPMG measurements. Magnetic field inhomogeneity was also considered an important factor by Hansen and Lawson (24), whose studies of the liquid crystalline dimethyldodecylamine oxide–D_2O system revealed T_1 decays to be exponential and T_2 decays, using the Meiboom-Gill modification of the CP method, to be nonexponential. The value of T_{2eff} required for decay to $1/e$ of the original height was also found to depend on pulse spacing. The nonexponential decay and pulse spacing dependence are indicative of self-diffusion through magnetic field gradients or of chemical exchange effects (cf. Sections 2.1.3 and 5.7.2). No conceivable model for chemical exchange exists, so self-diffusion through magnetic field gradients originating from anisotropic magnetic susceptibilities in the liquid crystal was chosen as the explanation. Magnetic field gradients in liquid crystalline systems were therefore considered the cause of the broad lines in the CW spectra and of the field dependence of the linewidths.

Tiddy (25) subsequently showed, in the case of liquid crystalline sodium caprylate–decanol–D_2O, that the value of T_{2eff} measured with the Meiboom-Gill modification depended on the pulse spacing (τ) when $\tau \lesssim T_{2eff}$. However, no dependence of T_{2eff} on τ was found when the CP method without the Meiboom-Gill modification was employed. This clearly rules out the proposal by Hansen and Lawson (24) that the pulse interval effect is caused by diffusion through magnetic field gradients. A 90° phase shift between the 90° pulse and the following 180° pulses is used in the Meiboom-Gill modification. A 90° phase shift between the 90° pulse and the immediately following second pulse is also employed in $T_{1\rho}$ measurements. Tiddy (25) ascribed the T_{2eff} variations using the Meiboom-Gill modification to an increasing $T_{1\rho}$ contribution to the T_{2eff} measurement as the pulse spacing is decreased below T_{2eff} and as the second pulse overlaps the FID following the first pulse. In addition, no field dependence of T_{2eff} in the range 8–90 MHz was evident, in contrast to the report of Hansen and Lawson (24).

A pulsed NMR study of the liquid crystalline potassium laurate–D_2O system by Charvolin and Rigny (26) has implied the existence of two types of motion, one governing T_2 relaxation and the other governing T_1 relaxation. The observed T_1 decay displayed no field dependence and was described by a single exponential. The value of T_1 decreased with decreasing temperature (E_{act} = 5.5 kcal/mole), being 0.68 sec at 90°C. The single value for T_1 was attributed to spin diffusion, and the correlation time for T_1 relaxation was estimated to be on the order of 10^{-9} sec. In contrast, the proton NMR free induction decay was nonexponential and decayed much more rapidly than T_1. Careful analysis of the FID revealed that it can be approximated by a combination of two Gaussian (solidlike) components and a Lorentzian (liquidlike) component. A log plot of the free induction decay is shown in Fig. 8-9. Originally, the three components were regarded as suggesting three regions of mobility along the fatty acid chain, with protons near the head being less mobile than those near the methyl terminus. In subsequent studies, the complex shape of the FID was attributed to a distribution of local dipolar fields in the sample originating from (*a*) varying orientations of the microcrystalline domains in the polycrystalline sample and (*b*) a distribution of local fields along each chain (27). The observed FID would therefore be a superposition of the relaxation in each of the local domains. This interpretation was supported by the frequency and temperature dependence of pulsed experiments in the rotating frame (e.g., $T_{1\rho}$ measurements). Charvolin and Rigny (26), furthermore, found no field dependence of the FID in the range 7–30 MHz. They suggested that sample preparation may be an important factor in the linewidth and FID field dependence reported by others.

Charvolin and Rigny (28) have also indicated that the proton NMR spectral linewidth of unsonicated egg lecithin results largely from static dipolar broadening.

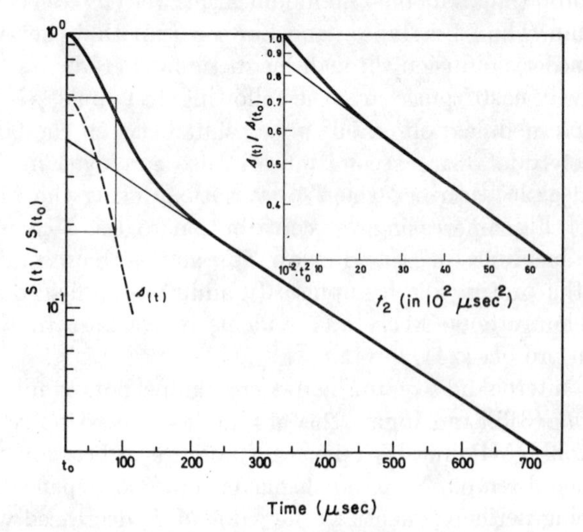

FIG. 8-9. Logarithmic plot of the free induction decay, $S(t)$, following a 90° pulse at at 30 MHz for potassium laurate–28% D_2O in the lamellar phase at 90°C. t_o is the sum of the pulse width and the recovery time of the instrument. $s(t)$, enlarged in the inset as a function of t^2, is the difference between the total decay $S(t)$ and its exponential part. The nonlinearity in the inset plot reveals two Gaussian components (26).

Chan *et al.* (29) examined the proton NMR of unsonicated egg lecithin bilayers in D_2O, making observations similar to those by Charvolin and Rigny (26) on another liquid crystalline system. It was found that the T_1 decay for lecithin is described by a single exponential, attributed to the effects of spin diffusion, with a time constant of 220 msec. The value of T_{2eff} obtained from the nonexponential FID was found to be 0.14 msec, independent of magnetic field in the range 16–60 MHz.

Chan *et al.* (29) also commented that the Carr-Purcell results (23, 24) could provide little insight into the T_2 relaxation mechanism. Because there is a distribution of T_2 relaxation times for the various protons in the lecithin bilayers (FID and CW results), with most being very short, it appears that a large part (caused by short T_2 protons) of the transverse magnetization following the initial 90° pulse would be lost before the first 180° pulse is applied even with the shortest pulse spacing employed. Consequently, T_2 measurements with the CP method will have a contribution from only that fraction of protons with longer T_2 values.

Oldfield *et al.* (30) investigated the liquid crystalline phase of dipalmitoyl lecithin, egg lecithin, and potassium laurate in D_2O using both pulsed and CW proton NMR. Other reports of nonexponential FID behavior independent of frequency in the range 10–40 MHz were confirmed. The use of selective isotope substitution was illustrated using di(perdeutero)palmitoyl lecithin. The proton NMR spectra of the deuterated lecithin in the gel and liquid crystalline phases are shown in Fig. 8-10. The appearance of the relatively narrow peak from the choline methyl protons in the liquid crystalline phase is clearly resolved.

A frequency dependence of the hydrocarbon chain proton linewidth in unsonicated egg lecithin dispersions above 60 MHz was reported by Finer *et al.* (22) (cf. Fig. 8-11). However, there are some difficulties in determining the true baseline, and therefore the linewidth, because a considerable fraction of the protons contribute to the low, very broad portion of the spectrum. The authors suggested that anisotropic motion of the hydrocarbon chain protons may account for the observed frequency dependence rather than internal magnetic field inhomogeneities.

Chan *et al.* (31, 31a) have been able to theoretically reproduce the methyl proton NMR lineshape in the spectrum of unsonicated lecithin. They obtained a relatively narrow peak on top of a very broad peak, the lineshape being neither Lorentzian nor Gaussian. The basis of the earlier treatment (31) was Woessner's theory for the dipolar relaxation of spins on groups undergoing anisotropic motion (6), which was discussed above. A more generalized treatment (31a) was based on a stochastic theory with boundary conditions. For the hydrocarbon chain methyl protons, free rotation about the methyl C_3 rotational axis was permitted. However, the orientation of the rotational axis with respect to the chain direction was restricted such that the orientation could vary only through an angle of 70°. The two motions (as discussed above) lead to two correlation times, τ_c and τ_{c1}, which are, respectively, the correlation times associated with reorientation about the rotational axis and reorientation of the rotational axis. The methyl resonance lineshape and relaxation were considered to be caused by dipolar interactions among the methyl protons and from an additional proton (e.g., from the adjacent methylene group).

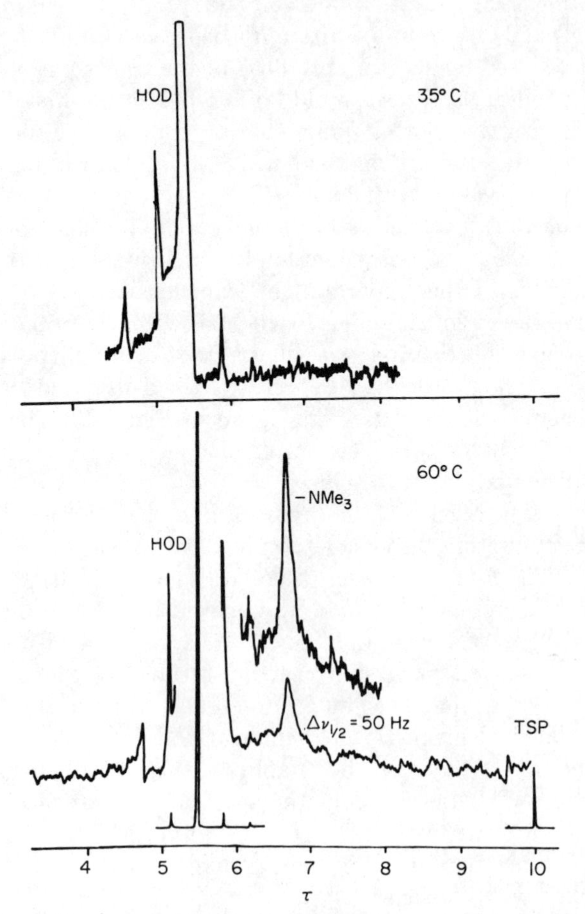

FIG. 8-10. 220 MHz proton NMR spectrum of di(perdeutero)palmitoyl lecithin (5%)–D_2O in the gel (35°C) and liquid crystalline (60°C) phases. The sharp lines appearing in the spectrum at 35°C arise from the residual HOD resonance and its spinning side bands. The reference is sodium 3-trimethylsilylpropionate-2,2,3,3-d_4 (TSP) (30).

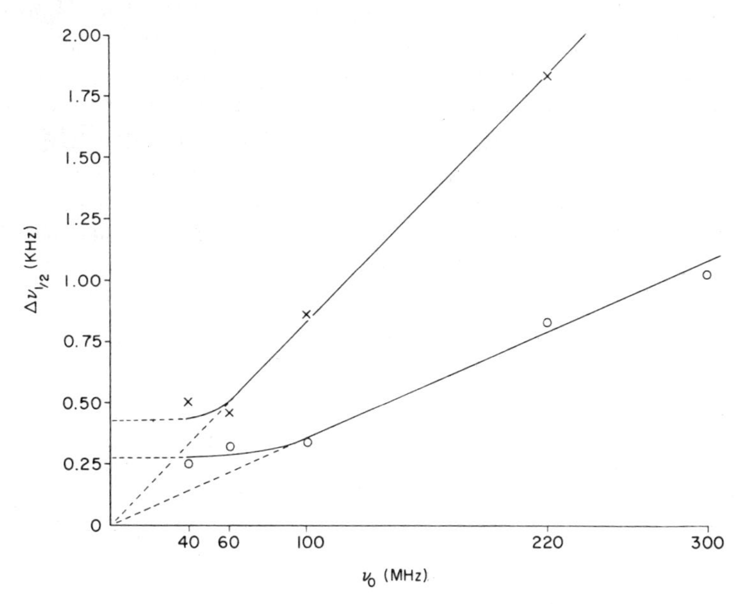

FIG. 8-11. Frequency dependence of the CH_2 proton NMR linewidth $\Delta\nu_{1/2}$ from unsonicated egg yolk lecithin dispersions at two concentrations: 80 wt % (×) and 60 wt % (○) (22).

It was pointed out by Chan *et al.* (31, 31a) that the existence of two correlation times could account for the apparent discrepancy between the observed linewidth and the spin–lattice relaxation time (cf. Eqs. 8-2 and 8-3). T_1 will be governed by the motion with the faster correlation time, namely τ_c, and may exhibit a temperature dependence characteristic of the short correlation time limit (i.e., increasing T_1 with increasing temperature). T_2, however, will be governed by the slower motion (τ_{c_1}) provided τ_{c_1} is less than the reciprocal of the rigid lattice linewidth.

The description by Chan *et al.* (31, 31a) would appear in general to explain the proton NMR spectral features satisfactorily. However, further refinements could take into consideration the interaction of the methyl protons with additional protons, both intramolecular and intermolecular. It is quite possible, for example, that intermolecular dipole–dipole relaxation may make a contribution if the translational correlation time is not much greater than $1/\omega_0$. In the case of relaxation in *n*-dodecane, the intermolecular contribution is about eight times larger than the intramolecular contribution (32). Moreover, NMR studies of sonicated egg lecithin have indicated that intermolecular relaxation is important for phospholipids as well (33).

It has been shown for solids that rapid rotation of the sample, aligned at the "magic angle" (54°44′) with respect to the direction of the external magnetic field, results in a dramatic reduction of the dipolar broadening (34, 35). If the sample tube rotation is more rapid than the dipolar linewidth of the sample, line narrowing may be achieved. Chapman *et al.* (36) have applied the sample rotation technique to lecithin in the gel and liquid crystalline states. The results for an egg lecithin dispersion in the liquid crystalline state are shown in Fig. 8-12. Clearly, a significant part of the linewidth is caused by a dipolar contribution (cf. Eq. 8-1), which can be affected by macroscopic rotation of the sample. The linewidth of the methyl proton resonance of $N(CH_3)_3^+$ occurring at 3.2δ (6.8τ) is reduced to 17 Hz, and the hydrocarbon chain methyl resonance at 0.9δ (9.1τ) is reduced to less than 50 Hz.

Carbon-13 NMR. The use of ^{13}C NMR would appear to offer some advantages over proton NMR, compensating for its lower sensitivity. First, the chemical shifts are typically large. Second, the smaller gyromagnetic

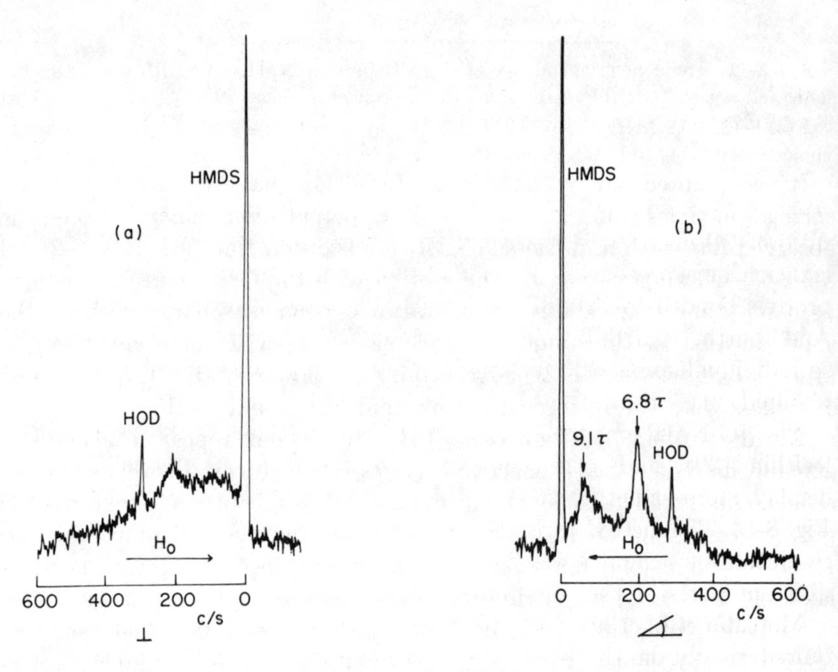

Fig. 8-12. Effect of sample rotation on the 60 MHz proton NMR spectrum of unsonicated egg yolk lecithin (33%)–D_2O, 25°C (liquid crystalline). Rotation frequency = 500 Hz. With respect to the external magnetic field, the sample tube rotation axis is (a) 90° and (b) $\cos^{-1} \sqrt{1/3}$. The reference is hexamethyldisiloxane (HMDS) (36).

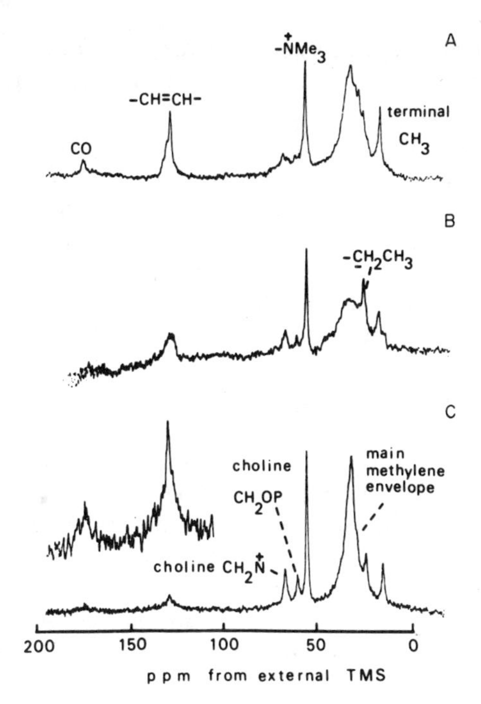

FIG. 8-13. 25.2 MHz ^{13}C NMR spectra of (A) egg lecithin (EYL)(B) egg lecithin–cholesterol (1:1), and (C) ox brain sphingomyelin (38).

ratio of ^{13}C makes it less susceptible to dipolar broadening. Third, the relaxation of a protonated ^{13}C nucleus is caused almost exclusively by the protons bonded to that ^{13}C, so translational motion may be disregarded. And fourth, spin diffusion does not occur with natural abundance ^{13}C, so individual spin–lattice relaxation times of various carbons in the phospholipid may be obtained.

The ^{13}C NMR spectra of unsonicated egg lecithin (37, 38), dipalmitoyl lecithin (39), and sphingomyelin (38) in the liquid crystalline phase display more structure than the proton NMR spectra, as illustrated in Fig. 8-13. The methyl carbon resonances may easily be distinguished. The resolution of several peaks makes it possible to consider the mobility in different parts of the phospholipid molecules.

Metcalfe *et al.* (39) noted that more accumulated transients were required to obtain the spectrum of unsonicated lecithin compared with sonicated lecithin, leading to the suggestion that only the outer layers of the liposomes in the unsonicated dispersions contribute to the observed narrow lines.

Phosphorus-31 NMR. The examination of ^{31}P NMR permits study of the polar head group in phospholipids without complications from nuclei on other parts of the molecule.

Davis and Inesi (40) obtained a ^{31}P NMR linewidth for an unsonicated egg yolk lecithin dispersion of 450 ± 50 Hz at 36.4 MHz and 1000 ± 50 Hz at 101 MHz. Davis (41) later noted that the ^{31}P linewidths of sonicated lecithin dispersions and membranes reported in the literature revealed that the field dependence of the linewidths was linear. Extrapolation to zero field gave a linewidth of 200 ± 120 Hz. It was suggested that the field dependence is caused by a chemical shift anisotropy of the lipid phosphate group.

C. *Sonicated Aqueous Lipid Dispersions.*

Proton NMR. It was observed some time ago that phosphatidylserine and phosphatidylcholine proton NMR linewidths are narrowed considerably on sonication of the aqueous dispersions (42). It is much easier to study the high-resolution spectra (cf. Fig. 8-5) of sonicated phospholipids than the broad lines generally observed with unsonicated phospholipids. However, why do the lines narrow on sonication? Are there significant structural changes in the bilayer, or can the line narrowing be ascribed simply to the existence of the bilayers in smaller particles (e.g., 250 Å vesicles)? There has been disagreement among researchers as to the cause of line narrowing on sonication of aqueous dispersions.

Sheard (43) proposed that sonication disrupts the bilayer structure, resulting in a high-resolution spectrum. However, using gel filtration, Finer *et al.* (44) showed that two fractions of sonicated egg lecithin particles exist with particle diameters smaller than that of the unsonicated particles (0.5–20 μm). Fraction I consists of fairly large-diameter, multilamellar liposomes (0.2–0.4 μm). Fraction II consists of spherical vesicles composed of a single bilayer, with the average diameter of the vesicle being 230 Å. Fraction II, as determined by analytical gel filtration, increases as sonication proceeds. The molecules in fraction II apparently give rise to the observed high-resolution proton NMR spectrum because the integrated $N(CH_3)_3^+$ proton signal intensity corresponds to the number of molecules in fraction II.

Experiments using the line broadening and pseudocontact shift capabilities of paramagnetic ions support the concept that the proton NMR spectra of sonicated phospholipids arise from small vesicles consisting of sealed shells formed from a single bilayer. Bystrov *et al.* (45) have examined the effect of Mn(II) and Eu(III) on the 100 MHz $N(CH_3)_3^+$ proton signal of sonicated egg lecithin and concluded the signal is composed of two components. Addition of Mn(II) after sonication results in a decrease in the intensity of the sharp $N(CH_3)_3^+$ signal. Addition of Eu(III) after sonication causes the $N(CH_3)_3^+$ to split into two components, one of which is shifted upfield relative to the normal choline methyl signal. The integrated intensity of the upfield component is 1.5 times that of the unshifted component. Addition of Mn(II) to this sample broadens only the upfield component. These observations show that only the cholines on the external surface of the lecithin vesicle are accessible to the broadening effects of Mn(II) and the shift effects of Eu(III); the choline methyl signal from the inner surface of the bilayer remains unperturbed. The metal ions are presumably complexed by phosphate groups in the head group region. The intensity ratio of 1.5 is in agreement with the expected molecular ratio of outer to inner layer lecithins for a spherical vesicle with the observed dimentions. Sonication in the presence of Mn (II) or resonication after addition of Mn(II) results in a single severely broadened $N(CH_3)_3^+$ signal.

Kostelnik and Castellano (46) confirmed these results and, in addition, used a paramagnetic anion $Fe(CN)_6^{3-}$ to produce an upfield shift of one component. Nd(III) and UO_2^{2+} have also been shown to promote shifts in the $N(CH_3)_3^+$ proton resonances of dipalmitoyl lecithin vesicles, permitting resolution of external and internal $N(CH_3)_3^+$ resonances (47).

The paramagnetic effect studies show that (*a*) the sonicated lecithins consist of closed sphere vesicles with a single bilayer for a shell, (*b*) the internal and external head groups may be distinguished, (*c*) the bilayer is impermeable to the paramagnetic ions, (*d*) the vesicle has sufficient integrity to prevent mixing of the internal and external aqueous solutions, and (*d*) sonication disrupts the bilayer, permitting mixing of internal and external solutions, with subsequent reformation of the vesicle.

By employing a 250 MHz proton NMR spectrometer, Kostelnik and Castellano (48) were able to show that the $N(CH_3)_3^+$ signal in a sonicated dispersion of egg lecithin in D_2O naturally consists of two peaks with an intensity ratio 1.85:1.00. The peaks are separated by about 5 Hz, the more intense peak being downfield. Progressive addition of Mn(II) to the vesicles results in broadening of the downfield peak, leaving only the weaker upfield resonance discernable in the spectrum with 1.0 m*M* Mn(II) present. This leads to the conclusion that the downfield peak is due to choline methyls in the external layer.

Although the evidence is good for a bilayer structure in a sonicated phospholipid system, there is still a possibility that the structure within the bilayer has been altered by sonication, leading to the high-resolution spectra. Whether or not this happens has been a source of some contention.

Electron paramagnetic resonance studies of spin labels in sonicated and unsonicated lecithin dispersions have indicated that there is little difference in the mobility of the spin label in the hydrocarbon moiety (49, 50).

Finer *et al.* (22) have maintained that the high-resolution proton NMR spectrum obtained after sonication is caused by an increased tumbling rate of the small vesicles (compared with unsonicated particles) that averages out the dipolar broadening. Using an equation similar to the second term of Eq. 8-1, describing the line narrowing caused by isotropic motion of the vesicle (51),

$$W_{1/2}^2 = \frac{4 \log 2}{\pi} W_{1/2}''^2 \tan^{-1} (W_{1/2}\tau_c/\pi) \tag{8-4}$$

it was shown that tumbling could produce the relatively narrow lines observed (22). For motional narrowing to occur, $1/\tau_c \ll W_{1/2}$ (cf. Eq. 8-4). The correlation time estimated from the Debye-Stokes equation (Eq. 2-40) for a vesicle of 115 Å radius is sufficiently small to produce motional narrowing. However, tumbling of unsonicated particles or even the smallest particle in fraction I (44) with a radius of 1000 Å would yield very little motional narrowing.

In contrast, Horwitz *et al.* (52) contend that vesicle tumbling is not an important cause of line narrowing in sonicated egg lecithin. They argue that the value of $W_{1/2}''^2$ used by Finer *et al.* (22) should have been the rigid lattice linewidth rather than the linewidth of the unsonicated lecithin. The point is equivocal because the rigid lattice linewidth $(W_{1/2}' + W_{1/2}'')$ may have a component $W_{1/2}'$ caused by anisotropic motion, which need not be considered for the isotropic tumbling of the vesicle.

It was also noted that the linewidths in sonicated egg lecithin are independent of the amount of added glycerol; i.e., they are independent of viscosity (52). This point argues against the importance of vesicle tumbling if the Debye-Stokes relation (eg. 2-40) containing a translational viscosity is valid for rotational (tumbling) motion.

The results found by Kohler *et al.* (14), using yeast and egg lecithin (cf. Fig. 8-5), also argue against vesicle tumbling as the cause of line narrowing on sonication. If vesicle tumbling were the major source of line narrowing, it would be expected that the unsonicated-to-sonicated linewidth ratio of yeast and egg lecithin should be the same. It is clearly seen in Fig. 8-5 that the linewidth ratios are not the same. In fact, the yeast lecithin linewidths narrow only a little on sonication, whereas the egg lecithin linewidths narrow considerably. At this time, the question of line narrowing in sonicated vesicles must be considered as unresolved.

Fourier transform NMR has been used by Horwitz *et al.* (53) to measure the individual T_1 and T_2 relaxation times of some protons in sonicated egg lecithin dispersions. (See Chapter 5 for a discussion of the T_1 and T_2 measurements using Fourier transform NMR.) The results are given in Table 8-4. Although the T_1 values are similar, they are not identical and are therefore not the result of spin diffusion to an energy sink, as apparently is the case with unsonicated lecithin. The T_2 values show more variability than the T_1 values. In addition, the T_2 decay of the methylene protons is nonexponential. About 80% decay rapidly and nonexponentially; the remaining 20% decay with a single T_2 of 56 msec.

As in the case of unsonicated lecithin, T_1 increases with increasing temperature, indicating that T_1 relaxation is in the short correlation time limit (53). In this limit, $T_1 = T_2$ if the motion is isotropic. Experimentally, $T_1 > T_2$, leading to the conclusion that the motion is anisotropic, with different correlation times governing the T_1 and T_2 relaxation processes. Two plausible types of motion were suggested to account for the T_1 and T_2 results. A high-frequency oscillation involving small rotational displacements of individual methylene carbon atoms may regulate the T_1 relaxation process. It is expected that the rotational oscillations will differ little from one end of the molecule to the other, in accord with the approximately equal T_1 values. Trans–gauche interconversion (cf. Fig. 8-2B) was proposed as the motion leading to T_2 relaxation. The large angular displacements involved in the trans–gauche interconversion entail progressively less steric hindrance as the methyl terminus is approached along the chain. This variation along the chain may account for the heterogeneity in the observed methylene T_2 relaxation. This interpretation of the T_1 and T_2 data was later amplified (52).

Lee *et al.* (54) have used Fourier transform NMR to obtain individual proton T_1 values in sonicated dipalmitoyl lecithin vesicles. The results were compared with those obtained on lecithin dissolved in CD_3OD. The temperature dependence of the observed $N(CH_3)_3^+$ T_1 values exhibits a well-defined inflection at the transition temperature. The T_1 values for sonicated dipalmitoyl lecithin (similar to those for protons on saturated carbons of egg lecithin given in Table 8-4) were interpreted to mean that motional freedom increases from the glycerol backbone toward both the methyl terminus of the alkyl chain and the $N(CH_3)_3^+$ group on the surface of the bilayer. The variation in T_1 values was ascribed to tighter packing of lecithin molecules in the glycerol region.

Proton T_1 and linewidth measurements for a 1:10 mixture of dipalmitoyl lecithin with di(perdeuteropalmitoyl)lecithin in sonicated vesicles above the transition temperature were obtained by Lee *et al.* (33). The larger proton NMR linewidths and longer T_1 values (cf. Table 8-5) in the sample diluted with the perdeutero species illustrate the importance of an inter-

TABLE 8-4

SPIN–LATTICE (T_1) AND SPIN–SPIN (T_2) RELAXATION TIMES AND ACTIVATION ENERGIES FOR SOME PROTON RESONANCES OF SONICATED EGG LECITHIN[a]

	—N(CH₃)₃⁺	—CH₂—C(=O)—O—	—CH₂—	—CH₂—C(H)=C(H)—	—HC=CH—	—CH₃	^{31}P
T_1 (sec)	0.41 ± 0.02	0.34 ± 0.02	0.47 ± 0.03	0.41 ± 0.04	0.54 ± 0.03	0.76 ± 0.06	1.4 ± 0.1 (*a*) 8.5 ± 0.7 (*b*)
T_2 (sec)	0.075	0.008	0.056 (20%) <0.02 (80%)	0.015	0.020	0.036	0.110
E_a (kcal/mole)	4.3 ± 0.3	2.8 ± 0.4	3.0 ± 0.2	2.7 ± 0.2	3.2 ± 0.3	4.2 ± 0.3	—

[a] The T_1 values were determined at 40°C. For a given experiment the estimated error was within 10%, as indicated; however, for experiments performed on different days with different samples, the error sometimes exceeded this limit. The estimates of T_2 were made at 20°C; the text contains an explanation of the two relaxation times for the methylene protons. The phosphorus nuclear relaxation times were measured at 34°C. (*a*) Refers to dimyristoyl-L-α-lecithin and (*b*) refers to egg yolk lecithin. Horwitz *et al.* (53).

TABLE 8-5

COMPARISON OF THE 100 MHz PROTON NMR DATA FOR SONICATED VESICLES (I) OF DIPALMITOYL LECITHIN AND (II) A 1:10 MIXTURE OF DIPALMITOYL LECITHIN–DI(PERDEUTEROPALMITOYL) LECITHIN AT 54°C (I) AND 52°C (II)[a]

	T_1 (SEC)		$W_{1/2}$ (Hz)	
	$(CH_2)_n$[b]	CH_3	$(CH_2)_n$[b]	CH_3
I	0.53 ± 0.04	0.84 ± 0.08	25	16[c]
II	0.8 ± 0.2	2.0 ± 0.5	14	13[c]

[a] Lee *et al.* (33, 54).
[b] Weighted average for observed composite resonance.
[c] The CH_3 resonance in CD_3OD, as solvent, appears as a distorted triplet of total linewidth ~11 Hz, caused by ^1H–^1H coupling.

molecular contribution to relaxation. This observation, at the very least, necessitates a modification of the interpretation by Horwitz *et al.* (52, 53), who described relaxation solely in terms of intramolecular motions. In fact, Lee *et al.* believed the intermolecular contribution to be predominant for T_2 relaxation.

With a series of assumptions, the intermolecular contribution to relaxation was used to calculate a self-diffusion coefficient for the lecithin molecule in the bilayer (33). It was in reasonable agreement with values estimated for lateral diffusion from electron paramagnetic resonance spin label studies (55).

A means of directly measuring the self-diffusion coefficient of lipids has been described by Roberts (56). The value for T_2 in lipid–water systems is generally too small to permit a direct measurement of the self-diffusion coefficient using the conventional 90°–τ–180° pulse sequence (or the pulsed gradient modification) described in Section 5.8.1. The problem of T_2 being too small has been circumvented by using a sequence of three 90° pulses, producing a stimulated echo. Tanner (57) has developed the stimulated echo sequence for use in diffusion coefficient measurements. Utilizing Tanner's development, Roberts (56) obtained values for the self-diffusion coefficient of potassium laurate in heavy water for both the hexagonal phase (50% D_2O) and the lamellar phase (28% D_2O) at 80°C. The values were $2.3 \pm 0.3 \times 10^{-6}$ in the hexagonal phase and $2.4 \pm 0.35 \times 10^{-6}$ cm² sec⁻¹ in the lamellar phase. The large values indicate that the individual lipid molecules maintain considerable translational freedom.

Carbon-13 NMR. Carbon-13 spin–lattice relaxation time measurements of some of the individual carbon nuclei in lecithin have provided information about the relative mobility of different parts of the molecule.

Levine *et al.* (58) examined the ^{13}C NMR spectra of sonicated dipalmitoyl lecithin as a function of temperature. As shown in Fig. 8-14, sharp

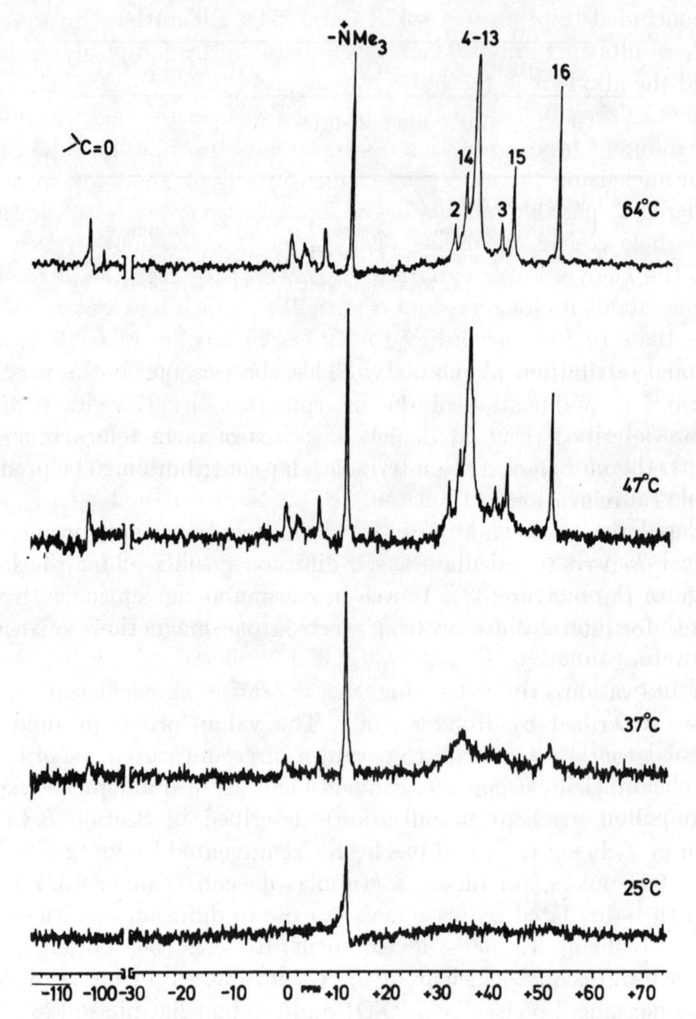

FIG. 8-14. 25 MHz ^{13}C NMR spectra of sonicated dipalmitoyl lecithin (230 mM) in D_2O buffer (pD 7.4) as a function of temperature. The numbers refer to the position on the hydrocarbon chain. The three peaks just downfield of the choline are caused by CH_2OP choline, (CH_2OP plus CH_2O) glycerol, and CH_2N choline, in order of decreasing field. Chemical shifts are expressed relative to dioxane as internal reference (58).

lines are observed above the transition temperature (~40°C). T_1 relaxation times for the lecithin vesicles at 52°C increase progressively from the glycerol carbon nuclei toward the methyl carbon of the hydrocarbon chain and toward the $N(CH_3)_3^+$ carbon nucleus, with the exception of the nonprotonated carbonyl nuclei. The trend is illustrated in Fig. 8-15. As was also concluded from proton NMR data (54), the authors interpreted the ^{13}C T_1 results in terms of increasing mobility from the glycerol moiety toward the alkyl chain methyl terminus and toward the $N(CH_3)_3^+$ group.

The ^{13}C T_1 results for sonicated lecithins were further examined in terms of the different types of motion, with estimated correlation times, for each carbon nucleus on the fatty acid chain (59). Three modes of motion were considered as possible modulators of T_1 relaxation: (*a*) isotropic tumbling of the whole vesicle, which has a correlation time of about 10^{-6} sec according to the Debye-Stokes equation (Eq. 2-40); (*b*) rotation of the lecithin molecule about its long axis (cf. Fig. 8-2D), which has a correlation time of less than 10^{-7} sec according to EPR spin label studies (50); and (*c*) rotational oscillations about individual carbon–carbon bonds in the alkyl chain.

It was observed that T_1 increases with increasing temperature for all carbons; therefore $(\omega_H + \omega_c)\tau_c < 1$, and vesicle tumbling is eliminated as a viable cause of T_1 relaxation (59). Rotation about the long axis was also considered an unlikely motional contributor because it would predict identical T_1 values for all methylene carbons in the chain. The motion governing T_1 relaxation was therefore attributed to rotational oscillations about the carbon–carbon bonds. Correlation times for these internal rotations were estimated using the theory of Wallach (60), which was mentioned in Section 2.3.1. That theory will not explain the observed abrupt increase in T_1 toward the terminal methyl group (cf. Fig. 8-15), and no physical basis for the observed increase was offered.

Batchelor *et al.* (12) have examined the temperature dependence (above the transition temperature) of the ^{13}C chemical shifts in sonicated egg

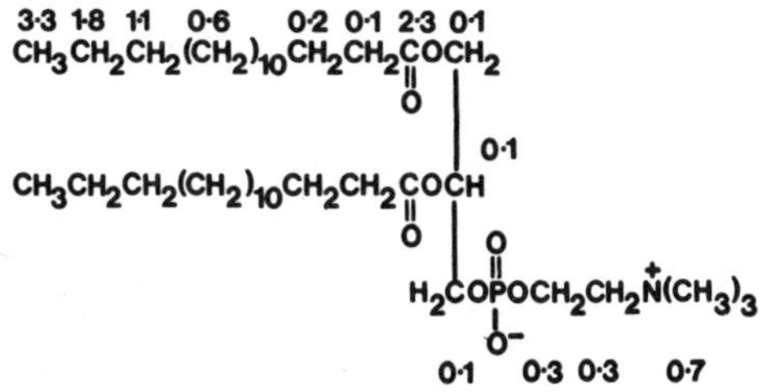

3·3 1·8 1·1 0·6 0·2 0·1 2·3 0·1
$CH_3CH_2CH_2(CH_2)_{10}CH_2CH_2\overset{O}{\overset{\|}{C}}OCH_2$
0·1
$CH_3CH_2CH_2(CH_2)_{10}CH_2CH_2\overset{O}{\overset{\|}{C}}OCH$
$H_2COPOCH_2CH_2\overset{+}{N}(CH_3)_3$
O^-
0·1 0·3 0·3 0·7

FIG. 8-15. Carbon-13 (sec) T_1 values for dipalmitoyl lecithin in D_2O at 52°C. (58).

lecithin. The observed chemical shift dependence was interpreted in terms of changes in the relative populations of trans and gauche conformations along the fatty acid chains. The basis for interpretation is that the ^{13}C resonances of alkanes shift upfield about 5 ppm when the carbon nucleus is involved in a carbon–carbon bond in the gauche conformation rather than the trans conformation. With rapid exchange between the two conformations for any carbon–carbon bond, changes in the ^{13}C chemical shift for any carbon nucleus on the alkyl chain will be indicative of an alteration in the population of trans and gauche conformers at that site. Although quantitation of the conformer population was not possible, the general observed shift to lower field with decreasing temperature was ascribed to an increasing population of lower energy trans conformer. Laser raman studies (61) indicate that there are about ten gauche conformers per lecithin molecule.

Fluorine-19 NMR. To overcome the problem of peak overlap in the case of ^{1}H and ^{13}C NMR, for resonances originating from nuclei in the middle region of the fatty acid chain, ^{19}F NMR of monofluoro compounds has been utilized.

Birdsall *et al.* (62) examined the ^{19}F NMR linewidth of monofluorostearic acid in egg lecithin vesicles as a function of temperature and fluoro substituent position. Presumably, the fluorostearic acid will be oriented in the vesicle in the same manner as the lecithin molecules. The three monofluorostearic acids employed (substituted at positions 4, 7, and 12, respectively) revealed a sharply decreased ^{19}F linewidth, with the position of the fluoro substitution approaching the methyl terminus. The ^{19}F results were attributed to increased motional freedom in the chain progressing toward the terminal methyl group.

Phosphorus-31 NMR. Phosphorus-31 spin–lattice and spin–spin relaxation time measurements, performed by Barker *et al.* (63), enabled a preliminary study of the lipid head group. T_2 values, determined from the free induction decay following a 90° pulse, are summarized for several phospholipids in Table 8-6. It will be noted from the table that addition of water to anhydrous egg lecithin increases T_2; the interpretation made was that the water facilitates motion in the polar head group. However, it also appears plausible that the increase in T_2 is caused by the tumbling motion of the vesicles. The T_2 difference between egg lecithin and dipalmitoyl lecithin at 21°C simply reflects the fact that egg lecithin is in the liquid crystalline state at that temperature and dipalmitoyl lecithin is in the gel state.

It was noted earlier in this section that the inner and outer surfaces of

TABLE 8-6

^{31}P (84.5 MHz) RELAXATION TIMES IN PHOSPHOLIPID DISPERSIONS[a]

PHOSPHOLIPID	T_2 (μSEC)	TEMPERATURE (°C)
Anhydrous egg lecithin	<50[b]	21
33% (w/w) egg lecithin–H_2O	190	21
33% (w/w) egg lecithin–D_2O	180	21
30% (w/w) phosphatidylserine–H_2O	180	28
30% (w/w) phosphatidylserine–D_2O	230	28
50% (w/w) dipalmitoyl lecithin–D_2O	90	21
50% (w/w) dipalmitoyl lecithin–D_2O	210	71
50% (w/w) dipalmitoyl lecithin–cholesterol–D_2O (dipalmitoyllecithin:cholesterol 70:30 mole %)	240	22

[a] Barker *et al.* (63).
[b] Too short to be observable because of the recovery time of the instrument.

lecithin vesicles could be distinguished by the effect of paramagnetic ions on the proton resonance of the outer surface $N(CH_3)_3^+$ groups. It is expected that the paramagnetic metal ions will be coordinated to the lecithin phosphate groups. On that basis, the internal and external head groups should also be distinguishable by ^{31}P NMR. Bystrov *et al.* (64) showed that the ^{31}P signals of the inner and outer surfaces of egg lecithin vesicles could be completely resolved by adding Pr(III). Adding 0.01 *M* $Pr(NO_3)_3$ to a 20% sonicated egg lecithin dispersion in D_2O resulted in a shift of 11.5 ppm downfield for the ^{31}P peak of the external phosphate groups, demonstrating that ^{31}P NMR of lecithin vesicles is much more sensitive to paramagnetic metal ions than ^{1}H NMR. The linewidth in the absence of added metal ions was observed to be only 50 Hz, implying a value of 6–7 msec for T_2; this is more than an order of magnitude larger than the values reported by Barker *et al.* (63) (cf. Table 8-6). The apparent discrepancy between the results from linewidth measurements and free induction decay measurements remains to be reconciled.

D. *Presence of Cholesterol or Other Membrane Constituents*

The next step in a consideration of model membrane systems is to add some of the natural membrane constituents to a phospholipid bilayer. Both cholesterol and proteins comprise a large fraction of many membranes, and it is therefore of interest to investigate the effect of adding these constituents on the structure and mobility of phospholipid molecules in the bilayer.

Cholesterol. Early studies by Chapman and Penkett (65) on the effect of cholesterol addition to sonicated egg lecithin (liquid crystalline) revealed the relatively sharp alkyl chain methyl and methylene proton NMR resonances to broaden considerably and the $N(CH_3)_3^+$ proton resonances to remain fairly sharp. This observation implies that the cholesterol molecules affect the fatty acid chain motions without disturbing the polar head group. No proton resonances of cholesterol could be detected in the 1:1 lecithin–cholesterol dispersion.

Further investigation of the effect of cholesterol on lecithin was reported by Darke *et al.* (66), who obtained proton NMR spectra of both unsonicated and sonicated dispersions. The results are shown in Fig. 8-16 for sonicated egg lecithin–cholesterol in the liquid crystalline phase. The observable resonances are caused only by lecithin, not by cholesterol. This was demonstrated by using lecithin incorporating deuterated alkyl chains; the alkyl chain proton resonances disappeared with no evidence of cholesterol proton resonances. Cholesterol clearly has a differential effect on the lecithin mobility (cf. Fig 8-16). The choline methyl resonance is little changed and the chain terminal methyl resonance is affected much less than the methylene resonance. Magnetic anisotropic chemical shift effects (cf. Section 2.4.6) as a partial cause of line broadening was suggested from the observation that the methylene proton linewidth was 770 $\pm$ 50 Hz at 220 MHz and 550 $\pm$ 70 Hz at 60 MHz. This relatively small variation is nevertheless not sufficient to account for much of the selective line-broadening effect of cholesterol, which is therefore ascribed to a considerable diminution in mobility of most of the fatty acid chain.

Variation of the amount of cholesterol present, molar ratio from 10:1 to 1:1, in the lecithin dispersion results in a decrease in amplitude of the methylene peak, with no observable line broadening on progressive addition of cholesterol (66). This observation leads to the conclusions that (*a*) a specific molecular interaction of cholesterol and the lecithin fatty acid chains occur; (*b*) the lifetime of the "complex" is long, being >30 msec (estimated from line broadening in the 2:1 mixture); and (*c*) the interaction is probably equimolar because no more than one cholesterol per lecithin molecule may be incorporated in the lecithin dispersion.

Darke *et al.* (66) also examined the wide-line proton NMR spectrum of unsonicated dipalmitoyl lecithin–cholesterol dispersions below the transition temperature, noting that four broad lines of width 5.2, 3.4, 1.2, and 0.2 Gauss were present in the equimolar dispersion. Tentative assignments

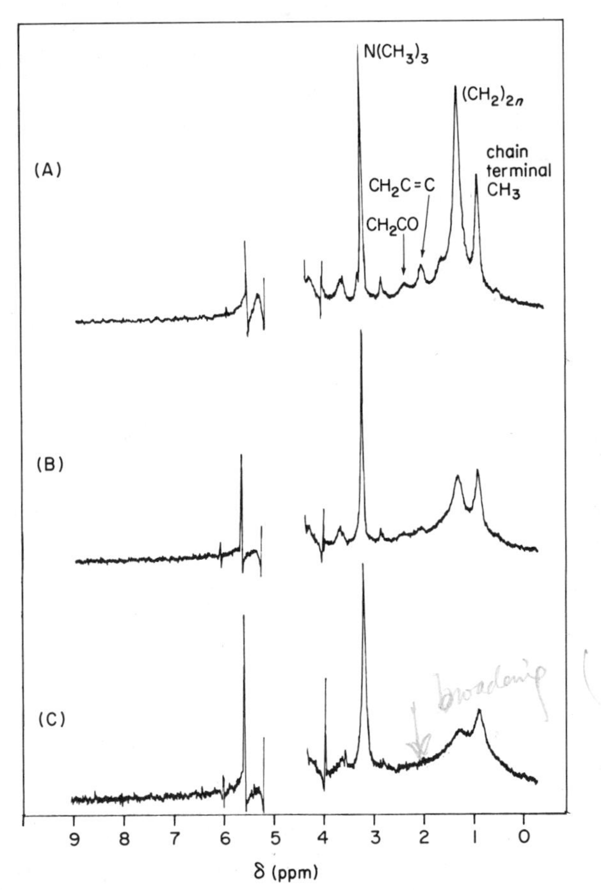

FIG. 8-16. Effect of cholesterol on the 220 MHz proton NMR spectra of sonicated egg lecithin dispersions (5% lecithin in D_2O). (A) egg lecithin; (B) 2:1 (molar ratio) egg lecithin–cholesterol; (C) 1:1 (molar ratio) egg lecithin–cholesterol (66).

for the lines arising both from cholesterol and lecithin were made. These assignments and the spectral features are listed in Table 8-7. The reason the peaks could be observed in the lecithin–cholesterol dispersion below the transition temperature is that cholesterol promotes line narrowing below the transition temperature. This is illustrated in Fig. 8-17 for unsonicated egg lecithin and dipalmitoyl lecithin at 18°C. Some line broadening effected by cholesterol is evident by the loss of resolution of the terminal methyl and methylene peaks in the case of egg lecithin (cf. Fig. 8-17, A and B). However, the presence of cholesterol promotes line narrowing (cf. Fig. 8-17, C and D) in the case of dipalmitoyl lecithin. At 18°C, egg lecithin is in the liquid crystalline state, characterized by greater motional freedom

TABLE 8-7

ASSIGNMENT OF FOUR WIDE LINES OBSERVED IN UNSONICATED EQUIMOLAR LECITHIN–CHOLESTEROL DISPERSIONS[a]

LINE	LINEWIDTH AT 30°C FOR 1:1 MIXTURE OF DIPALMITOYL LECITHIN AND CHOLESTEROL (GAUSS)	ASSIGNMENT	LINEWIDTH IN UNMIXED STATE (GAUSS)		INTENSITIES[b]			
					DIPALMITOYL LECITHIN AND CHOLESTEROL, 50°C		DEUTERATED DIMYRISTOYL LECITHIN AND CHOLESTEROL, 25°C	
			30°C	ABOVE LECITHIN CHAIN MELTING TEMPERATURE	CALC.	OBS.	CALC.	OBS.
a	5.2 }	Cholesterol backbone and chain	11.5	—	30		30	
	}	CH_3's on cholesterol ring system	4.5	—	6		6	
b	3.4 }	10 CH_2's on each lecithin chain	4.0	~0.05	40		0	
					76	77	36	33
c	1.2	Other lecithin chain CH_2's	4.0	~0.05	16		0	
		Cholesterol chain CH_3's	4.5	—	9		9	
					25	25	9	11
		Lecithin $N(CH_3)_3$	0.4	~0.05	9		9	
d	0.2	Lecithin chain CH_3's	4.0	~0.05	6		0	
		Lecithin polar backbone	4.0	1.4	9		9	
					24	23	18	19
Sharp component		Total spectrum from small particles tumbling rapidly				9%		10%

[a] Darke *et al.* (66).

[b] Normalized and expressed in protons per molecule calculated from the known composition of each mixture. Estimated errors in intensity measurements ±10%.

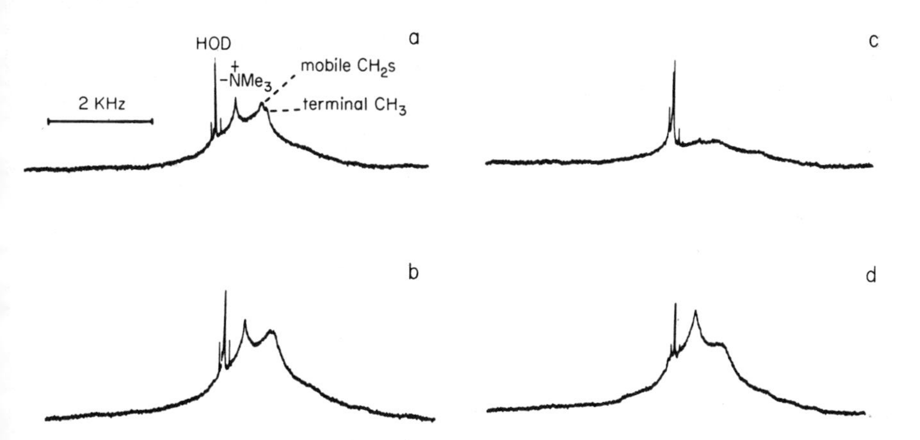

FIG. 8-17. 270 MHz proton NMR spectra at 18°C in D_2O of unsonicated dispersions of (a) egg lecithin, (b) egg lecithin–cholesterol (1:1), (c) dipalmitoyl lecithin, and (d) dipalmitoyl lecithin–cholesterol (1:1) (38).

of the lecithin molecules, but dipalmitoyl lecithin is in the relatively rigid gel state. The decreased linewidth effected by cholesterol was attributed to a state of "intermediate fluidity" in the lecithin–cholesterol system below the transition temperature. The proton NMR results of Oldfield and Chapman (67) on unsonicated sphingomyelin–cholesterol dispersions also indicate that cholesterol "mobilizes" the phospholipid below the transition temperature but "rigidizes" it above the transition temperature.

The ^{13}C NMR spectra of unsonicated egg lecithin above the transition temperature, shown in Fig. 8-13, indicate that cholesterol selectively interacts with part of the lecithin alkyl chain because the terminal methyl carbon and adjacent methylene carbon resonances remain sharp (38). The olefinic carbon resonances (from carbons 9 and 10 of oleyl chains) are appreciably broadened in the presence of cholesterol.

Dipalmitoyl lecithin–cholesterol multilayers containing 16% D_2O were examined by Hemminga and Berendsen (68). The multilayers were oriented between glass plates, and the dependence of the gel-state proton NMR spectrum on the angle of orientation of the normal to the glass plates with respect to the external magnetic field direction was investigated. Just as previously noted for oriented potassium oleate–D_2O (20), it was found that the spectrum has an angular dependence, with line narrowing observed at an orientation angle of 55°. The observed narrowing indicates that the lecithin chains and cholesterol are oriented perpendicular to the plane of the glass slide and the linewidth is a result of dipolar broadening.

Deuterium NMR of unsonicated di(perdeutero)myristoyl lecithin with equimolar cholesterol was investigated as a function of temperature (69). Below the transition temperature, only a very broad pooly defined resonance was observed for lecithin. Above the transition temperature, quadrupole splitting of the deuteron ($I = 1$) resonance was observed. The splitting was temperature dependent, being 27 ± 1 kHz at 30°C. The splitting (at 30°C) increased to 49.4 ± 1.5 kHz in the presence of cholesterol. It follows from Abragam (5, p. 234 ff.) that the quadrupole splitting of a spin 1 nucleus located at a site of axial symmetry in a crystalline sample is

$$\Delta\nu = \frac{3}{4}\left(\frac{e^2qQ}{h}\right)\overline{(3\cos^2\theta - 1)} \tag{8-5}$$

where e^2qQ/h is the quadrupole coupling constant and θ is the angle between the principal component of the electric field gradient tensor and the applied magnetic field. The bar in $\overline{(3\cos^2\theta - 1)}$ denotes a time average. The value of e^2qQ/h for a static C—D bond is about 170 kHz. The increase in $\Delta\nu$ found on addition of cholesterol to lecithin was attributed to a greater restriction of movement of deuterons in the presence of cholesterol, leading to a larger value of $(\cos^2\theta - 1)$ (69).

The spin–lattice relaxation times of the different proton resonances of sonicated lecithin were examined via Fourier transform NMR to determine the effect of cholesterol (54). The results, summarized in Table 8-8, show that the alkyl chain proton resonances experience a greater reduction in T_1 than the $N(CH_3)_3^+$ resonances when cholesterol is present. It was noted, however, that the change in T_1 was not commensurate with the change in linewidth caused by addition of cholesterol.

Protein. Certain proteins play an important functional and structural role in the membrane mosaic. Berger *et al.* (70) isolated the apoprotein of human erythrocyte membrane and studied the interaction of the apoprotein with lysolecithin. The results are given in Fig. 8-18. It is evident that the methylene proton resonances of lysolecithin are selectively broadened compared with the choline methyl resonance. It is also clear that the membrane apoprotein has a greater effect on lysolecithin than does a nonfunctional protein, acid casein. The NMR results imply that the apoprotein is in the interior of the membrane and interacts via hydrophobic forces with the alkyl chains, reducing the motional freedom of those chains.

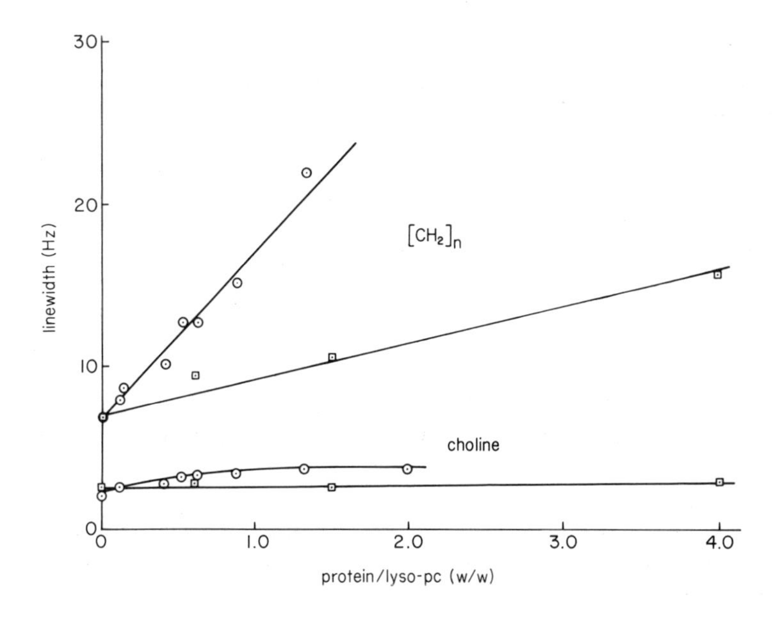

FIG. 8-18. Linewidths of $(CH_2)_n$ and $N(CH_3)_3^+$ signals of lysolecithin as a function of protein concentration for the apoprotein (⊙) of human erythrocytes and for acid casein (⊡) (70).

E. *Presence of Antibiotics or Anesthetics*

Several cyclic molecules induce transport of specific ions across membranes. Finer *et al.* (71, 72) examined the effect of the cyclic antibiotics alamethicin, valinomycin, and gramicidin S dihydrochloride on the proton NMR spectra of sonicated ox brain phosphatidylserine and egg lecithin dispersions. Alamethicin and gramicidin S are cyclic polypeptides containing 19 and 10 amino acid residues, respectively, and valinomycin is a cyclic molecule composed of six amino acid and six hydroxy acid residues.

Alamethicin and valinomycin cause the proton resonances of the alkyl chains to become broadened beyond detection in both sonicated phosphatidylserine and lecithin dispersions. In contrast, gramicidin S does not affect the alkyl chain proton resonances but does produce some broadening of the choline methyl resonance. The interpretation of these results is that valinomycin and alamethicin interact with the hydrophobic parts of the phospholipid bilayer whereas gramicidin S interacts with the polar head group.

Although NMR signals from the alkyl chain protons were observable

TABLE 8-8

T_1 VALUES FOR SONICATED EGG LECITHIN SYSTEMS[a]

SYSTEMS	TEMP. (°C)	T_1 (SEC)		
		$—N(CH_3)_3^+$	$—(CH_2)_n—$	$—CH_3$
Sonicated egg lecithin	48	0.45 ± 0.01	0.53 ± 0.02[b]	0.72 ± 0.02
	40	0.39 ± 0.01	0.47 ± 0.02[c]	0.54 ± 0.02
	28	0.31 ± 0.02	0.38 ± 0.02	0.36 ± 0.02
	16	0.24 ± 0.05	0.25 ± 0.06	0.28 ± 0.06
Sonicated egg lecithin + 4:1 cholesterol	48	0.45 ± 0.07	0.35 ± 0.05	0.40 ± 0.03
	40	0.40 ± 0.01	0.31 ± 0.01	0.31 ± 0.01
	28	0.29 ± 0.01	0.31 ± 0.02	0.34 ± 0.01
Sonicated egg lecithin + 2:1 cholesterol	48	0.41 ± 0.02	0.18 ± 0.01	0.22 ± 0.02
	40	0.35 ± 0.01	0.14 ± 0.03	0.19 ± 0.02
	28	0.21 ± 0.01	0.08 ± 0.02	0.12 ± 0.01

[a] Lee *et al.* (54).
[b] Average of 0.43 ± 0.02 and 0.57 ± 0.01.
[c] Average of 0.42 ± 0.02 and 0.53 ± 0.04.

above the transition temperature, no lipid proton resonances were detectable below the transition temperature in gramicidin S–dipalmitoyl lecithin (73). Surprisingly, proton signals from the antibiotic were observed below the transition temperature.

Studies of the effect of valinomycin (2 mole %) on the delayed Fourier transform proton NMR spectrum of unsonicated egg lecithin revealed the linewidth of the choline methyl resonance to be halved, whereas the terminal methyl resonance was unchanged on addition of valinomycin (18). It was concluded that valinomycin interacts with the polar head groups of the bilayer, in contrast to the conclusion of Finer *et al.* (66, 67).

The effect of antibiotics on the lipid bilayer is presently not entirely clear. The apparent discrepancies in the different investigations remain to be reconciled.

Cerbón (74) has observed differential broadening of the proton resonances of local anesthetics interacting with lecithin and phosphatidylserine vesicles. Little or no effect on the proton resonances of the phospholipids was found, however, implying that the anesthetics place no restrictions on the motions of the phospholipid molecules. The differential broadening of the drug proton resonances was taken as evidence for a specific site of interaction with the phospholipid bilayer on the basis that greater peak broadening for a nonpolar group, as opposed to a polar group, indicates that the drug molecule's site of interaction is in the hydrophobic interior of the bilayer. It was noted that the strength of the hydrophobic interaction of the anesthetics with the lipid bilayer occurred in the order dibucaine > tetracaine > butacaine > procaine, in direct correlation with the potency of the anesthetics.

8.1.2. Biological Membranes

The application of NMR to the study of biological membranes has relied to a large extent on comparisons with the lipid membrane models discussed in the preceding section. The dynamic state of the membrane, as well as the detailed structure, will depend on the lipid, lipopolysaccharide, and protein composition of the membrane. For example, on the basis of the model membrane studies, increasing amounts of cholesterol or membrane protein might be expected to result in a decrease in motional freedom of the liquid crystalline phospholipids in the biological membrane.

A. *Sarcoplasmic Reticulum Membrane*

The membrane of the sarcoplasmic reticulum is especially suitable for NMR studies because it contains only a small amount of cholesterol and more than 80% of the fatty acid chains are unsaturated, both factors favorable to narrow lines in the room temperature spectrum.

Another favorable feature of the sarcoplasmic reticulum membrane is that purified vesicular fragments may be isolated with retention of some functional properties, such as permeability and ion transport, enabling structure–function correlations to be made. Davis and Inesi (75) performed such a study, examining the proton NMR spectra and the ATP-dependent Ca^{2+} accumulation of the vesicular fragments.

The 90 MHz proton NMR spectra of unsonicated and sonicated suspensions of sarcoplasmic reticulum membrane are compared with sonicated egg lecithin in Fig. 8-19. The resonances occurring at −0.90, −1.26, and −2.00 ppm were attributed to $C\underline{H}_3$, $(C\underline{H}_3)_n$, and $C\underline{H}_2$—CH=C protons, respectively. The spectrum of the unsonicated suspension exhibiting the relatively narrow lines of Fig. 8-19A is actually superimposed on a broad line (~500 Hz), the well-resolved peaks accounting for about 20% of the phospholipid present. Sonication results in an increase in intensity of the narrow line spectrum and the appearance of a resonance at −3.3 ppm due to the $N(CH_3)_3^+$ protons (cf. Fig. 8-19B).

Raising the temperature of the unsonicated suspension also results in the appearance of the $N(CH_3)_3^+$ resonance, with the intensity gradually increasing with temperature until a plateau between 40°C and 50°C is

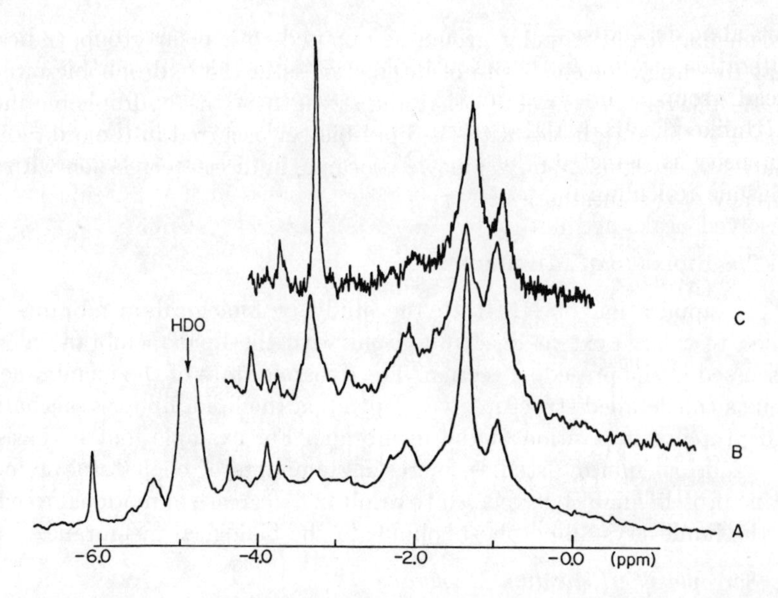

FIG. 8-19. Portions of the 90 MHz proton NMR spectra of (A) 6% (protein wt) sarcoplasmic reticulum suspension in D_2O, (B) 0.6% sonicated suspension of sarcoplasmic reticulum vesicles, and (C) 0.5% sonicated dispersion of egg lecithin. All spectra were taken at 25°C. Note the appearance of the choline line at ≈ -3.3 ppm and the intensity increase of the methyl line at ≈ -0.9 ppm in the sonicated sarcoplasmic reticulum suspension (75).

reached. The intensity increase correlates with the Ca^{2+} efflux from the microsomes. No change is observed in the methylene resonance over the same temperature range. Raising the temperature above 50°C results in a further increase in the choline methyl intensity. This is apparently a result of irreversible heat denaturation of membrane ATPase, because the temperature dependence of the choline methyl signal intensity displays a hysteresis effect and the efflux of Ca^{2+} from the microsomal membrane is increased.

The authors (75) also found other alterations of protein structure to simultaneously affect the choline methyl proton resonance intensity and the Ca^{2+} efflux. For example, treatment of the sarcoplasmic reticulum suspension with trypsin to hydrolyze protein resulted in an increase in the fraction of mobile choline methyl groups.

The implication of these studies is that the ability of the sarcoplasmic membrane to maintain a calcium ion concentration gradient depends on the integrity of the protein structure and on the protein–lipid interactions involving the polar head groups of the phospholipids.

In contrast, ^{31}P NMR spectra of sarcoplasmic membrane suspensions reveal no dependence of the membrane protein on temperature or chemical alteration, leading to the speculation that only a selective part of the polar head group is involved in lipid–protein interactions (75).

Unlike the ^{1}H NMR spectra, in which only 20% of the phospholipid intensity is detected, the ^{13}C NMR spectrum of an unsonicated sarcoplasmic reticulum membrane suspension, shown in Fig. 8-20, yields well-resolved peaks accounting for 75 ± 15% of the phospholipids (76). The

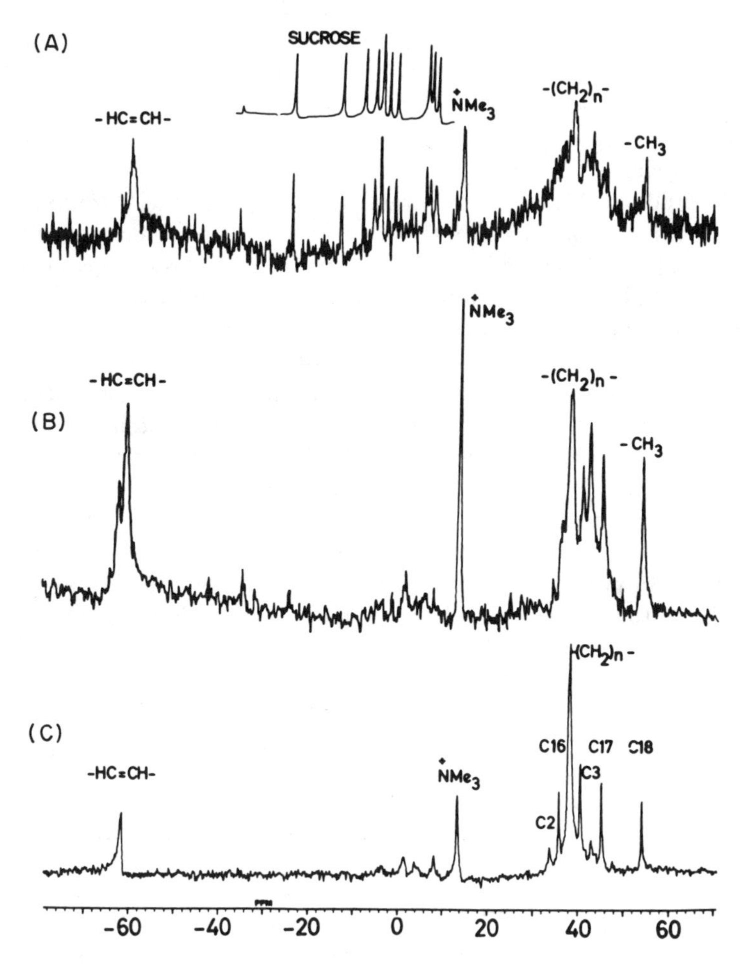

FIG. 8-20. 25.2 MHz ^{13}C NMR spectra of (A) an unsonicated sarcoplasmic reticulum dispersion (15% w/w) at 31°C (the inset sucrose spectrum indicates the resonances that arise from sucrose trapped in the vesicles during preparation); (B) sonicated vesicles of lipids extracted from sarcoplasmic reticulum (20% w/w) at 31°C; and (C) sonicated dioleyl lecithin (20% w/w) at 52°C. All samples are in D_2O (76).

values of the ^{13}C T_1 relaxation times for the membrane suspension are compared in Table 8-9 with the T_1 values of lipid vesicles extracted from the membrane. The methylene and terminal methyl T_1 values reveal no significant difference between the membrane and the lipid extract. However, the lower T_1 value in the case of the choline methyl carbon was tentatively attributed to interaction of the head group with membrane protein.

B. *Erythrocyte Membrane*

In spite of the larger amount of cholesterol in erythrocyte membranes (42 mole % in the human variety), successful NMR studies have been carried out. Broad-line proton NMR studies of intact human erythrocyte membranes revealed the spectrum to be quite similar to that of lipid in the presence of cholesterol (13).

Early studies (77) of the interactions in erythrocyte membranes entailed treatment with detergents to separate lipid from protein. The detergent treatment resulted in the appearance of proton resonances from the phospholipid alkyl chains. However, it was noted that the choline groups of lecithin and sphingomyelin are relatively unrestricted because the choline methyl resonance intensity is not affected by detergent treatment. It was suggested from these results that the choline groups of lecithin and sphingomyelin are not involved to any extent in lipid–protein interactions.

In a continuation of earlier research, Sheetz and Chan (78) described proton NMR studies of human erythrocyte membranes as a function of temperature, protein solubilization, and sonication. Below 41°C, only a few broad peaks are discernable. Above 41°C, several peaks become observable as shown in Fig. 8-21A for the intact membrane at 75°C. The spectrum of the intact membrane is compared with those of lecithin plus cholesterol and the isolated membrane proteins in Figs. 8-21B and C,

TABLE 8-9

^{13}C SPIN–LATTICE RELAXATION TIMES OF LIPID NUCLEI IN SUSPENSIONS OF SARCOPLASMIC RETICULUM MEMBRANES AND IN EXTRACTED LIPID VESICLES AT 31 ± 1°C[a]

	T_1 (SEC)		
	$N(CH_3)_3^+$	$(CH_2)_n$	CH_3
Membrane	0.36 ± 0.03	0.42 ± 0.03	3.1 ± 0.6
Lipid vesicles	0.55 ± 0.06	0.37 ± 0.01	3.7 ± 0.3

[a] Robinson *et al.* (76).

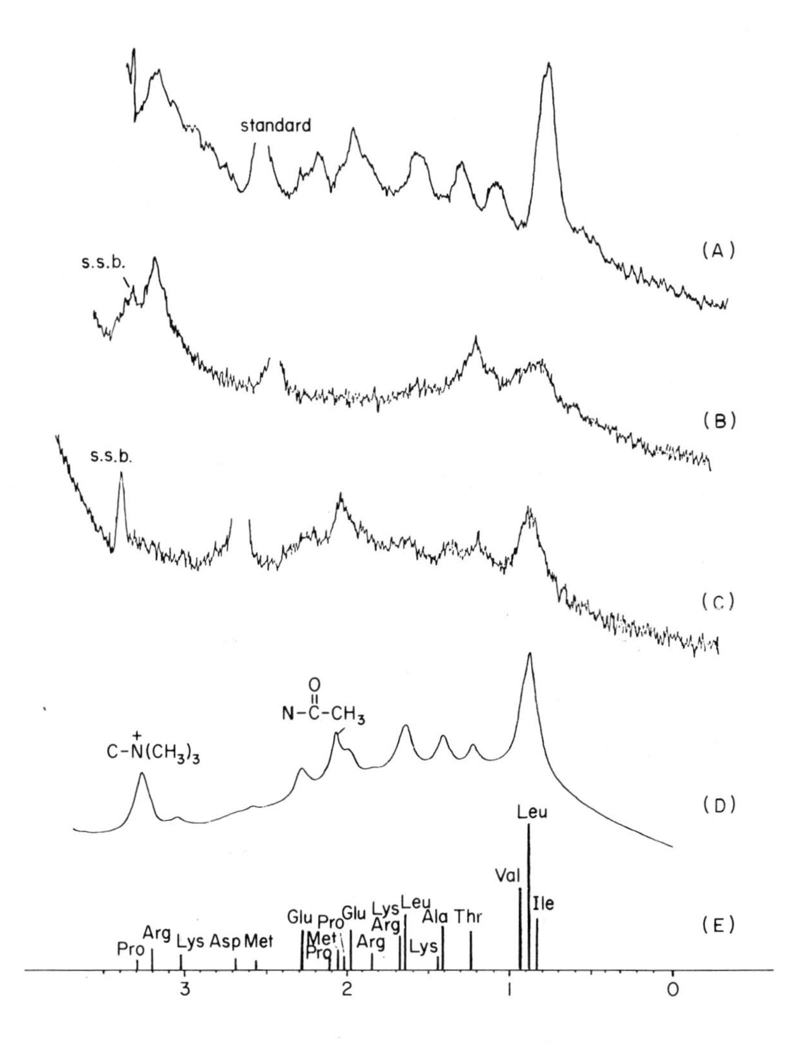

FIG. 8-21. 220 MHz proton NMR spectra of (A) whole human erythrocyte membranes at 75°C; (B) egg lecithin and cholesterol, 1:1 (20 mg of lecithin/ml) at 72°C; (C) delipidated membrane protein from human erythrocyte membranes in D_2O at 75°C (5 mg/ml); (D) computed spectrum for human red cell membrane; (E) assignment of the amino acid resonances in computed spectrum (78).

respectively. The comparison led to the assignment of the peaks to 20% of the amino acid residues, to 20% of the choline methyl groups of the phospholipids, and to the acetamide groups of the sugars, notably neuraminic acid. The spectrum calculated on this basis is shown in Fig. 8-21D to represent the erythrocyte membrane spectrum rather well.

Treatment of the erythrocyte with EDTA and β-mercaptoethanol to remove spectrin resulted in solubilization of about 12% of the membrane proteins and a large increase in the observed intensity of the choline methyl resonance from about 15–20% to about 50% of the total choline groups present (78).

Studies of the linewidth of the phenyl proton resonance of benzyl alcohol, as a function of benzyl alcohol concentration, employed the alcohol as a structural probe as well as a cause for lysis of the whole erythrocytes (79). The experiments showed that the reaggregated membrane structure was altered from that of the native membrane, probably because of faulty reassembly of some membrane proteins.

The ^{13}C NMR spectrum of erythrocyte membranes exhibits relatively sharp lines for the $N(CH_3)_3^+$, terminal CH_3, and penultimate CH_2 carbon resonances of phospholipids (38).

C. *Myelin*

The myelin sheath surrounds most of the nerve fibers in vertebrates. It contains relatively little protein, but approximately 40% of the lipid is cholesterol. The fatty acid acyl chains are largely unsaturated.

Jenkinson *et al.* (80) have compared the broad-line proton NMR spectrum of the myelin membrane with that of the extracted lipids. The two spectra were quite similar, consistent with the idea that the lipid in the membrane is in the form of bilayers, as it is in the extracted lipid–cholesterol system.

Lecar *et al.* (81) measured proton NMR linewidths of bovine, rat, and guinea pig myelin as a function of temperature. The broad line caused by phospholipid protons narrowed, revealing two transition temperatures, as shown in Fig. 8-22 for the bovine species. The transition occurring in the physiological range (25°–40°C) corresponds to the transitions observed in phospholipid–cholesterol systems with a liquid crystalline phase above 40°C.

D. *Rabbit Sciatic Nerve*

The proton NMR spectrum of an isolated rabbit sciatic nerve in its native state has been observed and resonances assigned to water and the phospholipids of the nerve. The spectrum of the whole nerve obtained by Dea *et al.* (82) is shown in Fig. 8-23 together with spectra computed on the basis of the lipid composition of the nerve. It would appear that the computed spectrum of phospholipids, omitting cholesterol (Fig. 8-23B), more nearly matches the experimental spectrum (Fig. 8-23A) than the computed spectrum of phospholipids plus cholesterol (Fig. 8-23C). The difference between the two are not sufficiently pronounced to rule out the appearance of cholesterol resonances, however.

The spectrum does lead to the conclusion that the sciatic nerve contains fluidlike hydrocarbon regions. It was not determined whether the fluidlike phospholipid spectrum is caused by excitable membranes in the nerve fiber or to myelin membranes.

E. *Mitochondrial and Chloroplast Membranes*

The 270 MHz proton NMR spectrum of rat liver mitochondrial membranes exhibits well-resolved $N(CH_3)_3^+$ and CH_2 phospholipid peaks, as might be expected from the low cholesterol (~6% of total lipids) content (38).

The 25.2 MHz ^{13}C NMR spectrum of mitochondrial membranes has several well-resolved peaks, with at least three peaks clearly defined in the methylene region (38). It was demonstrated that the observable carbon resonances arise from lipids rather than proteins in the membrane, suggesting that the lipids have considerably greater mobility than the proteins in the membrane.

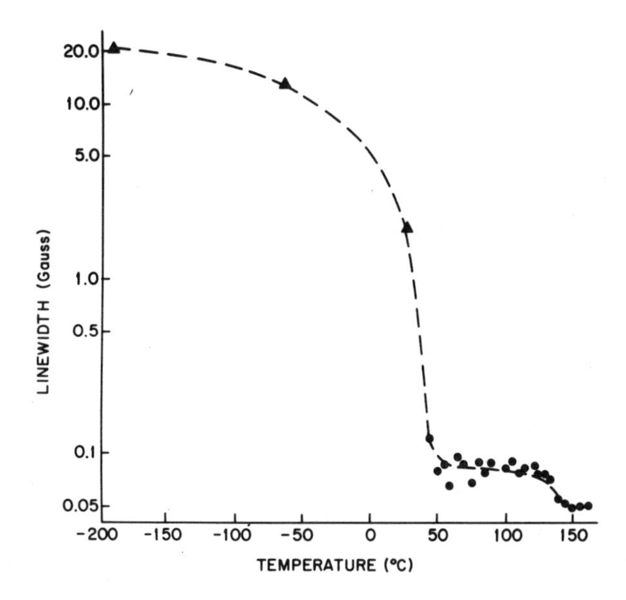

FIG. 8-22. Peak-to-peak linewidth (derivative spectrum) plotted as a function of temperature for lyophilized bovine myelin. Triangles taken with wide-line spectrometer, circles with high-resolution instrument. The major line narrowing occurs between 23° and 40°C. A second linewidth transition coincides with the melting of the sample at 140°C (81).

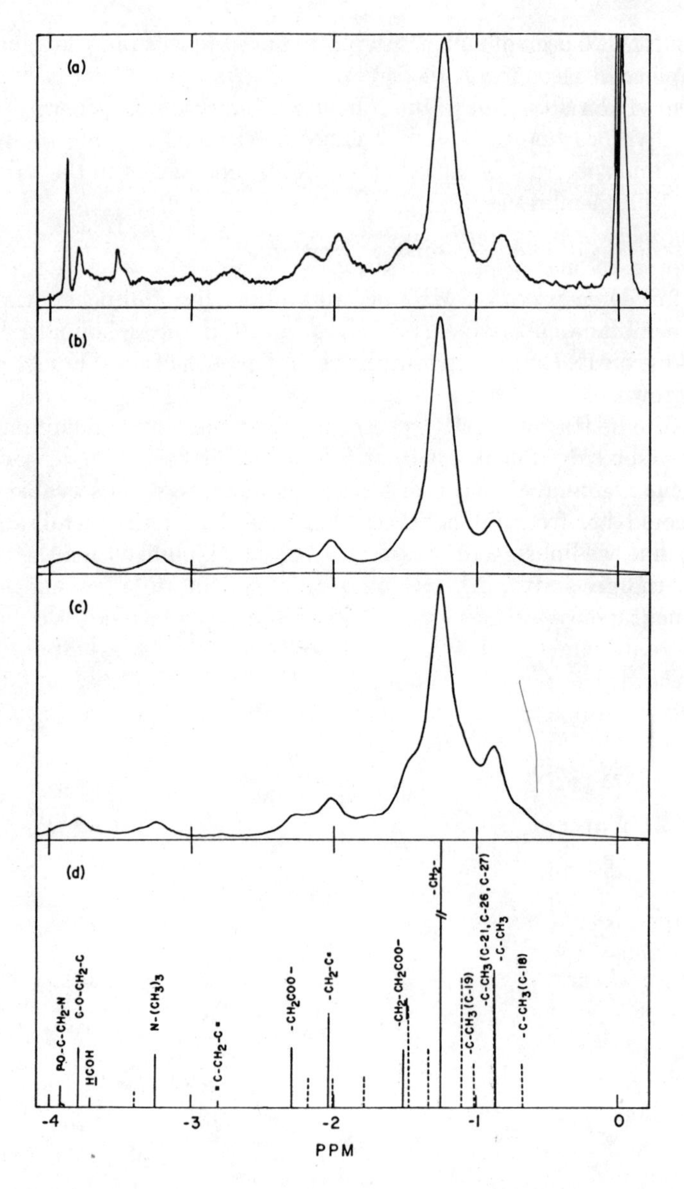

FIG. 8-23. (a) Proton magnetic resonance spectrum (220 MHz) of an intact sciatic nerve in the region from 0 to 4 ppm downfield from TMS. (b) A computed spectrum of the phospholipids in the rabbit sciatic nerve. (c) A computed spectrum of the phospholipids and cholesterol in the rabbit sciatic nerve. (d) Proton spectral assignments and expected intensities. The phospholipid protons are indicated by solid lines; the cholesterol protons are depicted by dotted lines. Except for the methylene peak, all intensities are drawn to scale (82).

The ^{13}C NMR spectrum of chloroplast membranes (from lettuce) is more complicated than that of mitochondrial membranes because of the appearance of chlorophyll carbon resonances in the alkyl chain region (38).

F. *Mycoplasma laidlawii and Acholeplasma laidlawii Membranes*

The *Mycoplasma* membranes have been studied by examination of the phenyl proton resonance of benzyl alcohol, which was used as a probe molecule for detecting structural differences between native and reaggregated membranes (79).

Carbon-13 NMR studies of membranes isolated from *Mycoplasma laidlawii* grown on a medium containing [^{13}C]-labeled palmitic acid has been shown to be feasible (83). Deuterium NMR has been employed with membranes isolated from *Acholeplasma laidlawii B* grown on a medium containing perdeuterated lauric and palmitic acids (84). No quadrupole splitting could be observed up to 37°C. Instead, a broad (~65 kHz) deuterium line was observed, similar to that of di(perdeutero)myristoyl lecithin in the gel state, rather than a doublet deuterium resonance exhibiting quadrupole splitting, as with the deuterated lecithin in the liquid crystalline state. On that basis, it was concluded that the hydrocarbon region of the *Acholeplasma* membrane is "relatively rigid" and, at physiological temperatures, the membrane lipids are predominantly in the gel state.

8.2. Water in Biochemical and Biological Systems

The state of water in biological tissue may play an important role in cellular structure and function. The nature of that water may potentially influence cellular reaction equilibria and kinetics, ionic and molecular transport, preservation of chemical gradients and osmotic pressure, and macromolecular conformation.

The state of water in biological samples has recently been a subject of considerable controversy in the scientific literature. At one time, cellular water was thought to be much like water in a "water balloon," differing only in containing some ions and proteins in otherwise ordinary liquid water. About a decade ago, however, this concept was challenged. The basic challenge came from Ling's association–induction theory (85), although modifications of that theory have been presented. The association theories maintain that the fixed charges on macromolecules and their associated counterions constrain much of the cellular water to form a matrix of polarized multilayers having restricted motion compared with pure water.

The concept of "ordered" water is plausible according to quantum mechanical calculations, which imply that hydrogen bonding of water molecules is a cooperative process (86). In the context of the association theories, this cooperativity may be viewed in the following manner. Hydrophilic groups on a macromolecule (e.g., $-NH_2$ and $-OH$) may form hydrogen bonds with nearby water molecules, which in turn will be more inclined to form hydrogen bonds with water molecules in the next layer, and so on. The stimulus for the formation of the water multilayer is the polarization of the water molecules caused by hydrogen bonding. A charged group on a macromolecule could also serve as the origin of polarized water multilayers. The possible existence of these multilayers is viewed as a short-lived phenomenon, having a lifetime on the order of 10^{-11} sec. It has been suggested that hydrophobic moieties of macromolecules may also induce short-lived structure (87) by forcing the water into a clathrate cage structure incorporating the hydrophobic group (88). Whether such "ordered" water exists, and if it does exist whether it differs significantly from pure liquid water, are questions still open to argument. The majority viewpoint would appear to be the classical conception (with possible modifications); the onus is largely on the minority viewpoint to show that cellular water differs from ordinary water. If the minority concept is ultimately proved, it may have a tremendous impact on currently held ideas concerning bioenergetics.

Elucidation of the nature of water in biological systems is a difficult task for two reasons. First, the nature of pure water is not entirely understood (89); and second, experimental observations must be made on a heterogeneous system, the cell. NMR may be the most useful experimental technique for studying the nature of water in biological material. NMR data have, in fact, been used as ammunition by both sides in the current controversy. Recent reviews have covered certain aspects of the subject of water in biological systems (88, 90, 91). In particular, Walter and Hope (90) have reviewed the earlier NMR studies.

Part of the problem with the subject of biological water has been the imprecise use of such terms as "bound," "free," "ordered," "structured," "icelike," and "water of hydration." The lack of exact definitions has resulted partially from the complexity of the system and partially from the generally qualitative nature of the experimental observations. Quantitative estimations have varied widely depending on the experimental technique and interpretations of observations. Therefore, unfortunately, it is not quite possible to define these terms exactly. In general, it will be necessary to use the terms loosely with the operational definitions implied in the individual studies. However, the following comments concerning certain terms may be appropriate.

The spin–lattice relaxation time for water protons may be expressed as (cf. Eqs. 2-45 and 2-52)

$$1/T_1 = K\overline{H^2}J(\omega) \tag{8-6}$$

where K is a constant, $\overline{H^2}$ is the magnitude of the local fields effecting relaxation, and $J(\omega)$ is the spectral density. $J(\omega)$ is the Fourier transform of a correlation function $G(\tau_c)$, which denotes the rate of fluctuation of the interaction between the nuclear spin and the dipolar field. $J(\omega)$ is therefore a measure of the mobility of water. Generally, it is assumed that $J(\omega) = \tau_c/(1 + 2.5\omega^2\tau_c^2)$, where τ_c is the correlation time for the motion. The linewidth and spin–spin (T_2) relaxation time may be expressed in an equation similar to Eq. 8-6 (cf. Eq. 2-47). In the case of biologial water, it is possible to distinguish between "motional" features, which are manifest in $J(\omega)$, and "structural" features, which may be manifest in $\overline{H^2}$. Most NMR studies of water in biological material have entailed T_1, T_2, and linewidth measurements. The changes in these parameters, relative to pure water, have often been interpreted as an indication of "structuring" or "ordering." In fact, it is conceivable that only $J(\omega)$ changed, and not $\overline{H^2}$. As pointed out by Hertz (92), changes in "structure" and changes in molecular motion in a fluid may be independent. However, it may be incorrect to attribute the variations in relaxation times solely to motional changes. The interaction factor $\overline{H^2}$ may vary for at least part of the water in biological systems. Combination of the two variables $\overline{H^2}$ and $J(\omega)$ complicates matters considerably for relaxation time or linewidth measurements. T_2 and linewidth measurements may have an additional contribution from exchange between bulk water and "water of hydration."

Recently, however, measurements of the self-diffusion coefficient D have been made via pulsed NMR. The self-diffusion coefficient is a measure only of the translational motion of the water molecules and, therefore, would appear to be a more direct indication of the mobility. Once the extent of mobility is ascertained, comparison of D with T_1 or T_2 may be used to provide information about the possible causes for alterations in water mobility.

"Bound" water or "water of hydration" may differ from bulk water in both relaxation times and self-diffusion coefficient. If a water molecule is bound to a macromolecule or membrane, it may be expected to take on slower motional characteristics. If the "bound" water molecules are exchanging rapidly with "free" water molecules, observed NMR parameters will be a weighted average of those in the "bound" and "free" states. If the concept of polarized multilayers of water molecules around a macromolecular is correct, it would be expected that the effect would be attenuated with distance from the macromolecule. The point of distinction between "free" and "bound" water in that case is arbitrary.

Some investigations have been concerned with cellular systems, whereas others have considered molecular systems as a basis for understanding the more complicated cellular systems. Some studies have employed very low water contents in attempts to study only the water of hydration.

8.2.1. Molecular Systems

For ease in presentation, water studies in molecular systems will be divided into three parts. The first is concerned with solutions of macromolecules that contain a large amount of water. The second considers macromolecular systems with a sufficiently small amount of water that the individual water molecules are forced to interact with the macromolecule. The third part covers studies made with aqueous lipid dispersions.

A. *Solutions of Macromolecules*

There have been two central questions considered in most studies. (*a*) How much water is "bound"? (*b*) How does bound water differ from pure water? Water proton peak intensities, linewidths, free induction decays, T_1, T_2, and self-diffusion coefficients have been measured, as well as some of the same parameters for deuterium in D_2O solutions. Peripheral questions have been concerned with the effect of the biopolymer's composition, size, and conformation on the bound water.

The conclusion from the first studies of water in biopolymer solutions was that 20–25% of the water in solutions of DNA (93) and tobacco mosaic virus (94) is in an "icelike" state. That conclusion was based on low water peak intensities. Later investigations did not uphold that conclusion, however (95–97). The water proton resonance intensities in the later studies, (and most subsequent ones on other biopolymer solutions) were able to account for all the water in solution. However, the linewidth was observed to be greater than that of pure water. Balazs *et al.* (95) attributed the broad line to a chemical shift anisotropy arising from the π electrons of the stacked bases in DNA affecting nearby water molecules. The adjacent water molecules were presumed to exchange rapidly with the remainder of the water molecules in solution, giving rise to an averaged water proton resonance. Depireau and Williams (96) suggested that the presence of DNA contributed to a decrease in motion for all water molecules in solution. Douglass *et al.* (97) believed that paramagnetic ions produced water proton line broadening in tobacco mosaic virus solutions.

Experiments with other macromolecule solutions have led many investigators (98–100) to conclude that line broadening results from fast exchange between bulk water and bound water, viz,

$$W_{1/2} = X_f W_{1/2}(\text{free}) + X_b W_{1/2}(\text{bound}) \tag{8-7}$$

where X_f and X_b are the respective mole fractions of free and bound water, and $W_{1/2}$(free) and $W_{1/2}$(bound) are the respective proton linewidths for free and bound water molecules. In the past, it has been assumed that rapid exchange conditions hold and Eq. 8-7 may be used. More thorough studies may consider the effect of exchange on line broadening (cf. Eq. 6-7) and may therefore include more extensive experiments to justify the assumption of rapid exchange.

An interesting method of estimating the amount of water bound to a macromolecule was presented by Kuntz *et al.* (101). The amount of bound water was simply determined from the intensity of the relatively narrow water proton resonance caused by the water in macromolecule solutions that remained unfrozen at temperatures as low as −60°C (cf. Fig. 8-24). Bulk water freezes near 0°C, leaving a proton signal from water interacting

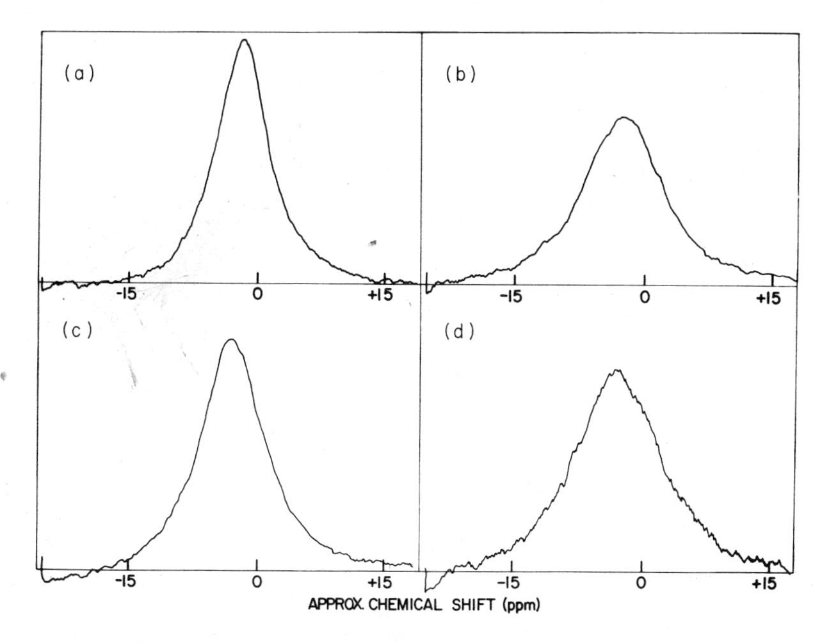

FIG. 8-24. Water proton NMR spectra (60 MHz): (a) lysozyme, 75 mg/ml, −35°C; (b) egg albumen, 75 mg/ml, −35°C; (c) native serum albumin pH 5.03, −35°C; and (d) pH "denatured" serum albumin, pH 2.45, −35°C. Experimental conditions were: −35°C, 0.001 *M* KCl, all solutions approximately 75 mg/ml of protein except (d), which was 50 mg/ml but whose spectrum was obtained at twice the spectrum amplitude of the others. All spectra are averages of four repetitions, using a CAT (101).

in some manner with the ions and macromolecules in solution. The studies were carried out with 5–10% solutions of macromolecules (from a salt-free source) containing 1 m*M* KCl. Evidence was cited for attributing the peak to water bound to a macromolecule (101). (*a*) Ice has a resonance line much broader than those observed, as witnessed by the fact that water and salt solutions gave no observable signal below their freezing (or eutectic) points. (*b*) The peaks are diminished by >90% when D_2O is used. (*c*) The peak intensity varies linearly with macromolecule concentration. (*d*) The linewidths are greater than those for supercooled water at the same temperature. (*e*) The linewidth and intensity of the signal are dependent on the macromolecular conformation. (*f*) The estimated amounts of bound water agree fairly well with the amounts determined by other methods. An obvious conclusion from the freezing studies is that the bound water is not "icelike" as some have described it.

It was found that proteins contained 0.3–0.5 g of water per gram of protein but nucleic acids contained three to five times as much if the unfrozen water criterion (101) is used. A combination of two sources for the unfrozen water was suggested. The first is water physically or chemically absorbed on the surface of macromolecules. The second is water trapped in pores throughout the molecule. Both types of water may be expected to have lower freezing points.

Further investigation of the nonfreezing phenomenon was carried out by Kuntz (102) on a series of high molecular weight polypeptides. The results for the amount of unfrozen water are summarized in Table 8-10. The generalities apparent from study of Table 8-10 are that the polypeptides with ionized side chains are highly hydrated and polypeptides composed of uncharged basic amino acids or proline also contain much water. Polypeptides with hydrophobic side chains contain much less water. In addition, it was found that the hydration per mole of polypeptide is independent of polymer concentration, and the hydration is independent of molecular weight. It was also observed that the amount of bound water is not affected by polypeptide conformation.

On the basis of the peak intensity observations on unfrozen water, the number of moles of water hydrating each amino acid residue was calculated; the results are tabulated in Table 8-11. From amino acid compositions, the amount of hydration was calculated for several proteins using the results of Table 8-11. The agreement with the observed hydration of the proteins was fairly good; however, the calculated values were higher than the experimental values because it was assumed in the calculation that all amino acid residues are exposed. For example, the hydration (grams of water per gram of protein) calculated for bovine serum albumin was 0.445, whereas the observed value was 0.40. Denaturing BSA with

TABLE 8-10

POLYPEPTIDE HYDRATION FROM NMR PEAK INTENSITY OF NONFREEZING WATER[a,b]

POLYPEPTIDE	PH	HYDRATION[c] −25°C	−35°C	−45°C	ERROR[c]	CONFORMATION[d]
L-Glu	7–12	8.3	7.7	6.3	±1	Coil
L-Glu	4.5		1.8		±0.5	Helix[e]
L-Asp	8–12	8.1	6.0	4.8	±1	Coil
L-Asp	4.5	2.1			±0.5	Helix[e]
L-Tyr	11.5–12		8.5	6.5	±1.5	Coil
L-Tyr	11.3		5.5	5.1	±1	Helix[e]
DL- or L-Lys	3–9	5.0	4.3	3.8	±1	Coil
DL- or L-Lys	10–12	5.0	4.5	3.7	±1	Helix[e,f]
L-Orn	1.5–9	4.0	3.4	3.5	±1	Coil
L-Orn	10–12	4.5	3.7	3.5	±1	Helix[e]
L-Arg	3–8	3.1	2.7		±1	Coil
L-Arg	10	3			±1	Helix[e,h]
L-Pro		3.1	2.8		±1	e
L-Asn		2.0			±0.5	?
DL-Ala		1.4			±0.5	Helix[f]
L-Val		0.9			±0.5	?
Gly		0.9			±0.5	?
Copolymers						
$Lys^{40}Glu^{60}$	2–4	2.5	2.4		±0.5	Helix
$Lys^{40}Glu^{60}$	11–12	7.8	7.5		±1	Coil
$Lys^{50}Phe^{50}$	2–9	2.6	3.8		±0.5	h
$Lys^{33}Phe^{67}$		1.2			±0.5	?

[a] Kuntz (102).

[b] The macromolecular concentrations are 5–10% by weight and the solutions also contain 10 m*M* KCl + KOH to bring to pH.

[c] As moles of water per mole of amino acid. Includes estimates of error in area and concentration measurements.

[d] Unless otherwise indicated, these assignments are made from literature data.

[e] Solubility is very low at extremes of pH. Probably some mixture of helix and coil exists at the pH shown.

[f] DL polymers probably contain considerable sequences of D and L residues, permitting helix formation under appropriate conditions.

[g] Polyproline II.

[h] Assumed similar to polylysine.

urea to expose all residues resulted in an experimental value of 0.44, in good agreement with the calculated value. For lysozyme, the hydration was observed to be 0.34. The value calculated assuming full exposure of all residues was 0.36, but correcting for buried residues using an x-ray model produced a calculated value of 0.335.

TABLE 8-11

AMINO ACID HYDRATIONS AT −35°C AND pH 6–8 UNLESS OTHERWISE INDICATED[a]

AMINO ACID[b]	HYDRATION[c]	BASIS OF ASSIGNMENT[d]
	Acidic Groups	
Asp^-	6.0	M
Glu^-	7.5	M
Tyr (uncharged)	3	E
Asp (pH 4)	2	E
Glu (pH 4)	2	M, E[e]
Tyr^- (pH 12)	7.5	M
	Basic Groups	
Arg^+	3.0	M
His^+	4	As Lys^+
Lys^+	4.5	M
Arg (pH 10)	3	M
Lys (pH 10–11)	4.5	M
	Hydrophilic Groups	
Asn	2.0	M
Gln	2	As Asn
Pro-OH	4	*g*
Pro	3.0	M
Ser, Thr	2	*f*
Trp	2	*h*
	Hydrophobic Groups	
Ala	1.5	M
Cys, Met	1	As Val
Gly	1	M
Ile, Leu	1	As Val
Phe	0	M[i]
Val	1	M

[a] Kuntz (102).

[b] Arranged by functional group, alphabetically within group.

[c] Moles of water per mole of amino acid.

[d] M, measured; E, extrapolated from portion of titration curve.

[e] Measured from Lys–Glu copolymer assuming Lys value as 4.5.

[f] Assuming one water per peptide, one water per hydroxyl.

[g] As Pro plus one water per hydroxyl.

[h] Assuming one water per peptide, one water per ring nitrogen.

[i] Measured using Lys–Phe copolymers, range of values −0.5 to +0.5.

The linewidth of the unfrozen water was examined for several polypeptide solutions and was found to vary from 200 Hz to 7600 Hz (103). It was found that most, but not all, of the charged polypeptides had narrower linewidths. The linewidths also appeared to be independent of the amount of hydration. In addition, it was noted that the polypeptides that maintain a helix conformation at room temperature exhibited broader water peaks at −35°C than the random-coil polypeptides.

It has been found by several observers that water proton relaxation times are decreased in the presence of macromolecules. Daszkiewicz *et al.* (99) first measured the water proton relaxation time as a function of protein concentration. For protein concentrations less than 100 mg/ml, the relationship could be expressed by the empirical linear equation

$$1/T_1 = 1/T_1(w) + k_1 c \tag{8-8}$$

where T_1 is the measured relaxation time, $T_1(w)$ is the relaxation time for pure water, c is the concentration (weight/weight) of protein, and k_1 is an experimentally determined parameter. T_2 exhibits a similar behavior, having an empirical constant k_2. Assuming rapid exchange, k_1 and k_2 were estimated to be 3.07 and 6.80, respectively, for egg albumen at 20°C and 14 MHz. Denaturation resulted in k_1 and k_2 increasing to 12.4 and 100.4, respectively. Further developments along this line indicated that k_1 (and therefore the amount of bound water) depends on the molecular weight in the case of DNA (104). It was also observed that denaturation of DNA resulted in an increase in water T_1.

It has often been assumed that biopolymer solutions could be described by a two-state model, i.e., a model in which water molecules rapidly exchange between free and bound sites with $k_{ex} > 1/T_1$ (bound) (e.g., 105, 106). On the basis of the frequency and temperature dependence of the water proton T_1 in solutions of apotransferrin, Koenig and Schillinger (107) estimated that the correlation time for the protons of the bound water moiety is of the order of 10^{-7} sec, which is, in fact, the correlation time for rotational reorientation of the protein. The implication of that result is that the water may be strongly, perhaps irrotationally, bound to the protein.

Outhred and George (108) examined the proton free induction decay (as an indication of water content) and the frequency dependence of T_1 and T_2 for water with agarose and gelatin. For agarose, the T_1 and T_2 decays were exponential and the height of the FID extrapolated to zero time was observed to be within 10% of the height expected for the water protons alone. However, in the case of gelatin containing less than 50% water, T_2 relaxation was nonexponential, displaying a fast relaxing component ($T_2 < 4$ msec), and the height of the FID was 20 ± 10% larger than expected. It was proposed that the additional fast relaxing component was caused by exchangeable gelatin protons. The authors also considered that a simple two-state model was not sufficient for agarose. It was hypothesized, therefore, that the water is in two states (as before), but that there is a continuous distribution of correlation times for the water protons in the bound state, ranging up to 10^{-6} sec. Specifically, the correlation times were fit to a log distribution model with the distribution heavily weighted in favor of short correlation times, only 10% of the protons having $\tau_c > 10^{-9}$ sec in the sample of 50% water–50% agarose.

Self-diffusion coefficient (D) measurements (cf. Sections 2.1.3,C and 5.8) provide information about the translational mobility of water molecules. The early measurements by Douglass *et al.* (97) revealed the self-diffusion coefficient of water containing 4% tobacco mosaic virus to be only a few percent less than that of pure water. On that basis, it was concluded that no more than 3% of the water (amounting to four to five water molecules per amino acid residue) is bound to the tobacco mosaic virus, in direct contrast to the earlier conclusions of 20–25% bound (93, 94). Later self-diffusion coefficient measurements in solutions of egg albumen and bovine serum albumin led to the similar conclusion that only a small amount of water differed significantly from pure water (106).

Woessner *et al.* (109) examined samples of agar–water and found the proton and deuterium T_2 values to exhibit temperature hysteresis, as shown in Fig. 8-25. The break in the curves corresponds to the sol–gel transition. It was proposed that the cause of the hysteresis may be that the water molecules act as links for inter- and intrachain bonding of the agar molecules in the gel state. In the sol state, there is sufficient thermal energy to break the bonding links, freeing the water molecules and leading to an increase in T_2. The hysteresis presumably reflects a dependence of the chain linking on the conformation of the macromolecule in addition to the arrangement of water molecules. The self-diffusion coefficient for 0.0785 gm agar per cubic centimeter of water was found to be 1.95×10^{-5} cm^2/sec, compared to a value of 2.25×10^{-5} cm^2/sec for pure water obtained under identical conditions. These observations imply that the majority of water in the system is unaffected by the agar macromolecules.

Woessner and Snowden (110) investigated further the temperature and frequency dependence of the water relaxation times. The dependence of the relaxation rates on concentration were expressed in terms of k_1 (and k_2 for T_2 measurements) as given by Eq. 8-8. The results, shown in Table 8-12, reveal that $T_1 > T_2$, which implies that $\tau_c \gtrsim 0.6/\omega$ (cf. Eqs. 2-52 and 2-53, as well as Fig. 2-16). Assuming a single correlation time for bound

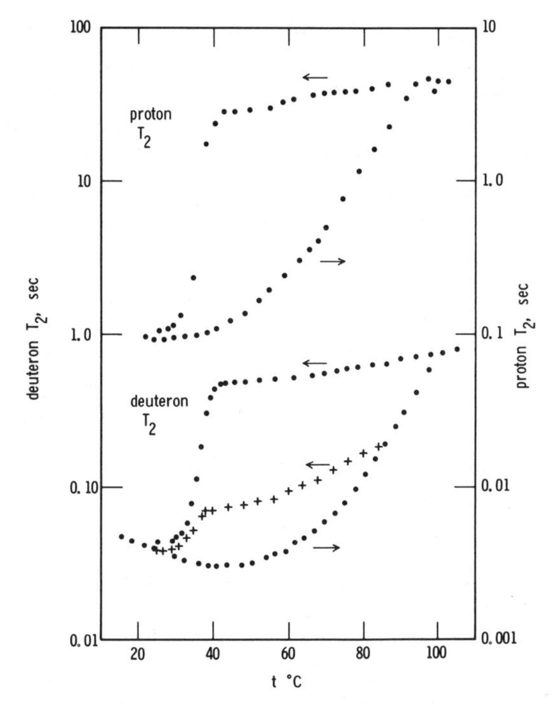

FIG. 8-25. The temperature dependences at 8 MHz of the aqueous proton and deuteron T_2 values observed for a sample of 0.0238 gm of dry agar per cubic centimeter of water. The composition of the water is 30 vol % H_2O and 70 vol % D_2O. The arrows denote the direction of temperature change (109).

water, τ_c for deuterium may be determined from the frequency dependence of T_1 via Eq. 2-52. The value for τ_c was found to be 6.23×10^{-9} sec (110). Using the values of k_1 and k_2 in Table 8-12 for a frequency of 8 MHz, this correlation time predicts (Eqs. 2-52 and 2-53) a $T_1(\text{bound})/T_2(\text{bound})$ ratio of 1.19, in distinct contrast to the observed ratio of 60.7. The temperature dependence of T_1 supported the value of 1.19, which implies the correlation time is on the low τ_c side of the T_1(bound) maximum in Fig. 2-16. The explanation for the observed ratio of 60.7 is that there are at least two pertinent correlation times. The smallest correlation time, 6.23×10^{-9} sec, is appropriate for T_1 relaxation, and a correlation time at least 100 times larger determines T_2. The discussion concerning Eqs. 8-2 and 8-3 presented earlier in this chapter should elucidate that point.

Glasel (111) has also used deuterium NMR to study the nature of D_2O in solutions with some biopolymers, a decrease in deuterium T_1 being taken as an indication of D_2O–polymer interaction. On that basis, it was con-

TABLE 8-12

THE CONCENTRATION DEPENDENCE PARAMETERS FOR THE EXCHANGEABLE HYDROGEN NUCLEI IN THE AGAR–WATER SYSTEM AT 25°C[a]

NUCLEUS	MOLE FRACTION OF H_2O IN WATER	FREQUENCY (MHz)	k_1 (SEC^{-1})	k_2 (SEC^{-1})
D	0.30	8.0	77.3	1033
D	0.30	13.75	53.7	—
H	1.00	8.00	11.20	661
H	0.30	8.00	9.58	415
H	1.00	13.75	8.22	—
H	0.30	13.75	7.01	—

[a] Woessner and Snowden (110).

cluded that the following functional groups interact weakly with D_2O:

$$>C{=}O, \quad -\overset{O}{\overset{\|}{C}}-O^-M^+, \quad >NH, \quad \text{and} \quad -C-NH_3^+X^-$$

However, the T_1 evidence suggested a strong interaction with $-COOH$ and $-NH_2$. It was presumed that counterions effectively shielded the charged groups on the biopolymers from interactions with water. It was also concluded that polyuridylic acid bound no water molecules, whereas polyadenylic acid bound only slightly. This conclusion may be compared with the study of frozen biopolymers by Kuntz *et al.* (101), who demonstrated that the nucleic acids bound three to five times as much water as proteins. It was also reported by Kuntz (102) that the amount of hydration is independent of molecular weight, whereas the deuterium T_1 measurements by Glasel (111) reveal a molecular weight dependence for polylysine but not for poly-L-glutamic acid. The apparent inconsistencies between the two techniques are probably caused by the differing nature of the observables.

Evidence for the possible existence of three fractions of water was obtained from deuterium NMR linewidth and intensity measurements on starch (112). It was found that not all of the D_2O present in the sample could be accounted for by the peak intensity. The difference between the expected and observed intensity was attributed to "solidlike" or "irrotationally bound" water. The observable signal could still conceivably arise from two mobile fractions, one free and the other bound. The criterion for the bound mobile species was taken to be the existence of an observable deuterium NMR signal at temperatures below the freezing point. As shown in Fig. 8-26, there is no discontinuity at the freezing point for starch containing 20% D_2O, implying the absence of free water. The amount of "solidlike" water is in good agreement with the amount of water required to form a monolayer. In curious contrast to the intensity measurements, Tait *et al.* (112) noted a break at 4°C (freezing point of D_2O) for the temperature dependence of the linewidth as shown in Fig. 8-27. The low-temperature activation energy of 6.5 kcal/mole is between the values of 13.3 kcal/mole for molecular rotation in ice and 4 kcal/mole for rotational motion of liquid D_2O. It was estimated that exchange contributes no more than 10% to the observed linewidths. The lack of temperature dependence above 4°C is inexplicable. If bulk water is present above 4°C, the linewidth would be expected to continue to decrease (albeit with a different slope) as the temperature is increased. The break may provide evidence for a phase change in the mobile bound water at 4°C.

B. *Water Adsorbed on Macromolecules.*

In Section 8.1.1, B, it was mentioned that lipids could be oriented and the NMR spectrum studied as a function of the angle between the applied magnetic field and the lipid chain axis (e.g., 20). In a similar manner, the proton and deuterium NMR of water molecules in oriented biopolymers, such as collagen, DNA, cellulose, silk fibroin, and keratin, have been examined (113–121).

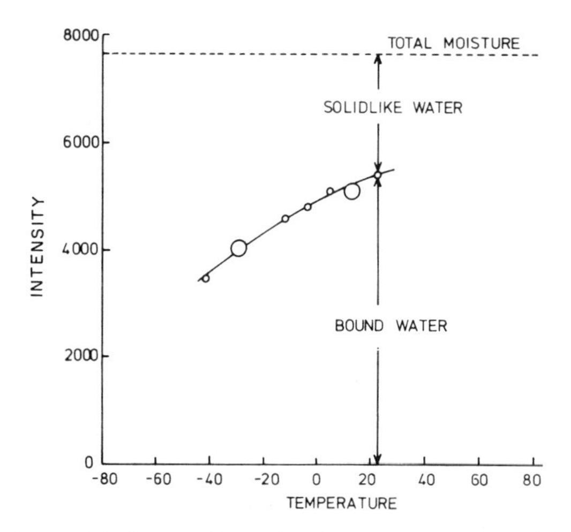

FIG. 8-26. Deuterium NMR signal intensity (arbitrary units) as a function of temperature for a sample of starch containing 20% D_2O (112).

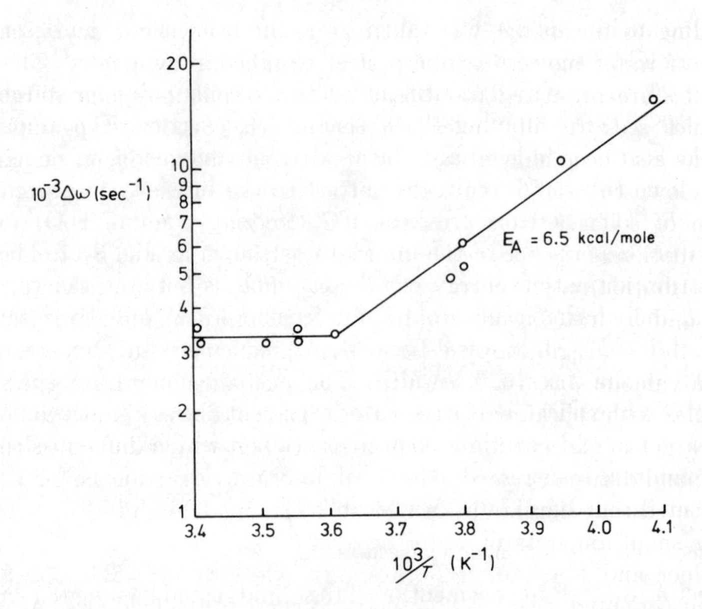

FIG. 8-27. Deuterium NMR linewidth (log scale) as a function of reciprocal temperature for a sample of starch containing 20% D_2O (112).

Berendsen and Migchelsen (113, 114) found the width of the water proton line for wet collagen to depend on the fiber axis orientation, with the anisotropic rotational axis for the water molecules being parallel to the direction of the fiber axis. It was proposed that the collagen molecules stabilize chainlike water structures along the fiber axis by hydrogen bonding of polar groups with water molecules. It is very interesting that a chain of six water molecules fits the fiber repeat distance almost exactly (within 5%). Similarly, oriented water of hydration was found for several other macromolecules.

Although they considered models in which water exhibits a periodic structure with either an extended chain or a helix with a high pitch, Chapman and McLauchlan (119) also suggested the possibility that the water of hydration may not be oriented by interaction with the biopolymer, but that the water may be constrained to assume a particular structure if it is encased in a channel of molecular dimensions. Such a possibility exists for the triple-stranded polypeptide chain of collagen.

Narrow doublets have been observed over a limited temperature range for the proton or deuterium signal of the water of hydration on some oriented biopolymers (116, 117). The splitting depends on orientation. Khanagov (120) developed a model to explain the observed splittings.

According to the model, the splittings result from slight deviations of a lattice of water molecules from perfect tetrahedral symmetry. The water molecules are presumed to be in a state of rapid molecular diffusion as molecules seek to fill "holes" in the lattice (Schottki-type diffusion).

It has also been shown that the water hydrating collagen remains unfrozen down to $-50°C$ (118, 121). The amount of unfrozen water is 0.55 gm/gm of collagen from kangaroo tail tendon.

The motion of the water of hydration has been described by a continuous log distribution for the correlation times of water in hydrated human hair (122) and hydrated wool keratin (123). Clifford and Sheard (122) assumed the τ_c distribution to range from about 10^{-11} to 10^{-8} sec, with a median value of 3×10^{-10} sec, with rapid exchange only among the water molecules with short correlation times. Studies of small amounts of water in silica gel (124) and ion-exchange resins (125) have indicated that exchange and diffusion can contribute to observed relaxation times. It might be expected that similar observations may be made with biopolymers containing small amounts of water of hydration.

Lindner and Forslind (126) examined the proton NMR of water in collagen (oriented with the fiber perpendicular to the field axis) as a function of water content and tension on the fiber. As shown in Fig. 8-28, both water content and tension affect the linewidth. It was found from independent tension measurements as a function of water content that the dehydration of collagen proceeds in two steps. The first stage (to 20% H_2O) has little effect on the linewidth (cf. Fig. 8-28), but the second stage ($<15\%$ H_2O) results in increasing linewidth with decreasing water content. The water proton line also shifts with varying water content, as shown in Fig. 8-29. The results were interpreted in terms of a model in which the tropocollagen triple helix conformation is largely determined by water bridges between various hydrophilic groups along the polypeptide chain. The two stages of dehydration referred to above were identified with different types of water bridges. Water trapped in interfibrillar channels was also considered a constituent, as was shown by the appearance of a separate peak at higher water contents.

C. *Aqueous Lipid Dispersions.*

It is quite apparent that water plays a role in maintaining the liposome and vesicle structures in dispersions; in certain organic solvents the lipids are soluble. However, the details of water interaction with the polar head group in lipid dispersions largely remains to be delineated. Deuterium NMR has been employed to provide some insight, however, it was found in the dipalmitoyl lecithin–D_2O system that the deuterium quadrupole coupling constant decreased with increasing temperature until the gel–liquid crystal transition temperature was attained (127). Above the transition temperature, the quadrupole splitting increased with temperature. In the transition region, the spectrum gave evidence for the existence of at least two phases. Studies with increasing D_2O content revealed that the temperature for which a minimum splitting is observed decreases until the D_2O-to-lecithin ratio is ~10. The spectrum of the unsonicated lecithin dispersion also provided evidence for two phases with D_2O present in excess of a maximum hydration of 40%. The "additional" phase was presumed to arise from unbound D_2O.

The observed quadrupole splittings were less than 3 kHz, in contrast to previously observed splittings of 200 kHz for simple crystalline hydrates. The diminution was attributed to rapid anisotropic rotation of D_2O molecules associated with the lecithin head groups.

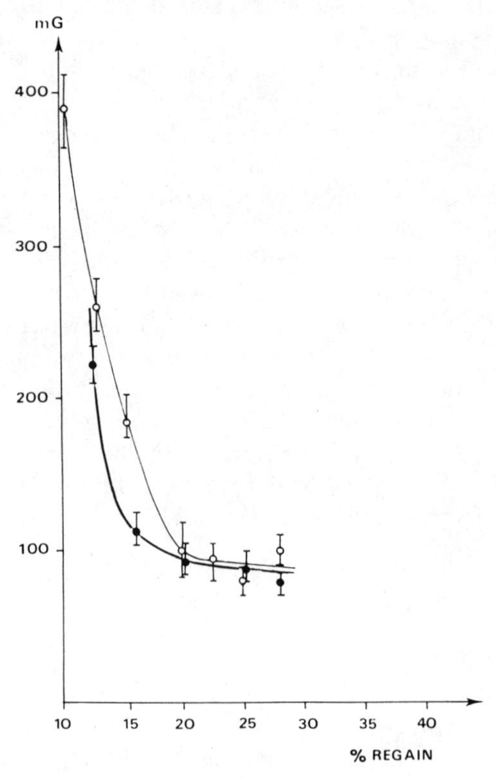

FIG. 8-28. Linewidth (milliGauss) of the water proton signal in collagen as a function of water content (% regain) for tensions of 1050 gm (●) and 1780 gm (○) (126).

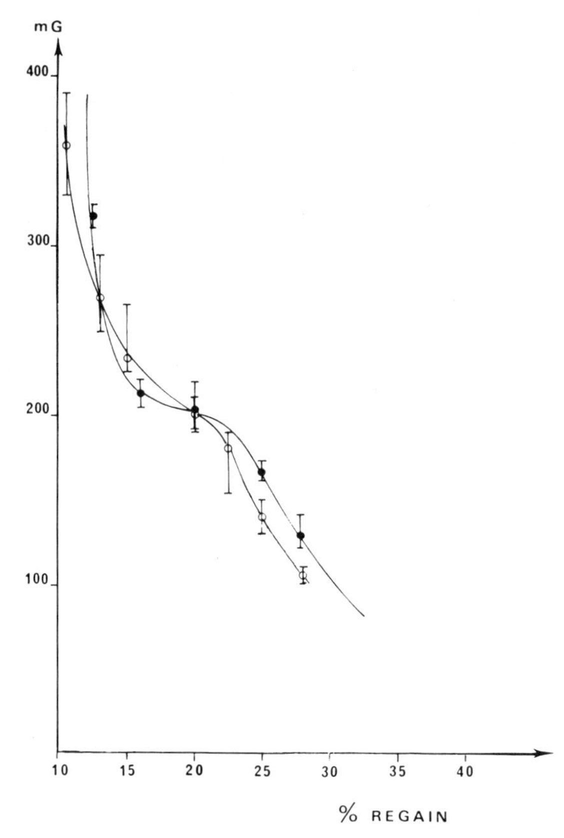

FIG. 8-29. Chemical shift (milliGauss) of the water proton signal in collagen as a function of water content (% regain) for tensions 1050 gm (●) and 1780 gm (○) (126).

Deuterium NMR has been used to study water molecules in the potassium laurate–D_2O system as well (128, 129). Two types of water were evident in the spectrum of the liquid crystalline lamellar phase: (*a*) a doublet from quadrupole splitting, ascribed to bound molecules, and (*b*) a singlet ascribed to unbound molecules. It was suggested that the adsorbed water molecules were free to reorient rapidly while diffusing along the lipid–D_2O interface.

8.2.2. Cellular Systems

NMR has been used for nearly two decades as a means for quantitatively determining water content in various materials, including biological samples. Various NMR techniques have also been applied to the question of the nature of water in biological material.

A. *Continuous Wave NMR*

Early studies of human erthyrocytes (130) and human vaginal cells (131) illustrated the exchange of intracellular water with extracellular water. Odeblad (132) later observed seven separate proton NMR lines in vaginal epithelial cell sediments and attributed four of them to different types of cellular water, although the assignments were not supported by further experiments.

Fritz and Swift (133) demonstrated the existence of intracellular and extracellular water in frog sciatic nerves by doping with paramagnetic ions. Co(II) caused a shift and Mn(II) caused a broadening of the spectral component caused by extracellular water in the 60 MHz proton NMR spectrum. The results were later verified at 220 MHz, as shown in Fig. 8-30 (82). Peak area measurements revealed that 65% of the water could be attributed to the intracellular component in the case of polarized nerve, but the percentage diminished to 34% when the nerve was depolarized by (*a*) applying a constant current through the axons or (*b*) immersing the axons in a Ringer's solution containing 133 m*M* excess KCl; both depolarizing methods yielded the same result (133).

Study of the temperature dependence of the linewidth $W_{1/2}$ of the intracellular component (cf. Eq. 2-78),

$$\pi W_{1/2} = 1/T_2 + k_{ex} \tag{8-9}$$

permitted determination of k_{ex}, the rate of exchange between intracellular

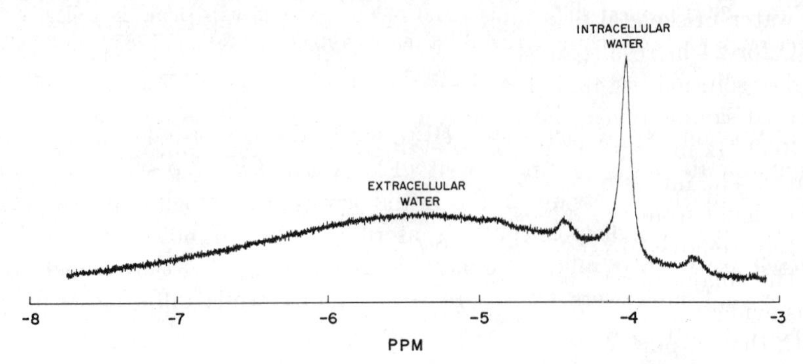

FIG. 8-30. The 220 MHz proton NMR spectrum of intracellular and extracellular water for a sciatic nerve trunk of rabbit immersed in a bathing solution doped with 0.002 *M* Mn(II). The two smaller peaks adjacent to the intracellular water resonance are spinning side bands (82).

and extracellular fractions. For polarized nerve, the value of k_{ex} is 6.8 sec^{-1} at 20°C, with an enthalpy and entropy of activation of 11.1 kcal/mole and −17.1 cal/degree-mole, respectively (133). These values were shown to be compatible with literature values for permeability coefficients determined for similar cells by other methods.

Further studies showed that use of cevadine, a depolarizing drug, produced the same effects on frog nerves as electrical or KCl depolarization, whereas acetylcholine (another depolarizer) did not alter the amount of intracellular water (134). Procaine and *d*-tubocuraine, which are blocking agents, affected the nerves by increasing the rate of exchange between intracellular and extracellular water.

The splitting of a doublet in the water proton NMR spectrum of rabbit sciatic nerve was found to depend on the orientation of the nerve axis with respect to the applied external field (135). Such an observation is consistent with oriented water molecules undergoing dipolar relaxation by anisotropic rotation. However, Klein and Phelps (136) presented evidence that the observation of splitting in the proton NMR is an artifact resulting from spectra of short lengths of slightly dried nerves, which do not uniformly fill the capillary within the sample coil. Short pieces of moistened twine also exhibited the orientation-dependent doublet. Using deuterium NMR, which should be about ten times more sensitive to orientation, no quadrupole splitting was observed with rat phrenic nerve isolated from rats given deuterium-enriched drinking water (136), providing evidence against orientation of water in the phrenic nerve.

Hazlewood *et al.* (137) used high-resolution and wide-line proton NMR spectra to demonstrate the existence of (at least) two "ordered" phases of water in skeletal muscle of rats and mice. Placing a muscle sample in D_2O for 24 hr resulted in loss of 90% of the water protons and all of the high-resolution signal. However, the wide-line spectrum of the D_2O-treated sample revealed, in addition to a very broad protein peak, a peak ~1000 Hz in width that was attributed to a "minor phase" of adsorbed water. The minor phase accounted for ~10% of the water in the muscle. The signal from the minor phase was diminished but could not be completely removed by evacuating the sample.

The "major phase" for water in muscle, accounting for 90% of the water, was evident in a line of width 14.5 ± 2.5 Hz (137). Comparing this width with that of pure water obtained under identical conditions, 0.75–1.0 Hz, led the authors to the conclusion that a second "ordered" phase of water was present in the muscle tissue. The ordering was attributed to interaction of the water molecules with cellular macromolecules, resulting in a loss of mobility. The linewidth of the "major phase" signal was found to decrease irreversibly from 14.5 Hz to 3 Hz following heat denaturation. This led to the conclusion that the water molecules interacted more extensively with the biopolymers in their native conformation.

It has been argued by some (24, 138) that the water proton linewidth in such complicated systems as biological tissue probably results from diffusion of water molecules through local magnetic field gradients in the heterogeneous sample. Packer (139) has developed an expression for the contribution of diffusion through internal magnetic field gradients to the measured transverse relaxation time and, consequently, the linewidth. It was demonstrated that this heterogeneity effect makes a negligible contribution to T_2 in striated muscle, even with a very conservative choice of parameters.

Swift and Barr (140) examined the ^{17}O NMR spectrum of bullfrog muscle that had been bathed in Ringer's solution containing 10% enrichment in $H_2{}^{17}O$. All of the ^{17}O present in the sample was found to contribute to the signal, eliminating the existence of a slowly exchanging water fraction. Oxygen-17 has a nuclear spin of 5/2; therefore, electric quadrupolar relaxation is the dominant relaxation mechanism. T_2 was determined from the linewidth and T_1 was determined by progressive saturation. Over the range 10°–30°C, it was found that the ratio of T_1 in pure water to T_1 in muscle did not change, but that ratio was about 2.6. That observation implies that there are not two fractions of water, but only one, which has slightly reduced mobility compared with pure water. However, that interpretation is not consistent with the additional observation that $T_2 < T_1$, which implies the existence of a fraction with a long τ_c.

B. *Pulsed NMR*

Water in muscle tissue has received the most attention. Bratton *et al.* (105) observed that T_2 increased with stimulation of frog muscle from a value of 40 msec in the relaxed state to greater than 60 msec in the completely exhausted state. The increase in T_2 was also reflected in a narrowing of the CW resonance line, but no change in T_1 was observed on contraction of the muscle. T_1 however, depended on the field strength, implying that at least one pertinent correlation time is $\gtrsim 0.6/\omega$. Although other models were considered, the model favored by Bratton *et al.* (105) involved rapid exchange between two phases, a "liquid" phase and a "solidlike" phase. The effect of contraction on T_2 was explained as a loss of part of the "solidlike" water, but T_1 was taken to be independent of the effects of the "solidlike" phase.

Deuterium NMR was utilized in a study of water in muscle and brain excised from rats raised on deuterium-enriched drinking water (141).

The values of T_1 for D_2O in muscle and brain were 92 and 131 msec. respectively, compared with a value of 470 msec for 10% D_2O–90% H_2O. T_2 values in muscle and brain were 9 and 22 msec, compared with 450 msec for 10% D_2O–90% H_2O. The decreased values of T_1 and T_2 in muscle and brain were interpreted to mean that water in those tissues differs in structure from that of liquid water. The observed free induction decay was described by a single exponential down to 2 msec, the limit imposed by the instrument. However, the FID extraplated to zero time could not account for all of the D_2O in the samples, implying the existence of a fraction of water with $T_2 < 2$ msec that exchanges slowly with the remainder of the water. This "slower fraction" amounted to 27% for muscle and 13% for brain tissue.

Hansen (142) repeated the deuterium NMR T_1 and T_2 measurements on rat brain and muscle and obtained the values shown in Table 8-13. The T_1 and T_2 decay curves were found to be simple exponential functions. Additionally, the observed FID, extrapolated to zero time, was able to account for all the D_2O, eliminating the existence of the "slower fraction" reported by Cope (141). It was also found that the calculated T_2 values increase with decreasing pulse spacing, using the Meiboom-Gill modification of the Carr-Purcell T_2 pulse sequence. The pulse spacing dependence was attributed to diffusion of water molecules across local magnetic fields caused by the microscopic heterogeneity of the samples. Such diffusion could also contribute to the observed linewidth. However, as noted in Section 8.1.1, B, Tiddy (25) demonstrated that the Meiboom-Gill modification was susceptible to contributions from $T_{1\rho}$ for short pulse spacing, which could lead to the observations made by Hansen (142). As already mentioned, the calculations made by Packer (139) also indicate contributions from diffusion to be negligible.

TABLE 8-13

DEUTERIUM NMR RELAXATION TIMES FOR D_2O IN RAT BRAIN AND MUSCLE AT 30°C[a,b]

	T_1 (MSEC)	T_2 (MSEC)
Brain	181 ± 5 (131)	33 ± 1.0 (22)
Muscle	112 ± 2 (90)	18 ± 0.5 (9)
Bulk water	480	450

[a] Hansen (142).

[b] The values in parentheses are those reported by Cope (141).

TABLE 8-14

DEPENDENCE OF WATER PROTON SPIN-LATTICE RELAXATION RATE ($1/T_1$) ON THE PROTEIN CONCENTRATION USING EQ. 8-8[a]

SAMPLE	k_1
Myosin A, 0.6 *M* KCl	3.4 ± 0.2
Myosin A, 0.08 *M* KCl	2.5 ± 0.2
G-Actin	1.6 ± 0.2
F-Actin	1.6 ± 0.2
Actomyosin, 0.08 *M* KCl	2.3 ± 0.2
Muscle fiber	1.8 ± 0.2

[a] Cooke and Wien (138).

Cooke and Wien (138, 143) used proton NMR T_1 and T_2 relaxation time measurements to investigate water in glycerinated rabbit muscle, suspensions of glycerinated myofibrils, and solutions of muscle proteins. Equation 8-8 was used to express the concentration dependence of T_1. As shown in Table 8-14, the values of k_1 for the fibers are approximately the same as for the muscle protein solutions, implying that the relaxation mechanisms operable in the live muscle fibers are similar to those of the *in vitro* protein solutions. It was found that the muscle relaxation time data could be fit satisfactorily with a two-state model in which there is rapid exchange between 96% of the water in the bulk phase and 4% of the water in the bound phase. The values of T_1(bound) and T_2(bound), necessary to fit the model to the data, were assumed to be the same as those obtained from a sample of partially hydrated muscle fibers containing 0.2 gm H_2O per gram protein.

A frequency dependence study of T_1 for water protons in toad muscle was interpreted in terms of a log distribution of correlation times (144). The T_1 values at the three frequencies (2.3, 8.9, and 30 MHz) could be adequately described by weighting the distribution in favor of short correlation times such that no more than 2% of the water protons had a correlation time longer than 10^{-9} sec. The possibility that a significant fraction of water protons has correlation times in the range 10^{-11}–10^{-9} sec, i.e., longer than that of bulk water, was still left open.

Conlon and Outhred (145) have extended the paramagnetic doping technique developed by Fritz and Swift (133) to study the water permeability of erythrocytes. By adding Mn(II) to the extracellular water and measuring the T_2 decay, they determined the exchange time between intracellular and extracellular water. The exchange time is defined as the time necessary for a fraction $1/e$ of the intracellular water molecules to diffuse through the erythrocyte membrane into the plasma. Following the initial rf pulse, the protons in the plasma relax rapidly because of the presence of Mn(II). The relaxation in the interior of the erythrocyte is much slower. However, when water molecules from the interior of the cell diffuse out, they too will be subjected to the relaxing influence of Mn(II). Accounting for the spontaneous relaxation of the water protons and back flux of unrelaxed water molecules from the plasma, the observed T_2 decay will be determined by the movement of water molecules from the interior of the erythrocyte into the plasma. The water exchange time was determined to be 8.2 ± 0.4 msec at 37°C, comparing well with a published figure of 8.9 msec obtained by more tedious tracer techniques.

C. *Self-Diffusion Coefficient of Water*

Abecedarskaya *et al.* (106) have measured the self-diffusion coefficient D, as well as T_1 and T_2, for several plant and animal tissues. The data are listed in Table 8-15. It was observed that the measured values of D depended on pulse spacing, which is indicative of restricted diffusion, so the values reported in Table 8-15 are those obtained by extrapolation to zero pulse spacing. The plasmalemma in animal tissues and the tonoplast, in

TABLE 8-15

EXPERIMENTAL AND CALCULATED VALUES OF D, T_1, AND T_2 FOR VARIOUS LIVE TISSUES AT 25°C[a]

	C[b]	T_2 (MSEC) ±10%	T_1 (MSEC) ±10%	$D \times 10^5$ IN CELL (CM²/SEC) ±12%	$D \times 10^5$ IN CYTOPLASM[c] (CM²/SEC)
Water	—	2500	2500	2.4	—
Maize leaves	10.5	96	250	2.0	1.82–1.60
Cytoplasm + vacuolar sap from maize leaves	7.8	—	—	2.1	—
Bean leaves	9.5	110	300	1.81	1.56–1.20
Maize roots	9.1	133	885	1.76	1.49–1.12
Frog gastrocnemius muscle	19.5	47	658	1.56	1.56
Frog liver	22.1	46	270	1.02	1.02

[a] Abecedarskaya *et al.* (106).
[b] Concentration of dry matter, weight %.
[c] Corrected for presence of vacuoles.

addition to the plasmalemma in plant tissues, were suggested as possible barriers causing restricted diffusion. Hansen (142) has also noted restricted diffusion in muscle and brain tissue. For plant cells, corrections were made for vacuoles assuming they occupy 30–50% of the internal volume of the cell and contain bulk water. It is apparent in the table that the uncorrected self-diffusion coefficient values for plant cells are larger than those for animal cells because of the presence of vacuoles. The values of D, as shown in Table 8-15, are only a factor of 1.3 to 2.4 less than the values for pure water, and they are comparable to values obtained in protein solutions and gels.

Wang (146) has proposed two possible reasons for the smaller self-diffusion coefficients of water in protein solutions:

1. The obstruction effect. Because proteins are larger and less mobile than water molecules, they impede the translational motion of the water molecules. A water molecule must travel further to get from one side of a protein to the other. Therefore, the diffusion coefficient of water in protein solutions is less than in pure water.
2. The hydration effect. Water hydrating the protein may have either increased or decreased translational motion depending on the effect of the protein on the water structure in the vicinity of the protein.

Abecedarskaya *et al.* (106) considered the obstruction effect to be the predominant cause of the decreased self-diffusion coefficients. The decreased values of T_1 and T_2 were attributed to rapid exchange between a small amount of water of hydration and bulk water.

Self-diffusion coefficient and T_2 measurements of water protons in rat and frog muscles in the relaxed, contracted, and exhausted states were made by Finch *et al.* (147). No evidence for restricted diffusion was obtained, in contrast to the report by Abecedarskaya *et al.* (106). In reasonable agreement with the previously published value (106), the diffusion coefficient was found to be 45–50% of the value for pure water. The measured value was independent of the orientation of the muscle fiber in the sample coil and was independent of the state of stimulation of the muscle. Finch *et al.* (147) also considered the obstruction effect to be paramount and the amount of bound water to be very small. It was also pointed out that the water could hardly be described as "icelike" because the self-diffusion coefficient of ice is $\sim 10^{-10}$ sec, in contrast to the observed values of $\sim 10^{-5}$ for cellular water. Arguments against paramagnetic ions being a cause of relaxation in most biological materials were also detailed.

The T_1, T_2, and D measurements by Chang *et al.* (148) on gastrocnemius muscle of mature rats were in agreement with other published values

(106, 147) but the interpretation differed, as discussed more fully in later publications (149, 150). Using Robertson's equations (151) describing the effects of restricted diffusion, it was argued that the effect of macromolecular structures can account for less than 10% of the reduction in D for muscle water compared to pure water. It was therefore concluded that a considerable part of the reduction in D is caused by "a change in structure of the water molecules as a result of the presence of macromolecules."

In order to avoid the possibility of restricted diffusion reported earlier, James and Gillen (152) chose to study the contents of a single large cell, the chicken egg. The T_1, T_2 and D results are shown in Table 8-16. The

TABLE 8-16

RELAXATION TIME (30 MHz) AND SELF-DIFFUSION COEFFICIENT MEASUREMENTS OF WATER IN SAMPLES OF CHICKEN EGG YOLK, EGG WHITE, EGG ALBUMEN, AND PURE WATER AT 22°C[a]

SAMPLE[b]	T_1 (SEC)	T_2 (SEC)	D ($CM^2/SEC \times 10^5$)	FRACTION "IMMOBILIZED" (%)[c]
Egg yolk (1)	0.072	0.023	0.50	76
Egg yolk (2)	0.058	0.019	0.37	82
Egg yolk (3)	0.069	0.027	—	—
Egg yolk (4)	0.070	0.027	0.60	71
Egg white (1) ("thin" layer)	1.30	0.44	1.71	18
Egg white (1) (11 days old)	1.30	—	—	—
Egg white (2) ("thin" layer)	1.07	—	1.66	22
Egg white (2) ("thick" layer)	1.07	—	1.66	22
Egg white (3) ("thick" layer)	1.15	—	—	—
Egg albumen, 10% solution[d]	1.27	0.34	1.83	12
Egg albumen, 10% solution	1.26	0.30	—	—
Distilled water[d]	2.83	2.83[e]	2.07	0

[a] James and Gillen (152).
[b] All fresh except as noted.
[c] Using simple two-state model presented in text. Fraction $\sim (1 - D/D_o)$.
[d] Sealed at room temperature under house vacuum to remove most oxygen.
[e] Assumed to be the same as T_1.

values of the measured parameters vary from sample to sample, but it is clear that the values for water in egg yolk, egg white, egg albumen solution, and pure water differ from one another. Assuming a simple two-state model with the measured self-diffusion coefficient D determined as the average for water in two different states: $D = D_o X_o + D_i X_i$ for ordinary and "immobilized" water, where X_o and X_i are the mole fractions of ordinary and "immobilized" water, respectively. The fraction of "immobilized" water may be expressed as

$$X_i = \frac{1 - D/D_o}{1 - D_i/D_o} \approx 1 - D/D_o \qquad (8\text{-}10)$$

where the approximation is that $D_i \ll D_o$. The assumption therefore establishes X_i as a lower limit for the model. The values of X_i given in Table 8-16 reveal a considerable difference between water in egg yolk (76%) and egg white (20%). A good part of that difference is probably caused by the higher concentration of biopolymer in yolk (30%) as compared to white (10%), but another factor may be the nature of the biopolymers. The white contains primarily protein as such, whereas the yolk contains primarily phosphoproteins and lipoproteins.

Comparison of T_1 values with D values offers a means of qualitatively distinguishing between the obstruction and hydration effects as causes for the reduced translational mobility (152). The observed relaxation rate $1/T_1$ is a weighted average of the relaxation rates at the various possible sites, with rapid exchange among the sites. The water proton relaxation rate, paramagnetic effects having been excluded, has both intramolecular and intermolecular dipole–dipole contributions. In pure water, the two effects contribute approximately equally (153). The relaxation rate may therefore be expressed as

$$\begin{aligned} 1/T_1 &= 1/(T_1)^{dd}_{\text{intra}} + 1/(T_1)^{dd}_{\text{inter}} \\ &= B\tau_r + C(N/aD) \qquad (8\text{-}11) \end{aligned}$$

where B and C are constants, τ_r is the rotational correlation time, N is the number of water molecules per milliliter, D is the self-diffusion coefficient, and a is the distance of closest approach of the relaxing water proton to the other protons promoting relaxation. Assuming N/a does not change drastically, variations in T_1 should be a reflection of changes in τ_r and D. The obstruction effect will only cause a change in D. However, it is apparent that the decrease in D given in Table 8-16 will account for little of the decrease in T_1. In comparison, water of hydration may have a larger τ_r and a smaller D. The hydration effect is therefore consistent with the

T_1 results. It is interesting to note, however, that if the amount of "immobilized" water for egg albumen, 12%, is attributed entirely to water of hydration, it can be calculated that the number of monolayers of the water of hydration is only 2.6. That is certainly not a very long-range effect.

The self-diffusion coefficient of water in untreated frog ovarian eggs from *Rana pipiens* was 0.68×10^{-5} cm²/sec, compared to 2.07×10^{-5} cm²/sec for pure water at 22°C (154). Using the same approach as James and Gillen (152), it was shown that ~67% of the water in ovarian eggs is "relatively immobile." Consideration of the T_1 values, in conjunction with D, similarly led to the conclusion that water of hydration is an important factor in the decreased mobility. The value of 0.68×10^{-5} cm²/sec is lower than the literature values reported using isotopic exchange techniques on frog eggs chemically treated to remove the membrane. The value of D from NMR, however, does agree with the value required in a model of diffusion permeability in untreated frog eggs, implying that the chemical treatment had modified the cytoplasm.

D. *Liquid Water in Frozen Systems*

Sussman and Chin (155) observed a high-resolution proton NMR signal in frozen cod muscle that increased in linewidth from 15 Hz to 140 Hz and decreased in intensity as the temperature was lowered from 0°C to −20°C. The intensity of the unfrozen fraction slowly diminished with decreasing temperature, being no longer detectable at −70°C.

The study of freezing effects in conjunction with T_2 measurements has led Belton *et al.* (156) to propose three fractions of water in striated frog muscle. Above the freezing point, it was verified that all water protons contributed to the observed free induction decay. At around −8°C, the bulk of the water freezes, leaving a signal with only 20% of the original intensity. This unfrozen fraction was attributed to water strongly bound to protein. In contrast to previous studies (105, 138, 143, 144, 147, 148), a multiexponential T_2 decay using the Carr-Purcell/Meiboom-Gill technique was reported for muscle water above the freezing point. Graphical resolution of the transverse relaxation decay yielded T_2 values for three fractions: ~230 msec (15%), ~40 msec (65%), and ~10 msec (20%). The 20% fraction was attributed to protein-bound water, the 15% fraction to water in extracellular spaces, and the 65% fraction to water in the myofibrils and the sarcoplasmic reticulum. Measurements of the transverse relaxation of the unfrozen water at −10°C were carried out, yielding T_2 values of 1 msec and 6 msec for two components of approximately equal population.

Deuterium NMR of D_2O-enriched muscle and tropomyosin samples was used in a study of freezing effects (157). More than one component was evident in the spectra of both systems. Because deuterium has a nuclear spin of 1, electric quadrupole splitting of the resonance may be observed in solids, whereas the motions in liquids will average the splitting so only a single line will be observed. As shown in Fig. 8-31 for the deuterium NMR spectra of skeletal muscle, a central, narrow, "liquidlike" component decreases in intensity commensurate with an increase in intensity of the broad "icelike" component as the temperature is lowered. Below −40°C, the broad component is further split into two components with quadrupole coupling constants (e^2qQ/h) of 208 and 236 KHz. These values are comparable to those of ice I and may be associated with static D_2O molecules possessing hydrogen bonds differing in length by 0.06 Å. The "liquidlike" component was broader in the muscle sample than in the tropomyosin

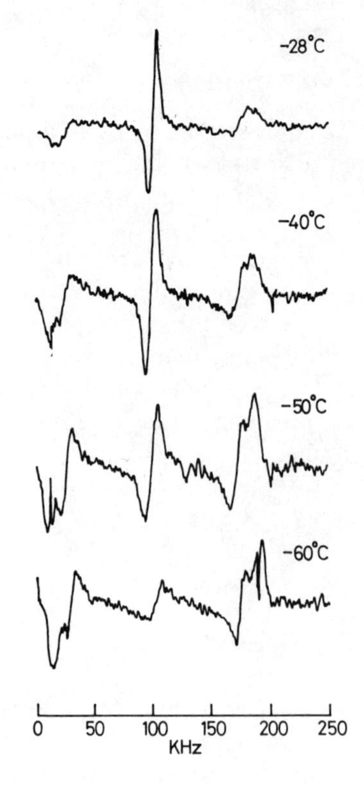

FIG. 8-31. Deuterium broad-line NMR spectra for skeletal muscle illustrating the temperature dependence of the "liquidlike" line (center) and ice peaks (sweep 500 gauss in 67 min) (157).

sample. The broader line was attributed to the D_2O molecules of this component undergoing a rapid but anisotropic motion.

Pearson *et al.* (158) examined the water proton NMR of frozen porcine muscle at various times *post mortem* to see if the effects of rigor might be detected. A significant fraction of water molecules remained unfrozen down to $-80°C$, but no dependence on the rigor process was found.

8.2.3. Water in Cancerous Tissue

Damadian (159) was the first to make the very interesting observation that water proton relaxation times in malignant tumors are greater than those of normal tissues in rats. The initial studies were followed by reports of other researchers generally substantiating the findings of Damadian. Several of the observations are summarized in Table 8-17. It is apparent that, although there is considerable variation among the several normal tissues, the proton relaxation times and self-diffusion coefficient of water in malignant tumors are generally larger than those of normal tissues but are not as large as the values for pure water.

Hazlewood *et al.* (160) made the important observation that normal, preneoplastic, and neoplastic tissue from the mouse mammary gland could be distinguished from one another. As shown in Table 8-17, the values for the parameters T_1, T_2, and D were found to increase in the order: normal tissue $<$ preneoplastic nodule $<$ neoplastic itssue $<$ pure water. It is apparent that the NMR parameters increase progressively as the cancerous tissue develops. The applications of such observations in diagnostic pathology are obvious.

Two Morris hepatomas having different growth rates and different degrees of deviation from normal tissue were excised from rats, and their T_1 values were measured (161). The measurements were not dependent on choice of surface or core samples from the nearly spherical tumors. The more rapidly growing tumor, No. 3924A, had a T_1 value of 0.812 ± 0.095 sec, clearly larger than values for normal tissues. However the more slowly growing tumor, No. 7800, had a T_1 value of 0.600 ± 0.0467 sec, overlapping with the T_1 values for some normal rat tissues.

In vivo malignant tissue has been distinguished from normal tissue using T_1 measurements on mouse tails (162). Weisman *et al.* (163) found the T_1 value to increase with growth of a malignant melanoma (Cloudman S91) on the tail of a mouse. Tails containing the developing melanoma had a T_1 value of $\sim$0.7 sec by day 31 following transplantation of the tumor, compared with $\sim$0.3 sec for normal tail tissue. The authors suggest that it may be useful to make *in vivo* T_1 measurements "on human limbs, breasts, or other protuberances" for diagnostic purposes.

TABLE 8-17

COMPARISON OF THE PROTON SPIN–LATTICE (T_1) AND SPIN–SPIN (T_2) RELAXATION TIMES AND THE SELF-DIFFUSION COEFFICIENT (D) OF WATER IN NORMAL, PRECANCEROUS, AND CANCEROUS TISSUE[a]

TISSUE	FREQ. (MHz)	T_1 (SEC)	T_2 (SEC)	D (CM²/SEC × 10⁵)	REF.
A. Normal					
Pure water	—	2.8 ± 0.3	1.4–1.5	2.07 (22°)	
Rat rectus muscle	24	0.538 ± 0.015	0.055 ± 0.005	—	159
Rat liver	24	0.293 ± 0.010	0.052 ± 0.003	—	159
Rat liver	24	0.356 ± 0.024	—	—	161
Rat stomach	24	0.270 ± 0.016	—	—	159
Rat small intestine	24	0.257 ± 0.030	—	—	159
Rat kidney	24	0.480 ± 0.026	—	—	159
Rat brain	24	0.595 ± 0.007	—	—	159
Rat brain	6	0.181 ± 0.005	0.033 ± 0.001	—	142
Rat thigh muscle	24	0.613 ± 0.031	—	—	161
Rat thigh muscle	30	—	0.052	1.12	147
Rat thigh muscle	6	0.112 ± 0.002	0.018 ± 0.0005	—	142
Rat skeletal muscle	30	0.72 ± 0.05	0.045 ± 0.002	1.43	148
Mouse mammary gland (pregnant)	30.3	0.380 ± 0.041	0.039 ± 0.002	0.34 ± 0.04	160
Mouse liver	24.3	0.350 ± 0.047	0.051	—	162
Mouse kidney	24.3	0.396 ± 0.043	0.052	—	162
Mouse spleen	24.3	0.571 ± 0.045	0.061	—	162
Mouse tail	18	0.330 ± 0.030	—	—	163
Mouse tail	8	0.230 ± 0.020	—	—	163
Rabbit liver	24.3	0.353 ± 0.045	—	—	162
Rabbit kidney	24.3	0.363	—	—	162
Rabbit spleen	24.3	0.614	—	—	162
Rabbit psoas (glycerinated)	51.6	0.560	0.066	—	138

Hamster liver	24.3	0.303 ± 0.045	—	—	162
Hamster kidney	24.3	0.398 ± 0.038	—	—	162
Hamster spleen	24.3	0.600	—	—	162
Dog liver	24.3	0.301	—	—	162
Dod kidney	24.3	0.534	—	—	162
Dog spleen	24.3	0.539	—	—	162
Human liver	24.3	0.383 ± 0.048	—	—	162
Human heart	24.3	0.873 ± 0.118	—	—	162
Human spleen	24.3	0.680 ± 0.177	—	—	162
Human pancreas	24.3	0.320 ± 0.015	—	—	162
Human skeletal muscle	24.3	0.807	—	—	162
Human kidney	24.3	0.77 ± 0.22	—	—	162
Human thymus	24.3	0.809	—	—	162
Human adrenal	24.3	0.585	—	—	162
Human prostate	24.3	0.767	—	—	162
Human thyroid	24.3	0.586	—	—	162
Human testis	24.3	1.01	—	—	162
Human stomach	24.3	0.585	—	—	162
Human placenta	24.3	1.00	—	—	162
B. Precancerous (or benign)					
Rat fibroadenoma	24	0.492	—	—	159
Mouse preneoplastic nodule (mammary)	30.3	0.451 ± 0.021	0.053 ± 0.001	0.44 ± 0.03	160
C. Cancerous					
Rat Walker sarcoma	24	0.736 ± 0.022	0.100	—	159
Rat Novikoff hepatoma	24	0.826 ± 0.013	0.118 ± 0.002	—	159
Rat Morris hepatoma (tumor No. 3924A)	24	0.812 ± 0.095	—	—	161
Rat Morris hepatoma (tumor No. 7800)	24	0.600 ± 0.046	—	—	161

(continued)

TABLE 8-17—(*Continued*)

TISSUE	FREQ. (MHz)	T_1 (SEC)	T_2 (SEC)	D (CM²/SEC × 10^5)	REF.
Mouse neoplastic tumor (mammary)	30.3	0.920 ± 0.047	0.091 ± 0.008	0.78 ± 0.05	160
Mouse melanoma Cloudman S91 (tail)	18	0.750 ± 0.050	—	—	163
Human osteogenic sarcoma	24.3	1.14	—	—	162
Human Wilm's tumor	24.3	1.05	—	—	162
Human lobular carcinoma (breast)	24.3	0.658	—	—	162
Human leiomyosarcoma	24.3	0.793	—	—	162
Human anaplastic thyroid carcinoma (formalin fixed)	24.3	0.505	—	—	162

[a] Measurements made at room temperature.

Cottam *et al.* (162) have measured the relaxation times for several tissues from different animals, including human normal and tumor tissue obtained at autopsy. A wide variation in T_1 was found for the different tissues. However, as listed in Table 8-17, the distinction is not as clear in the case of certain human tumors, e.g., the lobular carcinoma of the breast and the leiomyosarcoma. This would appear to be partially because many normal human tissues have larger T_1 values than the corresponding tissues from other animal species. It is obvious that more detailed comparisons must be made between tumors and their host tissues, e.g., between the lobular carcinoma and normal breast tissue. In this context, Frey *et al.* (163a) have examined the nonmalignant tissues from tumorous mice. It was observed that the T_1 values for spleen, kidney, liver, and heart were longer in mice with a tumor on the hind leg than for the corresponding organs of healthy mice, but the T_1 value for the tumor itssue itself was longer than for any other tissue.

Various reasons for the increased relaxation times and self-diffusion coefficient of water in malignant tissues have been considered. Damadian (159) believed the cause to be a result of increased potassium ion content in the malignant neoplasms, the potassium ion being a "water structure-breaking" cation. Hazlewood *et al.* (160) attributed the increased values to alterations in water–macromolecule interactions. Cottam *et al.* (162) also considered changes in interactions of water molecules with macromolecular structures to be important. However, Weisman *et al.* (163) suggested that the increase in the NMR parameters may be a result of higher water content in malignant cells.

8.3. State of Sodium and Potassium Ions in Biological Tissue

The intracellular concentration of K^+ is greater and the intracellular concentration of Na^+ is lower for most biological cells than the extracellular fluid concentrations. Most researchers have attributed the maintenance of such chemical gradients to an energy-requiring (from ATP) active transport of these ions. However, according to the association theories, the ionic distribution can be explained if a significant fraction of the intracellular ions are bound to macromolecules. Sodium-23 NMR spin–lattice relaxation time measurements have been used to determine the dissociation constants for Na^+ binding to macromolecules (164, 165), lending credence to the concept of alkali metal ion complexing in cells. Furthermore, ^{23}Na and ^{39}K linewidth studies by Magnuson *et al.* (166, 167) have implied that the alkali metal ions interact with membranes, specifically with membrane protein. The T_2 measurement of ^{39}K in the bacterium *Halobacterium halobium* has also implied K^+ interaction within macromolecules in the cell (168).

Cope (169) examined the ^{23}Na CW NMR signal of bullfrog muscle, rat muscle, rabbit brain, and rabbit kidney. It was noted that the ^{23}Na peak amplitudes for the intact biological tissues were much lower (~60%) than the peak amplitudes for the same samples following ashing. The "NMR-invisible" ^{23}Na was ascribed to Na^+ complexed to macromolecules in the tissue. The complexed Na^+ would be broadened beyond detection if there were no rapid exchange between complexed and free moieties because ^{23}Na ($I = 3/2$) has an electric quadrupole moment and the quadrupolar relaxation mechanism is very efficient (cf. Section 2.4.4).

Further studies, correctly using ^{23}Na intensity measurements and free induction decays rather than peak amplitudes, generally substantiated Cope's observations on several biological samples. The results are listed in Table 8-18 (169-177). The other authors also generally attributed the lower signal intensity to bound Na^+.

Shporer and Civan (178), however, noted that many of the ^{23}Na studies had found ~60% of the signal to be undetected (cf. Table 8-18), as they had in the ^{23}Na absorption spectrum of the mesophase composed of sodium linoleate in water (cf. Fig. 8-32). Use of the more sensitive derivative mode revealed the existence of broad satellite lines, as shown in Fig. 8-33. The

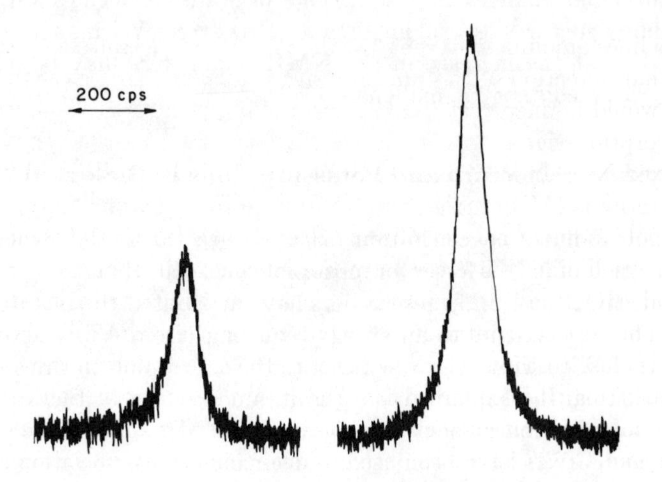

FIG. 8-32. Absorption spectrum of ^{23}Na obtained from sodium linoleate in water (left) and from a reference sample of NaOH in water and glycerol (right). The integrated intensity of the visible experimental signal is 39% of that for the reference signal, although both samples contained the same concentration and quantity of ^{23}Na nuclei (178).

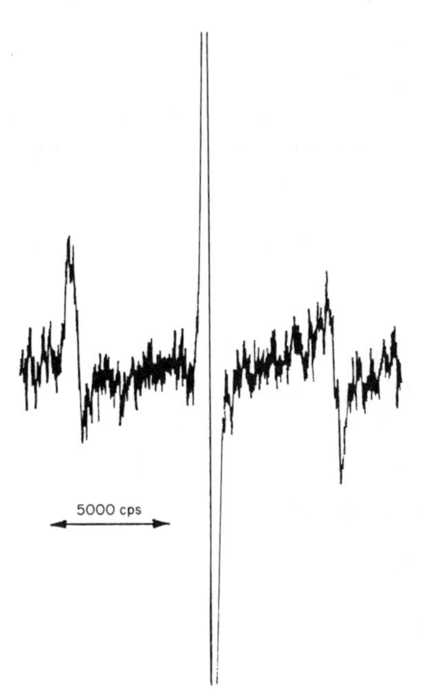

FIG. 8-33. Derivative of the ^{23}Na signal obtained from sodium linoleate in water. In addition to the central signal, satellite lines are easily distinguished (178).

satellite lines amount to 61–66% of the ^{23}Na resonance intensity but are so broad that detection of the full intensity is difficult. It is clear in this case that it would be incorrect to attribute the undetected 61–66% fraction in the absorption signal to complexed Na^+ and the observed signal to uncomplexed Na^+. Indeed, as already mentioned in connection with some of the deuterium NMR studies, the NMR spectrum of a nucleus possessing a quadrupole moment may exhibit quadrupole splitting if the nucleus is in a solid or "solidlike" environment. For nuclei with spin $I = 3/2$ (e.g., ^{23}Na and ^{39}K), the resonance will be a triplet with intensity ratios 3:4:3 (179). The observed intensity of 34–39% for the sodium linoleate–water system is close to the 40% expected for a ^{23}Na resonance possessing quadrupole splitting. It is apparent that the ^{23}Na resonance for liquid crystalline samples may exhibit quadrupole splitting. It was further suggested by Shporer and Civan)178) that this phenomenon may also be the cause of some of the observations with biological samples.

Lindblom (180) has found the ^{23}Na resonance to be split into three components for the lamellar liquid crystalline lecithin–sodium cholate–water system. The observation of quadrupole splitting was accepted as evidence for a strong interaction of Na^+ with the model membrane surfaces. Sodium-

TABLE 8-18

PERCENTAGE OF ^{23}NA NMR SIGNAL UNDETECTED IN SAMPLES OF BIOLOGICAL MATERIAL[a]

TISSUE	UNDETECTED ^{23}NA SIGNAL (%)	REF.
Frog muscle	63–74	169
	37	170
	58–65	171
	55–63	172
	53	173
Frog liver	66	170
Frog skin	57–64	174
	44	175
Rabbit muscle actomyosin	48	169
Rabbit kidney	58–70	169
Rabbit brain	60–70	169
Rabbit myelinated nerve	67	176
Rat muscle	65–72	169
Rat kidney	67	177
Rat brain	67	177
Rat testicle	24	175
Human erythrocyte	0–8	173

[a] The undetected signal is usually attributed to bound sodium ion.

23 quadrupole splitting has also been observed in the liquid crystalline system sodium decyl sulfate–decyl alcohol–water (181). The splitting was attributed to rapid exchange of mobile Na^+ with Na^+ interacting with the polar groups of the molecules, which are undergoing slow anisotropic tumbling. NMR spectra of ^{7}Li and ^{23}Na in hydrated oriented DNA fibers also exhibited splitting (182). The central component, accounting for 40% of the resonance intensity, was easily detectable but the two outer components were broader and more difficult to detect.

In a very important contribution, Berendsen and Edzes (183) presented a theoretical analysis of quadrupole splitting under various conditions. Their development suggested that the ^{23}Na NMR of most biological systems could be a result of "medium-range order," insofar as the electric field gradients interacting with the nuclear spin system are concerned. The "medium range" is viewed as extending over 100 Å or more. The field gradients giving rise to these medium-range domains are caused by heterogeneity in the biological samples. As the sodium ions diffuse through these local domains, the field gradient will vary slowly with time. In such cases, the theoretical development predicts two spectral components; the broad component may not be detectable. The narrow component retains 40% of the total resonance intensity, and T_2 for the narrow component is equal to T_1. It was pointed out that deviations from this idealized behavior may be observed for any particular system. Nevertheless, the conclusion by Berendsen and Edzes (193) is that the ^{23}Na NMR does not prove that Na^+ is complexed in cells. The width of the narrow-line component (the one observed) is an indication of mobility. On the basis that the width of the narrow-line component is less than five times as broad as that of aqueous NaCl, it would appear that the great majority of sodium ions in the cell are not complexed.

One other facet that has been generally ignored was pointed out by Yeh *et al.* (173); i.e., most of the biological material studied has a significant extracellular Na^+ concentration, although intracellular Na^+ has been the subject of study. It was also observed that, within experimental error, no "NMR-invisible" Na^+ occurs in human erythrocytes, but line broadening is observed. It was suggested that the line broadening may be caused by intraction of part of the sodium ions with membrane-bound protein (cf. 166, 167) or with hemoglobin.

References

1. R. W. Hendler, *Physiol. Rev.* **51,** 66 (1971).
2. M. H. F. Wilkins, A. E. Blaurock, and D. M. Engleman, *Nature* (*London*) **230,** 72 (1971).
3. D. Chapman and N. J. Salsbury, *Recent Progr. Surface Sci.* **3,** 121 (1970).
4. D. Chapman, *Ann. N.Y. Acad. Sci.* **195,** 179 (1972).
5. A. Abragam, "The Principles of Nuclear Magnetism," p. 456. Oxford Univ. Press, London and New York, 1961.
6. D. E. Woessner, *J. Chem. Phys.* **36,** 1 (1962).
7. T. B. Cobb and C. S. Johnson, Jr., *J. Chem. Phys.* **52,** 6224 (1970).
8. J. H. Noggle, *J. Phys. Chem.* **72,** 1324 (1968).
9. D. Chapman and N. J. Salsbury, *Trans. Faraday Soc.* **62,** 2607 (1966).
10. N. J. Salsbury and D. Chapman, *Biochim. Biophys. Acta* **163,** 314 (1968).
11. R. Haque, I. J. Tinsley, and D. Schmedding, *J. Biol. Chem.* **247,** 157 (1972).
12. J. G. Batchelor, J. H. Prestegard, R. J. Cushley, and S. R. Lipsky, *Biochem. Biophys. Res. Commun.* **48,** 70 (1972).
13. Z. Veksli, N. J. Salsbury, and D. Chapman, *Biochim. Biophys. Acta* **183,** 434 (1969).
14. S. J. Kohler, A. F. Horowitz, and M. P. Klein, *Biochem. Biophys. Res. Commun.* **49,** 1414 (1972).
15. J. T. Daycock, A. Darke, and D. Chapman, *Chem. Phys. Lipids* **6,** 205 (1971).
16. J. E. Anderson and W. P. Slichter, *J. Phys. Chem.* **69,** 3099 (1965).
17. N. J. Salsbury, D. Chapman, and G. P. Jones, *Trans. Faraday Soc.* **66,** 1554 (1969).

18. C. H. A. Seiter, G. W. Feigenson, S. I. Chan, and M.-C. Hsu, *J. Amer. Chem. Soc.* **94,** 2535 (1972).
19. S. A. Penkett, A. G. Flook, and D. Chapman, *Chem. Phys. Lipids* **2,** 273 (1968).
20. J. J. DeVries and H. J. C. Berendsen, *Nature (London)* **221,** 1139 (1969).
21. C. P. Slichter, "Principles of Magnetic Resonance," Chapter 3. Harper, New York, 1963.
22. E. G. Finer, A. G. Flook, and H. Hauser, *Biochim. Biophys. Acta* **260,** 59 (1972).
23. S. Kaufman, J. M. Steim, and J. H. Gibbs, *Nature (London)* **225,** 743 (1970).
24. J. R. Hansen and K. D. Lawson, *Nature (London)* **225,** 542 (1970).
25. G. J. T. Tiddy, *Nature (London)* **230,** 136 (1971).
26. J. Charvolin and P. Rigny, *Mol. Cryst. Liquid Cryst.* **15,** 211 (1971).
27. J. Charvolin and P. Rigny, *J. Chem. Phys.* **58,** 3999 (1973).
28. J. Charvolin and P. Rigny, *Nature (London)* **237,** 127 (1972).
29. S. I. Chan, G. W. Feigenson, and C. H. A. Seiter, *Nature (London)* **231,** 110 (1971).
30. E. Oldfield, J. Marsden, and D. Chapman, *Chem. Phys. Lipids* **7,** 1 (1971).
31. S. I. Chan, C. H. A. Seiter, and G. W. Feigenson, *Biochem. Biophys. Res. Commun.* **46,** 1488 (1972).
31a. C. H. A. Seiter and S. I. Chan, *J. Amer. Chem. Soc.* **95,** 7541 (1973).
32. D. E. Woessner, B. S. Snowden, Jr., R. A. McKay, and E. T. Strom, *J. Magn. Resonance* **1,** 105 (1969).
33. A. G. Lee, N. J. M. Birdsall, and J. C. Metcalfe, *Biochemistry* **12,** 1650 (1973).
34. E. R. Andrew, A. Bradbury, and R. G. Eades, *Nature (London)* **182,** 1659 (1958).
35. J. Lowe, *Phys. Rev. Lett.* **2,** 285 (1959).
36. D. Chapman, E. Oldfield, D. Doskocilova, and B. Schneider, *FEBS (Fed. Eur. Biochem. Soc.) Lett.* **25,** 261 (1972).
37. E. Oldfield and D. Chapman, *Biochem. Biophys. Res. Commun.* **43,** 949 (1971).
38. K. M. Keough, E. Oldfield, D. Chapman, and P. Beynon, *Chem. Phys. Lipids* **10,** 37 (1973).
39. J. C. Metcalfe, N. J. M. Birdsall, J. Feeney, A. G. Lee, Y. K. Levine, and P. Partington, *Nature (London)* **233,** 199 (1971).
40. D. G. Davis and G. Inesi, *Biochem. Biophys. Acta* **282,** 180 (1972).
41. D. G. Davis, *Biochem. Biophys. Res. Commun.* **49,** 1492 (1972).
42. D. Chapman and S. A. Penkett, *Nature (London)* **211,** 1304 (1966).
43. B. Sheard, *Nature (London)* **223,** 1057 (1969).
44. E. G. Finer, A. G. Flook, and H. Hauser, *Biochim. Biophys. Acta* **260,** 49 (1972).
45. V. F. Bystrov, N. I. Dubrovina, L. I. Barsukov, and L. D. Bergelson, *Chem. Phys. Lipids* **6,** 343 (1971).
46. R. J. Kostelnik and S. M. Castellano, *J. Magn. Resonance* **7,** 219 (1972).
47. Y. K. Levine, A. G. Lee, N. J. M. Birdsall, J. C. Metcalfe, and J. D. Robinson, *Biochim. Biophys. Acta* **291,** 592 (1973).
48. R. J. Kostelnik and S. M. Castellano, *J. Magn. Resonance* **9,** 291 (1973).
49. W. L. Hubbell and H. M. McConnell, *Proc. Nat. Acad. Sci. U.S.* **61,** 12 (1968).
50. W. L. Hubbell and H. M. McConnell, *J. Amer. Chem. Soc.* **93,** 314 (1971).
51. R. Kubo and K. Tomita, *J. Phys. Soc. Jap.* **9,** 888 (1954).
52. A. F. Horwitz, D. Michaelson, and M. P. Klein, *Biochim. Biophys. Acta* **298,** 1 (1973).
53. A. F. Horwitz, W. J. Horsley, and M. P. Klein, *Proc. Nat. Acad. Sci. U.S.* **69,** 590 (1972).
54. A. G. Lee, N. J. M. Birdsall, Y. K. Levine, and J. C. Metcalfe, *Biochim. Biophys. Acta* **255,** 43 (1972).

55. P. Devaux and H. M. McConnell, *J. Amer. Chem. Soc.* **94,** 4475 (1972).
56. R. T. Roberts, *Nature* (*London*) **242,** 348 (1973).
57. J. E. Tanner, *J. Chem. Phys.* **52,** 2523 (1970).
58. Y. K. Levine, N. J. M. Birdsall, A. G. Lee, and J. C. Metcalfe, *Biochemistry* **11,** 1416 (1972).
59. Y. K. Levine, P. Partington, G. C. K. Roberts, N. J. M. Birdsall, A. G. Lee, and J. C. Metcalfe, *FEBS* (*Fed. Eur. Biochem. Soc.*) *Lett.* **23,** 203 (1972).
60. D. Wallach, *J. Chem. Phys.* **47,** 5258 (1967).
61. J. L. Lippert and W. L. Peticolas, *Biochim. Biophys. Acta* **282,** 8 (1972).
62. N. J. M. Birdsall, A. G. Lee, Y. K. Levine, and J. C. Metcalfe, *Biochim. Biophys. Acta* **241,** 693 (1971).
63. R. W. Barker, J. D. Bell, G. K. Radda, and R. E. Richards, *Biochim. Biophys. Acta* **260,** 161 (1972).
64. V. F. Bystrov, Y. E. Shapiro, A. V. Viktorov, L. I. Barsukov, and L. D. Bergelson, *FEBS* (*Fed. Eur. Biochem. Soc.*) *Lett.* **25,** 337 (1972).
65. D. Chapman and S. A. Penkett, *Nature* (*London*) **211,** 1304 (1966)
66. A. Darke, E. G. Finer, A. G. Flook, and M. C. Phillips, *J. Mol. Biol.* **63,** 265 (1972).
67. E. Oldfield and D. Chapman, *FEBS* (*Fed. Eur. Biochem. Soc.*) *Lett.* **21,** 302 (1972).
68. M. A. Hemminga and H. J. C. Berendsen, *J. Magn. Resonance* **8,** 133 (1972).
69. E. Oldfield, D. Chapman, and W. Derbyshire, *FEBS* (*Fed. Eur. Biochem. Soc.*) *Lett.* **16,** 102 (1971).
70. K. U. Berger, M. D. Barratt, and V. B. Kamat, *Chem. Phys. Lipids* **6,** 351 (1971).
71. E. G. Finer, H. Hauser, and D. Chapman, *Chem. Phys. Lipids* **3,** 386 (1969).
72. H. Hauser, E. G. Finer, and D. Chapman, *J. Mol. Biol.* **53,** 419 (1970).
73. W. Pache, D. Chapman, and R. Hillaby, *Biochim. Biophys. Acta* **255,** 358 (1972).
74. J. Cerbón, *Biochim. Biophys. Acta* **290,** 51 (1972).
75. D. G. Davis and G. Inesi, *Biochim. Biophys. Acta* **241,** 1 (1971).
76. J. D. Robinson, N. J. M. Birdsall, A. G. Lee, and J. C. Metcalfe, *Biochemistry* **11,** 2903 (1972).
77. V. B. Kamat and D. Chapman, *Biochim. Biophys. Acta* **163,** 411 (1968).
78. M. P. Sheetz and S. I. Chan, *Biochemistry* **11,** 548 (1972).
79. J. C. Metcalfe, S. M. Metcalfe, and D. M. Engleman, *Biochim. Biophys. Acta* **241,** 412 (1971).
80. T. J. Jenkinson, V. B. Kamat, and D. Chapman, *Biochim. Biophys. Acta* **183,** 427 (1969).
81. H. Lecar, G. Ehrenstein, and I. Stillman, *Biophys. J.* **11,** 140 (1971).
82. P. Dea, S. I. Chan, and F. J. Dea, *Science* **175,** 206 (1972)
83. J. C. Metcalfe, N. J. M. Birdsall, and A. G. Lee, *FEBS* (*Fed. Eur. Biochem. Soc.*) *Lett.* **21,** 335 (1972).
84. E. Oldfield, D. Chapman, and W. Derbyshire, *Chem. Phys. Lipids* **9,** 69 (1972).
85. G. N. Ling, "A Physical Theory of the Living State." Ginn (Blaisdell), Boston, Massachusetts, 1962.
86. J. Del Bene and J. A. Pople, *J. Chem. Phys.* **52,** 4858 (1970).
87. H. A. Scheraga, *Ber. Bunsenges. Phys. Chem.* **68,** 838 (1964).
88. W. Drost-Hansen, *in* "Chemistry of the Cell Interface" (H. D. Brown, ed.), Chapter 6, Part B, p. 1. Academic Press, New York, 1971.
89. D. Eisenberg and W. Kauzmann, "The Structure and Properties of Water." Oxford Univ. Press, London and New York, 1969.
90. J. A. Walter and A. B. Hope, *Progr. Biophys. Mol. Biol.* **23,** 3 (1971).

91. M. J. Tait and F. Franks, *Nature* (*London*) **230,** 91 (1971).
92. H. G. Hertz, *Ber. Bunsenges. Phys. Chem.* **74,** 666 (1970).
93. B. W. Jacobson, W. A. Anderson, and J. J. Arnold, *Nature* (*London*) **173,** 772 (1954).
94. C. D. Jardetzky and O. Jardetzky, *Biochim. Biophys. Acta* **26,** 668 (1957).
95. E. A. Balazs, A. A. Bothner-By, and G. Gergely, *J. Mol. Biol.* **1,** 147 (1959).
96. J. Depireaux and D. Williams, *Nature* (*London*) **195,** 699 (1965).
97. D. C. Douglass, H. L. Frisch, and E. W. Anderson, *Biochim. Biophys. Acta* **44,** 401 (1960).
98. D. J. Blears and S. S. Danyluk, *Biopolymers* **3,** 585 (1965).
99. O. K. Daszkiewicz, J. W. Hennel, B. Lubas, and T. W. Szczepkowski, *Nature* (*London*) **200,** 1006 (1963).
100. R. Collinson and M. P. McDonald, *Nature* (*London*) **186,** 548 (1960).
101. I. D. Kuntz, Jr., T. S. Brassfield, G. D. Law, and G. V. Purcell, *Science* **163,** 1329 (1969).
102. I. D. Kuntz, *J. Amer. Chem. Soc.* **93,** 514 (1971).
103. I. D. Kuntz, *J. Amer. Chem. Soc.* **93,** 516 (1971).
104. B. Lubas and T. Wilczok, *Biochim. Biophys. Acta* **120,** 427 (1966).
105. C. A. Bratton, A. L. Hopkins, and J. W. Weinberg, *Science* **147,** 738 (1965).
106. L. A. Abecedarskaya, F. G. Miftahutdinova, and V. D. Fedotov, *Biofizika* **13,** 630 (1968); *Biophysics* (*USSR*) **13,** 750 (1968).
107. S. H. Koenig and W. E. Schillinger, *J. Biol. Chem.* **244,** 3283 (1969).
108. R. K. Outhred and E. P. George, *Biophys. J.* **13,** 83 (1973).
109. D. E. Woessner, B. S. Snowden, Jr., and Y.-C. Chiu, *J. Colloid Interface Sci.* **34,** 283 (1970).
110. D. E. Woessner and B. S. Snowden, Jr., *J. Colloid Interface Sci.* **34,** 290 (1970).
111. J. A. Glasel, *J. Amer. Chem. Soc.* **92,** 375 (1970).
112. M. J. Tait, S. Ablett, and F. W. Wood, *J. Colloid Interface Sci.* **41,** 594 (1972).
113. H. J. C. Berendsen, *J. Chem. Phys.* **36,** 3297 (1962).
114. H. J. C. Berendsen and C. Migchelsen, *Ann. N.Y. Acad. Sci.* **125,** 365 (1965).
115. A. Rupprecht, *Acta Chem. Scand.* **20,** 582 (1966).
116. C. Migchelsen, H. J. C. Berendsen, and A. Rupprecht, *J. Mol. Biol.* **37,** 235 (1968).
117. R. E. Dehl, *J. Chem. Phys.* **48,** 831 (1968).
118. R. E. Dehl and C. A. Hoeve, *J. Chem. Phys.* **50,** 3245 (1969).
119. G. E. Chapman and K. E. McLaughlin, *Proc. Roy. Soc. Ser., B* **173,** 233 (1969).
120. A. A. Khanagov, *Biopolymers* **10,** 789 (1971).
121. R. E. Dehl, *Science* **170,** 738 (1970).
122. J. Clifford and B. Sheard, *Biopolymers* **4,** 1057 (1966).
123. L. J. Lynch and K. H. Marsden, *J. Chem. Phys.* **51,** 5681 (1969).
124. J. R. Zimmerman and W. E. Brittin, *J. Phys. Chem.* **61,** 1328 (1957).
125. W. J. Blaedel, L. E. Brower, T. L. James, and J. H. Noggle, *Anal. Chem.* **44,** 982 (1972).
126. P. Lindner and E. Forslind, *Chem. Scripta* **3,** 57 (1973).
127. N. J. Salsbury, A. Darke, and D. Chapman, *Chem. Phys. Lipids* **8,** 142 (1972).
128. J. Charvolin and P. Rigny, *J. Phys.* **30,** C4-C76 (1969).
129. J. Charvolin and P. Rigny, *Chem. Phys. Lett.* **18,** 515 (1973).
130. E. Odeblad, B. N. Bhar, and G. Lindstrom, *Arch. Biochem. Biophys.* **63,** 221 (1956).
131. E. Odeblad, *Ann. N.Y. Acad. Sci.* **83,** 189 (1959).

132. E. Odeblad, *Ark. Kemi* **25,** 377 (1967).
133. O. G. Fritz and T. J. Swift, *Biophys. J.* **7,** 675 (1967).
134. O. G. Fritz, A. C. Scott, and T. J. Swift, *Nature (London)* **218,** 1051 (1968).
135. G. Chapman and K. A. McLauchlan, *Nature (London)* **215,** 391 (1967).
136. M. P. Klein and D. E. Phelps, *Nature (London)* **224,** 70 (1969).
137. C. F. Hazlewood, B. L. Nichols, and N. F. Chamberlain, *Nature (London)* **222,** 747 (1969).
138. R. Cooke and R. Wien, *Biophys. J.* **11,** 1002 (1971).
139. K. J. Packer, *J. Magn. Resonance* **9,** 438 (1973).
140. T. J. Swift and E. M. Barr, *Ann. N.Y. Acad. Sci.* **204,** 191 (1973).
141. F. W. Cope, *Biophys. J.* **9,** 304 (1969).
142. J. R. Hansen, *Biochim. Biophys. Acta* **230,** 482 (1971).
143. R. Cooke and R. Wien, *Ann. N.Y. Acad. Sci.* **204,** 197 (1973).
144. R. K. Outhred and E. P. George, *Biophys. J.* **13,** 97 (1973).
145. T. Conlon and R. Outhred, *Biochim. Biophys. Acta* **288,** 354 (1972).
146. J. H. Wang, *J. Amer. Chem. Soc.* **76,** 4755 (1954).
147. E. D. Finch, J. F. Harmon, and B. H. Miller, *Arch. Biochem. Biophys.* **147,** 299 (1971).
148. D. C. Chang, C. F. Hazelwood, B. L. Nichols, and H. E. Rorschach, *Nature (London)* **235,** 170 (1972).
149. D. C. Chang, H. E. Rorschach, B. L. Nichols, and C. F. Hazlewood, *Ann. N.Y. Acad. Sci.* **204,** 434 (1973).
150. H. E. Rorschach, D. C. Chang, C. F. Hazlewood, and B. L. Nichols, *Ann. N.Y. Acad. Sci.* **204,** 444 (1973).
151. B. Robertson, *Phys. Rev.* **151,** 273 (1966).
152. T. L. James and K. T. Gillen, *Biochim. Biophys. Acta* **286,** 10 (1972).
153. J. G. Powles and M. Rhodes, *Mol. Phys.* **11,** 515 (1966).
154. K. Hansson Mild, T. L. James, and K. T. Gillen, *J. Cell. Physiol.* **80,** 155 (1972).
155. M. V. Sussman and L. Chin, *Science* **151,** 324 (1966).
156. P. S. Belton, R. R. Jackson, and K. J. Packer, *Biochim. Biophys. Acta* **286,** 16 (1972).
157. W. Derbyshire and J. L. Parsons, *J. Magn. Resonance* **6,** 344 (1972).
158. R. T. Pearson, W. Derbyshire, and J. M. V. Blanshard, *Biochem. Biophys. Res. Commun.* **48,** 873 (1972).
159. R. Damadian, *Science* **171,** 1151 (1971).
160. C. F. Hazlewood, D. C. Chang, D. Medina, G. Cleveland, and B. L. Nichols, *Proc. Nat. Acad. Sci. U.S.* **69,** 1478 (1972).
161. D. P. Hollis, L. A. Saryan, and H. P. Morris, *Johns Hopkins Med. J.* **131,** 441 (1972).
162. G. L. Cottam, A. Vesek, and D. Lusted, *Res. Commun. Chem. Pathol. Pharmacol.* **4,** 495 (1972).
163. I. D. Weisman, L. H. Bennett, L. R. Maxwell, Sr., M. W. Woods, and D. Burk, *Science* **178,** 1290 (1972).
163a. H. E. Frey, R. R. Knispel, J. Kruuv, A. R. Sharp, R. T. Thompson, and M. M. Pintar, *J. Nat. Cancer Inst.* **49,** 903 (1972).
164. T. L. James and J. H. Noggle, *Proc. Nat. Acad. Sci. U.S.* **62,** 644 (1969).
165. F. Ostroy, T. James, J. Noggle, and L. E. Hokin, *Fed. Proc., Fed. Amer. Soc. Exp. Biol.* **31,** 1207 (1972).
166. J. A. Magnuson, D. S. Shelton, and N. S. Magnuson, *Biochem. Biophys. Res. Commun.* **39,** 279 (1970).

167. J. A. Magnuson and N. S. Magnuson, *Ann. N.Y. Acad. Sci.* **204,** 297 (1973).
168. F. W. Cope and R. Damadian, *Nature* (*London*) **228,** 76 (1970).
169. F. W. Cope, *J. Gen. Physiol.* **50,** 1353 (1967).
170. D. Martinez, A. A. Silvidi, and R. M. Stokes, *Biophys. J.* **9,** 1256 (1969).
171. G. N. Ling and F. W. Cope, *Science* **163,** 1335 (1969).
172. J. L. Czeisler, O. G. Fritz, Jr., and T. J. Swift, *Biophys. J.* **10,** 260 (1970).
173. H. J. C. Yeh, F. J. Brinley, Jr., and E. D. Becker, *Biophys. J.* **13,** 56 (1973).
174. C. A. Rotunno, V. Kowalewski, and M. Cereijido, *Biochim. Biophys. Acta* **135,** 170 (1967).
175. I. L. Reisen, C. A. Rotunno, L. Corchs, V. Kowalewski, and M. Cereijido, *Physiol. Chem. Phys.* **2,** 171 (1970).
176. F. W. Cope, *Physiol. Chem. Phys.* **2,** 545 (1970).
177. F. W. Cope, *Biophys. J.* **10,** 843 (1970).
178. M. Shporer and M. Civan, *Biophys. J.* **12,** 114 (1972).
179. A. Abragam, "The Principles of Nuclear Magnetism," p. 232ff. Oxford Univ. Press, London and New York, 1961.
180. G. Lindblom, *Acta Chem. Scand.* **25,** 2767 (1971).
181. D. M. Chen and L. W. Reeves, *J. Amer. Chem. Soc.* **94,** 4384 (1972).
182. H. T. Edzes, A. Rupprecht, and H. J. C. Berendsen, *Biochem. Biophys. Res. Commun.* **46,** 790 (1972).
183. H. J. C. Berendsen and H. T. Edzes, *Ann. N.Y. Acad. Sci.* **204,** 459 (1973).

APPENDIX 1

LIST OF SYMBOLS

The following symbols may appear without explanation in certain parts of the text.

A, A_i	hyperfine coupling constants
a	solute radius
a_s	solvent radius
C^2	squared average of spin-rotation tensor
D	self-diffusion coefficient
D_r	rotational diffusion coefficient
E	energy
eQ	electric quadrupole moment
eq	principal component of electric field gradient tensor
F_z	sum of magnetic quantum numbers
f_R	microviscosity factor
G	magnetic field gradient
G_A, G_B	complex magnetic moments
$G_{mn}(\tau)$	correlation function
g	electronic "g" factor
$g(\nu)$	lineshape function
H_0, H_1, H_2	intensities of the stationary magnetic field, observing rf field, and saturating rf field, respectively

H_{eff}	effective field
Hb	hemoglobin
$\mathcal{H}$, $\mathcal{H}_0$, $\mathcal{H}_1$, etc.	Hamiltonians
$\hbar$	Planck's constant divided by 2π
I	total nuclear spin quantum number
I'	molecular moment of inertia
i	$\sqrt{-1}$
$\vec{i}, \vec{j}, \vec{k}$	unit vectors along x, y, and z axes, respectively
J	scalar spin-spin coupling constant
J'	angular momentum
$J_{mn}(\omega)$	spectral density
K, K_1, etc.	equilibrium constants
k	Boltzmann constant
k_1, k_{ex}, etc.	rate constants
$\vec{M}$, $\vec{M}_{\perp}$, etc.	macroscopic magnetic moments (magnetization)
M_x, M_y, M_z	components of the macroscopic magnetic moment along x, y, and z axes, respectively
M_0	macroscopic magnetic moment for system at thermal equilibrium
Mb	myoglobin
m_I	magnetic quantum number
n	number of water molecules in the coordination sphere of a metal ion
P_A, P_B, etc.	populations
R	resolution
r	internuclear distance
S	total electron spin quantum number
SW	spectral width
T	absolute temperature
$\vec{T}$	torque
T_1	spin–lattice (longitudinal) relaxation time
T_2	spin–spin (transverse) relaxation time
T_{1M}, T_{2M}	relaxation times for nuclei in the sphere of influence of paramagnetic species
$T_{1\rho}$	relaxation time in the rotating frame
t_{ac}	signal acquisition time
t_w	pulse length
u	component of the transverse macroscopic magnetic moment in phase with the applied rf field
v	component of the transverse macroscopic magnetic moment out of phase with the applied rf field
$W_{1/2}$	linewidth at half-maximal signal amplitude
(x, y, z)	laboratory frame coordinate system
(x', y', z)	rotating frame coordinate system
α, β	single spin wave functions
β	Bohr magneton
γ	gyromagnetic ratio
δ	chemical shift (ppm)
δ_c	contact shift
δ_p	pseudocontact shift
ϵ	water proton longitudinal relaxation enhancement

η	viscosity: also assymetry parameter
θ	dihedral angle in sugars and peptides
$\vec{\mu}$	magnetic moment of individual spin
ν	frequency (Hertz)
σ	shielding constant
τ	chemical shift (ppm) with TMS at 10.0τ
τ_c	correlation time
τ_e	hyperfine correlation time
τ_{ex}	chemical exchange time
τ_j	spin-rotation correlation time
τ_M	lifetime of nucleus in the coordination sphere of a paramagnetic species
τ_r	rotational correlation time
τ_s	electron spin relaxation time
τ_{tr}	translational correlation time
ϕ, ψ	dihedral angles in peptides
χ	bulk volume susceptibility
ψ, ψ_A, etc.	wave functions
ω	angular frequency ($= 2\pi\nu$)
ω_0, ω_I	nuclear resonance frequency
ω_s	electron resonance frequency

APPENDIX 2

TABLE OF NUCLEAR PROPERTIES

TABLE OF NUCLEAR PROPERTIES[a,b]

ISOTOPE	NMR FREQUENCY IN MHz FOR A 10 KG FIELD	NATURAL ABUNDANCE (%)	RELATIVE SENSITIVITY FOR EQUAL NUMBER OF NUCLEI		MAGNETIC MOMENT, μ IN MULTIPLES OF THE NUCLEAR MAGNETON $(eh/4\pi Mc)$	SPIN I IN MULTIPLES OF $h/2\pi$	ELECTRIC QUADRUPOLE MOMENT, Q, IN MULTIPLES OF $e \times 10^{-24}$ CM^2
			AT CONSTANT FIELD	AT CONSTANT FREQUENCY			
^{1}n*	29.1670	—	0.322	0.685	−1.91315	1/2	—
● ^{1}H	42.5759	99.9844	1.000	1.000	2.79268	1/2	—
● ^{2}H	6.53566	1.56×10^{-2}	9.65×10^{-3}	0.409	0.857386	1	2.77×10^{-3}
● ^{3}H*	45.414	—	1.21	1.07	2.9788	1/2	—
● ^{3}He	32.435	10^{-5}–10^{-7}	0.442	0.762	−2.1274	1/2	—
● ^{6}Li	6.265	7.43	8.50×10^{-3}	0.392	0.82192	1	4.6×10^{-4}
● ^{7}Li	16.547	92.57	0.294	1.94	3.2560	3/2	−0.1
● ^{9}Be	5.983	100	1.39×10^{-2}	0.703	−1.1773	3/2	2×10^{-2}
● ^{10}B	4.575	18.83	1.99×10^{-2}	1.72	1.8005	3	7.4×10^{-2}
● ^{11}B	13.660	81.17	0.165	1.60	2.6880	3/2	3.55×10^{-2}
● ^{13}C	10.705	1.108	1.59×10^{-2}	0.251	0.70220	1/2	—
● ^{14}N	3.076	99.635	1.01×10^{-3}	0.193	0.40358	1	7.1×10^{-2}
● ^{15}N	4.315	0.365	1.04×10^{-3}	0.101	−0.28304	1/2	—
● ^{17}O	5.772	3.7×10^{-2}	2.91×10^{-2}	1.58	−1.8930	5/2	-4×10^{-3}
● ^{19}F	40.055	100	0.833	0.941	2.6273	1/2	—
^{21}Ne	3.363	0.257	2.46×10^{-3}	0.395	−0.66176	3/2	—
^{22}Na*	4.434	—	1.81×10^{-2}	1.67	1.746	3	—
● ^{23}Na	11.262	100	9.25×10^{-2}	1.32	2.2161	3/2	0.1
^{24}Na*	3.22	—	11.5×10^{-3}	2.02	1.69	4	—
● ^{25}Mg	2.606	10.05	2.68×10^{-3}	0.714	−0.85471	5/2	—

(continued)

Table of Nuclear Properties—(*Continued*)

Isotope	NMR frequency in MHz for a 10 kG field	Natural abundance (%)	Relative sensitivity for equal number of nuclei: at constant field	at constant frequency	Magnetic moment, μ, in multiples of the nuclear magneton $(eh/4\pi Mc)$	Spin I in multiples of $h/2\pi$	Electric quadrupole moment, Q, in multiples of $e \times 10^{-24}$ cm^2
● ^{27}Al	11.094	100	0.206	3.04	3.6385	5/2	0.149
● ^{29}Si	8.458	4.70	7.84×10^{-3}	0.199	−0.55477	1/2	—
● ^{31}P	17.236	100	6.63×10^{-2}	0.405	1.1305	1/2	—
● ^{33}S	3.266	0.74	2.26×10^{-3}	0.384	0.64274	3/2	−0.053
^{35}S*	5.08	—	8.50×10^{-3}	0.597	1.00	3/2	0.038
● ^{35}Cl	4.172	75.4	4.70×10^{-3}	0.490	0.82091	3/2	-7.9×10^{-2}
● ^{36}Cl*	4.893	—	1.21×10^{-2}	0.919	1.2839	2	−0.168
● ^{37}Cl	3.472	24.6	2.71×10^{-3}	0.408	0.68330	3/2	-6.21×10^{-2}
● ^{39}K	1.987	93.08	5.08×10^{-4}	0.233	0.39094	3/2	+0.07
^{40}K*	2.470	1.19×10^{-2}	5.21×10^{-3}	1.55	−1.296	4	—
● ^{41}K	1.092	6.91	8.44×10^{-5}	0.128	0.21488	3/2	—
^{42}K*	4.34	—	8.50×10^{-3}	0.816	−1.14	2	—
● ^{43}Ca	2.865	0.13	6.40×10^{-2}	1.41	−1.3153	7/2	—
● ^{45}Sc	10.344	100	0.301	5.10	4.7492	7/2	−0.22
● ^{47}Ti	2.400	7.75	2.09×10^{-3}	0.658	−0.78711	5/2	—
● ^{49}Ti	2.401	5.51	3.76×10^{-3}	1.18	−1.1022	7/2	—
^{49}V*	10.2	—	0.288	5.03	4.68	7/2	—
● ^{50}V	4.245	0.24	5.55×10^{-2}	5.58	3.3413	6	—
● ^{51}V	11.193	~100	0.382	5.52	5.1392	7/2	0.2
● ^{53}Cr	2.406	9.54	9.03×10^{-4}	0.28	−0.47354	3/2	—
^{53}Mn*	11.00	—	0.362	5.42	5.050	7/2	—
● ^{55}Mn	10.553	100	0.178	2.89	3.4611	5/2	0.6

● ^{57}Fe	1.38	2.245	3.38×10^{-5}	3.2×10^{-2}	0.0903	1/2	—
^{56}Co*	7.347	—	0.137	4.60	3.855	4	—
^{57}Co*	10.1	—	0.283	5.00	4.65	7/2	—
^{58}Co*	15.44	—	0.382	2.90	4.052	2	—
● ^{59}Co	10.103	100	0.281	4.98	4.6388	7/2	0.5
● ^{60}Co*	4.6	—	5×10^{-2}	4.3	3.0	5?	—
^{61}Ni	3.79	1.25	3.53×10^{-3}	0.445	0.746	3/2	—
● ^{63}Cu	11.285	69.09	9.31×10^{-2}	1.33	2.2206	3/2	−0.16
^{64}Cu*	3.0	—	9.79×10^{-4}	0.191	0.40	1	—
● ^{65}Cu	12.090	30.91	1.14	1.42	2.3790	3/2	−0.15
● ^{67}Zn	2.664	4.12	2.86×10^{-3}	0.730	0.87354	5/2	0.18
● ^{69}Ga	10.219	60.2	6.91×10^{-2}	1.200	2.0108	3/2	0.178
● ^{71}Ga	12.984	39.8	0.142	1.525	2.5549	3/2	0.112
● ^{73}Ge	1.485	7.61	1.40×10^{-3}	1.15	−0.87677	9/2	−0.2
● ^{75}As	7.292	100	2.51×10^{-2}	0.856	1.4349	3/2	0.3
^{75}Se*	—	—	—	—	—	5/2	1.1
● ^{77}Se	8.131	7.50	6.93×10^{-3}	0.191	0.5325	1/2	—
^{79}Se*	2.211	—	2.94×10^{-3}	1.09	−1.015	7/2	0.7
● ^{79}Br	10.667	50.57	7.86×10^{-2}	1.25	2.0991	3/2	0.34
● ^{81}Br	11.499	49.43	9.85×10^{-2}	1.35	2.2626	3/2	0.28
^{83}Kr	1.64	11.55	1.88×10^{-3}	1.27	−0.96705	9/2	0.15
^{85}Kr*	1.70	—	2.08×10^{-3}	1.31	1.00	9/2	2.5
^{81}Rb*	10.2	—	6.80×10^{-2}	1.19	2.00	3/2	—
^{82}Rb*	—	—	—	—	—	5	—
^{83}Rb*	—	—	—	—	—	5/2	—
^{84}Rb*	—	—	—	—	—	2	—
● ^{85}Rb	4.111	72.8	1.05×10^{-2}	1.13	1.3482	5/2	0.28
^{86}Rb*	6.5	—	2.82×10^{-2}	1.2	(−)1.7	2	—
● ^{87}Rb	13.932	27.2	0.175	1.64	2.7414	3/2	0.14
● ^{87}Sr	1.845	7.02	2.69×10^{-3}	1.43	−1.0893	9/2	—

(continued)

TABLE OF NUCLEAR PROPERTIES—(*Continued*)

ISOTOPE	NMR FREQUENCY IN MHz FOR A 10 kG FIELD	NATURAL ABUNDANCE (%)	RELATIVE SENSITIVITY FOR EQUAL NUMBER OF NUCLEI		MAGNETIC MOMENT, μ IN MULTIPLES OF THE NUCLEAR MAGNETON $(eh/4\pi Mc)$	SPIN I IN MULTIPLES OF $h/2\pi$	ELECTRIC QUADRUPOLE MOMENT, Q, IN MULTIPLES OF $e \times 10^{-24}$ CM²
			AT CONSTANT FIELD	AT CONSTANT FREQUENCY			
● ^{89}Y	2.086	100	1.18×10^{-4}	4.90×10^{-2}	−0.13682	1/2	—
^{91}Zr	3.958	11.23	9.4×10^{-3}	1.08	−1.298	5/2	—
● ^{93}Nb	10.407	100	0.482	8.07	6.1435	9/2	−0.16
● ^{95}Mo	2.774	15.78	3.23×10^{-3}	0.760	−0.9099	5/2	—
● ^{97}Mo	2.833	9.60	3.44×10^{-3}	0.776	−0.9290	5/2	—
● ^{99}Tc*	9.583	—	0.376	7.43	5.6572	9/2	0.3
^{99}Ru	1.9	12.81	1.07×10^{-3}	0.526	−0.63	5/2	—
^{101}Ru	2.1	16.98	1.41×10^{-3}	0.576	−0.69	5/2	—
● ^{103}Rh	1.340	100	3.12×10^{-5}	3.15×10^{-2}	−0.0879	1/2	—
^{105}Pd	1.74	22.23	7.94×10^{-4}	0.48	−0.57	5/2	—
^{105}Ag*	—	—	—	—	—	1/2	—
● ^{107}Ag	1.723	51.35	6.62×10^{-5}	4.05×10^{-2}	−0.1130	1/2	—
● ^{109}Ag	1.981	48.65	1.01×10^{-4}	4.65×10^{-2}	−0.1299	1/2	—
^{111}Ag*	2.21	—	1.40×10^{-4}	5.19×10^{-2}	−0.145	1/2	—
● ^{111}Cd	9.028	12.86	9.54×10^{-3}	0.212	−0.5922	1/2	—
● ^{113}Cd	9.444	12.34	1.09×10^{-2}	0.222	−0.6195	1/2	—
● ^{113}In	9.310	4.16	0.345	7.22	5.4960	9/2	0.750
● ^{115}In*	9.329	95.84	0.347	7.23	5.5073	9/2	0.761
● ^{115}Sn	13.92	0.35	3.50×10^{-2}	0.327	−0.9132	1/2	—
● ^{117}Sn	15.17	7.67	4.52×10^{-2}	0.356	−0.9949	1/2	—
● ^{119}Sn	15.87	8.68	5.18×10^{-2}	0.373	−1.0409	1/2	—
● ^{121}Sb	10.19	57.25	0.160	2.79	3.3417	5/2	−0.53

● ^{123}Sb	5.518	42.75	4.57×10^{-2}	2.72	2.5334	7/2	−0.68
● ^{123}Te	11.16	0.89	1.80×10^{-2}	0.262	−0.7319	1/2	—
● ^{125}Te	13.45	7.03	3.16×10^{-2}	0.316	−0.8824	1/2	—
^{125}I*	—	—	—	—	—	5/2	0.8
● ^{127}I	8.519	100	9.34×10^{-2}	2.33	2.7937	5/2	−0.75
● ^{129}I*	5.669	—	4.96×10^{-2}	2.80	2.6031	7/2	−0.53
^{131}I*	—	—	—	—	—	7/2	−0.41
● ^{129}Xe	11.78	26.24	2.12×10^{-2}	0.277	−0.77255	1/2	—
● ^{131}Xe	3.490	21.24	2.75×10^{-3}	0.410	0.68680	3/2	−0.12
^{127}Cs*	21.5	—	0.129	0.505	1.41	1/2	—
^{129}Cs*	22.4	—	0.146	0.526	1.47	1/2	—
^{130}Cs*	—	—	—	—	—	1	—
^{131}Cs*	10.6	—	0.181	2.91	3.48	5/2	—
^{132}Cs*	—	—	—	—	—	2	—
● ^{133}Cs	5.585	100	4.74×10^{-2}	2.75	2.5642	7/2	−0.004
^{134}Cs*	5.64	—	6.21×10^{-2}	3.53	2.96	4	—
^{135}Cs*	5.94	—	5.70×10^{-2}	2.93	2.727	7/2	—
^{137}Cs*	6.18	—	6.44×10^{-2}	3.05	2.84	7/2	—
● ^{135}Ba	4.230	6.59	4.90×10^{-3}	0.497	0.83229	3/2	—
● ^{137}Ba	4.732	11.32	6.86×10^{-3}	0.556	0.93107	3/2	—
● ^{138}La*	5.617	0.089	9.18×10^{-2}	5.28	3.684	5	2.7
● ^{139}La	6.014	99.911	5.92×10^{-2}	2.97	2.7615	7/2	0.5
^{141}Ce*	0.35	—	1.1×10^{-5}	0.17	0.16	7/2	—
^{141}Pr	11.95	100	0.258	3.28	3.92	5/2	-5.4×10^{-2}
^{143}Nd	2.72	12.20	5.49×10^{-3}	1.34	−1.25	7/2	−0.57
^{145}Nd	1.7	8.30	1.33×10^{-3}	0.838	−0.78	7/2	−0.30
^{147}Nd*	0.37	—	2.21×10^{-5}	0.289	0.22	9/2	—
^{147}Sm	1.5	15.07	8.8×10^{-4}	0.730	−0.68	7/2	0.72
^{149}Sm	1.2	13.84	4.7×10^{-4}	0.591	−0.55	7/2	0.72
● ^{151}Eu	10.49	47.77	0.175	2.87	3.441	5/2	—

(continued)

TABLE OF NUCLEAR PROPERTIES—(*Continued*)

ISOTOPE	NMR FREQUENCY IN MHz FOR A 10 KG FIELD	NATURAL ABUNDANCE (%)	RELATIVE SENSITIVITY FOR EQUAL NUMBER OF NUCLEI: AT CONSTANT FIELD	RELATIVE SENSITIVITY FOR EQUAL NUMBER OF NUCLEI: AT CONSTANT FREQUENCY	MAGNETIC MOMENT, μ IN MULTIPLES OF THE NUCLEAR MAGNETON ($eh/4\pi Mc$)	SPIN I IN MULTIPLES OF $h/2\pi$	ELECTRIC QUADRUPOLE MOMENT, Q, IN MULTIPLES OF $e \times 10^{-24}$ CM²
● ^{153}Eu	4.638	52.23	1.51×10^{-2}	1.27	1.521	5/2	—
^{154}Eu	5.1	—	2.72×10^{-2}	1.91	2.0	3	—
^{155}Gd	1.2	14.68	1.33×10^{-4}	0.149	−0.25	3/2	1.1
^{157}Gd	1.7	15.64	3.34×10^{-4}	0.203	−0.34	3/2	1.0
^{159}Tb	7.72	100	2.99×10^{-2}	0.907	1.52	3/2	—
^{161}Dy	1.2	18.73	2.35×10^{-4}	0.317	−0.38	5/2	—
^{163}Dy	1.6	24.97	6.38×10^{-4}	0.443	−0.53	5/2	—
^{165}Ho	7.22	100	0.102	3.56	3.31	7/2	2
^{167}Er	1.04	22.82	3.11×10^{-4}	0.516	0.48	7/2	(10)
^{169}Tm	3.49	100	5.51×10^{-4}	8.20×10^{-2}	−0.229	1/2	—
^{171}Yb	7.51	14.27	5.50×10^{-3}	0.176	0.4926	1/2	—
^{173}Yb	2.1	16.08	1.33×10^{-3}	0.566	−0.677	5/2	—
^{175}Lu	4.86	97.40	3.12×10^{-2}	2.40	2.230	7/2	5.7
● ^{176}Lu*	5.3	2.60	0.110	7.02	4.2	6	8
^{177}Hf	1.3	18.39	6.38×10^{-4}	0.655	0.61	7/2	3
^{179}Hf	0.80	13.78	2.16×10^{-4}	0.617	−0.47	9/2	3
● ^{181}Ta	5.09	100	3.60×10^{-2}	2.52	2.340	7/2	4.0
● ^{183}W	1.75	14.28	6.98×10^{-5}	4.12×10^{-2}	0.115	1/2	—
● ^{185}Re	9.586	37.07	0.133	2.63	3.1437	5/2	2.8
● ^{187}Re	9.684	62.93	0.137	2.65	3.1760	5/2	2.6
^{187}Os	1.8	—	7.93×10^{-5}	4.3×10^{-2}	0.12	1/2	—
● ^{189}Os	3.307	16.1	2.34×10^{-3}	0.388	0.6507	3/2	2.0

^{191}Ir	0.813	38.5	3.5×10^{-5}	9.5×10^{-2}	0.16	3/2	—
^{193}Ir	0.86	61.5	4.2×10^{-5}	0.101	0.17	3/2	1.5
● ^{195}Pt	9.153	33.7	9.94×10^{-3}	0.215	0.6004	1/2	—
● ^{197}Au	0.731	100	2.53×10^{-5}	8.59×10^{-2}	0.1439	3/2	0.56
^{198}Au*	1.9	—	8.75×10^{-4}	0.358	0.50	2	—
^{199}Au*	1.2	—	1.17×10^{-4}	0.143	0.24	3/2	—
^{197}Hg*	7.9	—	6.46×10^{-3}	0.186	0.52	1/2	—
● ^{199}Hg	7.60	16.86	5.67×10^{-3}	0.178	0.4979	1/2	—
● ^{201}Hg	2.80	13.24	1.42×10^{-3}	0.329	−0.5513	3/2	0.45
^{199}Tl*	—	—	—	—	—	1/2	—
● ^{203}Tl	24.33	29.52	0.187	0.571	1.5960	1/2	—
^{204}Tl*	—	—	—	—	—	2	—
● ^{205}Tl	24.57	70.48	0.192	0.577	1.6115	1/2	—
● ^{207}Pb	8.899	21.11	9.13×10^{-3}	0.209	0.5837	1/2	—
● ^{209}Bi	6.842	100	0.137	5.30	4.0389	9/2	−0.4
^{209}Po*	—	—	—	—	—	1/2	—
^{227}Ac*	5.6	—	1.13×10^{-2}	0.656	1.1	3/2	−1.7
^{231}Pa*	9.96	—	6.40×10^{-2}	1.17	1.96	3/2	—
^{233}U*	1.6	—	6.75×10^{-4}	0.451	0.54	5/2	3.4
^{235}U*	0.75	0.71	1.21×10^{-4}	0.376	0.35	7/2	3.8
^{237}Np*	8.3	—	6.70×10^{-2}	2.09	2.5	5/2	(15)
^{239}Np*	—	—	—	—	—	1/2	—
^{239}Pu*	6.1	—	2.9×10^{-3}	0.14	0.4	1/2	—
^{241}Pu*	4.3	—	1.18×10^{-2}	1.17	1.4	5/2	—
^{241}Am*	4.3	—	1.18×10^{-2}	1.17	1.4	5/2	4.9
^{243}Am*	4.3	—	1.18×10^{-2}	1.17	1.4	5/2	4.9
^{244}Cm*	—	—	—	—	—	7/2	—
Free electron	28024.6	—	2.84×10^{8}	657	−1836	1/2	—

[a] Reproduced by permission of Varian Associates.

[b] Asterisk indicates radioactive; dot preceding entry indicates magnetic moment determined by NMR.

SUBJECT INDEX

A

M

T

U

V

A 5
B 6
C 7
D 8
E 9
F 0
G 1
H 2
I 3
J 4